Construction of Prestressed Concrete Structures

Construction of Prestressed Concrete Structures

Ben C. Gerwick, Jr.
Executive Vice-President
Santa Fe-Pomeroy, Inc.
San Francisco, California

Wiley-Interscience
a division of John Wiley & Sons, Inc.
New York • London • Sydney • Toronto

Library of Congress Catalog Card Number: 71-140176
ISBN 0-471-29710-0

Printed in the United States of America
10 9 8 7 6 5 4 3 2 1

Series Preface

The construction industry in the United States and other advanced nations continues to grow at a phenomenal rate. In the United States alone construction in the near future will exceed ninety billion dollars a year. With the population explosion and continued demand for new building of all kinds, the need will be for more professional practitioners.

In the past, before science and technology seriously affected the concepts, approaches, methods, and financing of structures, most practitioners developed their know-how by direct experience in the field. Now that the construction industry has become more complex, there is a clear need for a more professional approach to new tools for learning and practice.

This series is intended to provide the construction practitioner with up-to-date guides which cover theory, design, and practice to help him approach his problems with more confidence. These books should be useful to all people working in construction: engineers, architects, specification experts, materials and equipment manufacturers, project superintendents, and all who contribute to the construction or engineering firm's success.

Although these books will offer a fuller explanation of the practical problems which face the construction industry, they will also serve the professional educator and student.

M. D. Morris, P.E.

Prestressed Concrete Cylinder Piles are representative of the new construction techniques available for bridge construction.

Preface

Prestressed concrete and prestressing have rapidly become of universal importance in construction.

Prestressed concrete has gone through the research and development phase of the 1930's and 1940's, and through the specialized design and specialized construction phases of 1955 to the present. Now it is a *tool* and *technique* which every structural designer must possess. It is also a *technique* in which every good construction engineer and contractor must be versed and competent.

Because of the historical nature of its development, there are many excellent texts on *design of prestressed concrete*. Unfortunately, few references exist to guide the construction engineer and contractor. Specialized subcontracting organizations do exist but for economical reasons have had to become over-specialized as, for example, in post-tensioning. Very large national construction concerns, recognizing both the opportunity and the need, have established prestressing capabilities within their own firms, but these are frequently competent in one phase of prestressed construction only; e.g., pre-tensioned pile manufacture or post-tensioned bridge girders.

Serious problems have arisen where lack of full knowledge of prestressed concrete construction engineering has led to disastrous errors, such as errors in installation practice, or errors in concrete technology.

This book provides a basic and inclusive exposition of prestressed concrete construction technology, and a practical guide to the contractor and construction engineer. It is impracticable to cover every detail of every phase, but a serious attempt has been made to set forth construction engineering principles that will be an adequate guide for the majority of prestressed concrete construction, and to indicate, by references, sources of greater detail in specific aspects.

Contractors must perform their work at a profit. Economics are inextricably interwoven with techniques; hence, the emphasis throughout the book on methods of minimizing the cost and time of construction in prestressed concrete.

The orientation of this book is principally to construction; but, in highly technical work, design and construction are increasingly intertwined. Thus, the emphasis throughout this book will be placed on the integration of design with construction. As the demands of society require the construction of structures of increasing complexity and sophistication, the construction engineer and contractor must have the technical reserves at his command to enable him to carry out the design, not only in a timely and economical manner, but to assure that the intent of the designer and need of the owner are fulfilled. Then the contractor will have earned his profit; hopefully, it will be a good one.

This book is dedicated to G. W. "Bill" Harker, of Australia, a pioneer in prestressing, and a man of indomitable will.

Ben C. Gerwick, Jr.

San Francisco, California
November 1970

Contents

Construction of Prestressed Concrete Structures

Introduction

A. Prestressed Concrete Construction Principles

Prestressing is the creation within a material of a state of stress and deformation that will enable it to better perform its intended function.

As most commonly applied, prestressing creates a compressive stress within concrete that will partially or wholly balance the tensile stresses that will occur in service. Concrete, having a reserve of strength in compression, is an ideal material for prestressing; it is universally available, cheap, easily molded to the desired form, and provides corrosion protection to steel.

However, prestressing can be and has been applied to other materials: to steel trusses, to stone and ceramics and brick, to timber, and to native rock and soils. Furthermore, prestressing can be utilized to overcome not only tensile stresses due to applied loads, as in a bridge girder, but also tensile stresses and deformations due to dynamic waves (as in piledriving or machinery vibrations), temperature stresses (as in pressure vessels), shrinkage, direct tension, and shear (diagonal tension).

The prestressing principle is not confined solely to pre-compression. There are some applications under study where the introduction of a tensile prestress at a particular stage is desirable to overcome excessive compression.

Prestressing is usually introduced by means of internal steel tendons which are stressed and then anchored. While high-strength steel is currently the universally used material for tendons, research and development continues on such promising tendon substances as fiberglass. Tendons need not necessarily be inside the concrete. They can be external to the concrete section, as in a tie rod or a tied arch. Prestressing can also be applied by external force, such as jacking applied at the crown of an arch or at the ends of a pavement slab.

In many applications, the aim is to create, through prestressing, a state of stress that will just balance the stress that will be imposed in the member in service. In other applications, the intent is to overcome deflections; for example, to produce a truly flat floor under normal service loadings.

The construction engineer can utilize prestressing effectively to overcome excessive temporary stresses or deflections that may occur during

1

construction; for example, giving him a technique of cantilevering in lieu of falsework, or a means of handling large and unstable elements.

Prestressing is not a fixed state of stress and deformation, but is time-dependent. Both concrete and steel deform plastically under continued stress. This plastic flow is greatly increased by high temperature and decreased by low temperature. Much of the continuing research and development is aimed at finding more stable materials: high strength steel that is "stabilized" against stress relaxation, and high strength concrete that has low creep, shrinkage, and thermal response.

About 1890, Henry Jackson, a San Francisco engineer, reportedly "invented" prestressed concrete. He built lintels in which the steel bar was prestressed. But after a year or so, the lintels cracked and collapsed. He did not understand the phenomenon of creep of concrete and steel which "used up" the prestress during that period.

Hence, the demand for high-quality steels and high-quality concretes. High strengths are desirable for efficiency and economy of performance, but other properties are essential to the long-term stability and performance. These other properties basically relate to the assurance of keeping the state of stress and deformation of the prestressed concrete within certain acceptable limits.

The provision of a "stable" tendon is basically the province of the steel manufacturer. The contractor can contribute to this by protecting it during construction operations, so that it gets installed in as nearly the same condition as it was furnished by the steel manufacturer, by the quality of his anchoring and stressing operations, and by the durability of his concrete encasement.

The provision of a "stable" concrete is very much under the influence of the contractor. It is the most critical and sensitive aspect of "prestressed concrete." It is strange that so much attention is given in prestressing specifications and literature to the steel and its stressing, and so little to the concrete! The contractor can control the aggregates and the water. He may have considerable to say about the cement and admixtures. He controls the batching, the mixing, the transportation, the placement, the consolidation, and the curing. He controls the forms and the accuracy with which tendons and reinforcing bars are placed. Finally, in post-tensioned work, he controls the placing (and maintenance of position during concreting) of ducts and anchorages, and the grouting or other corrosion protection.

With prefabricated members, he controls the handling, transportation, erection or installation, and the fixing. Small wonder that most of the difficulties that have occurred with prestressed concrete relate to the construction phase and especially to the concrete.

B.   Basic Definitions and Principles

Prestressing is the imposition of a state of stress on a structural body—prior to its being placed in service—that will enable it to better withstand the forces and loads imposed on it in service or to better perform its design functions.

Prestress requires a pre-strain. It is important to remember that prestress cannot be imparted into a member unless that member is able to shorten.

A beam may be prestressed by pre-compressing its lower flange, so that it can resist bending tensile stresses in service without cracking.

A pile may be prestressed (pre-compressed) so as to maintain itself crack-free under shrinkage, transportation, handling, and driving stresses.

A column may be prestressed (pre-compressed) so as to enable it to act as an uncracked homogeneous section under long-column buckling or eccentric loading.

A pressure vessel may be prestressed so as to resist the membrance tensile stresses due to thermal gradients and internal pressure.

A thin floor slab may be prestressed so as to remain truly flat under normal service loads.

Tendons are the stretched or tensioned elements which are used to impart pre-compression to the concrete. Tendons may be high-strength steel wire, strands made of high-strength steel wire, or high-strength alloy steel bars. Fiberglass tendons have been used experimentally.

Internal prestressing is prestress which is imposed by tendons contained in or immediately adjacent to (in contact with) the concrete member.

External prestressing is prestress which is imparted by forces such as jacks reacting against abutments at the ends of the concrete member, or at intermediate locations, where the ends are restrained by abutments. This is usually applied only to pavements, arch ribs, and pedestals.

Pretensioning is the imposition of prestress by stressing the tendons against external reactions before the hardening of the fresh concrete, then allowing the concrete to set and gain a substantial portion of its ultimate strength, then releasing the tendons so that the stress is transferred into the concrete.

In the most common cases high-strength strands are stretched between two abutments and jacked to about 3/4 of their ultimate strength; the concrete is then poured within forms around the tendons. The cure of the concrete is accelerated by low-pressure steam curing, and the strands are then released, so that the stress is transferred into the concrete by bond. The high-stretched wires shorten slightly, pre-compressing and shortening the concrete.

Pretensioning is most commonly applied to precast concrete elements manufactured in a factory or plant. Typical products produced by pretensioning are roof slabs, floor slabs, piles, poles, bridge girders, wall panels, and railroad ties. A limited amount of pretensioning has been applied to cast-in-place concrete in the form of pavements and floor slabs.

Post-tensioning is the imposition of prestress by stressing and anchoring tendons against already hardened concrete.

In the most usual case, ducts are formed in the concrete body by means of tubes. After the concrete has hardened and gained sufficient strength, the tendons are inserted and elongated by jacking, then anchors are seated so as to transfer the load from the jacks through the anchors to the ends of the concrete member.

Post-tensioning is most commonly applied to cast-in-place concrete members, and to those involving complex curvatures. Bridges, large girders, floor slabs, roofs, shells, pavements, and pressure vessels are among the constructions usually prestressed by post-tensioning. Post-tensioning is extremely versatile and quite free from limitations on size, length, degree of stress, etc. It may be used for factory-made products as well but, for mass-manufacture, is usually more expensive than pretensioning per unit of prestress force.

Stage-stressing is the application of prestressing force in a sequence of two or more steps. This is done so as to avoid overstressing or cracking the concrete during the construction phase, before further dead load is applied.

Internal tendons are tendons which are embedded within the cross section of the concrete body or member. They may be either pretensioned or post-tensioned tendons. Usually this refers to post-tensioned tendons located in ducts cast in the concrete.

External tendons are tendons which lie outside the cross section of the concrete body or member as cast. They may later be bonded to it by concreting or grout. External tendons may be in grooves or channels at the side of the concrete member. They may be located inside the open box of a box girder.

Bonded tendons are tendons which are substantially bonded to the concrete throughout their entire length. Pretensioned tendons are almost always bonded. Small and moderate size post-tensioned tendons which are in ducts which are grouted are considered as "bonded."

External tendons may be bonded if stirrups and grout encasement provide full shear transfer along the entire length.

Unbonded tendons are tendons whose force is applied to the concrete member only at the anchorages. Bond throughout the length is purposely prevented. When the post-tensioned tendon is in a duct, the duct may be filled with grease. Some post-tensioned tendons are coated with a bituminous compound and wrapped with paper, then placed in the forms, the concrete poured and cured, and the tendons stretched. In this case there may be some slight degree of mechanical bond at bends, etc., but the tendon is essentially unbonded.

External tendons, suspenders, prestressed tie rods, etc., are generally considered as unbonded, except when fastened to the concrete member at close intervals, as by grout encasement and stirrups.

Partial prestress is a design philosophy in which the degree of prestress is purposely kept low. The aim usually is to provide a residual compression (zero tension) under normal service loads, but to permit tension and even a minor degree of cracking under occasional overloads.

Creep is the plastic change in volume of concrete under sustained stress. It is most marked during the early ages of the life of a concrete member; an old rule of thumb used to say, "One-third in three days, the second one-third in three months, the last one-third in ten years." Creep occurs in both the aggregate and the paste. The higher the sustained stress, the greater the creep. If stress is applied at an early age, the creep will be greater. Creep is essentially irreversible; when the stress is removed, the creep stops but does not reverse to any appreciable extent.

Camber is the upward deflection of the member due to prestress.

Shrinkage is a volume change of the concrete due to chemical reaction and the drying-out of the contained water. Some shrinkage occurs at set, but the largest amount occurs during drying. Shrinkage is reversible, and sustained humidity or soaking will cause a swelling, offsetting the drying shrinkage.. The effect of shrinkage is to reduce the pre-compression in the concrete. Before the application of prestress, shrinkage may produce tensile cracking in the concrete which, although closing again under prestress, may still reduce the inherent tensile strength of the concrete and its ability to remain crack-free under load.

Stress-relaxation is an irreversible plastic flow in the steel under sustained high stress. Stress-relaxation, as the name implies, leads to a reduction in the degree of stress in a tendon, thus reducing the prestress in the concrete.

While the greatest portion occurs during a very short time of high stress, relaxation may continue for very long periods. Since it is a molecular phenomenon, various treatments of the steel can be used to reduce the stress relaxation.

Elastic shortening is the volume change occurring in the concrete as the pre-compression is applied to it during prestressing.

C. Construction Engineering

Construction engineering is the art of applying engineering approaches to construction operations. Within the scope of this book, it will encompass such matters as planning, scheduling, and control. Planning starts with an analysis of the work to be accomplished, a selection of methods and techniques, a layout of sequential operations, assignment of equipment and labor. Scheduling includes the inter-relation with the other operations at the site as well as with external aspects such as weather, floods, air temperature, and contract requirements. Control involves the assignment of supervision and inspection, the establishment of detailed procedural instructions, the provision for adequate inspection, and cost control.

Construction engineering analyses, properly applied, will probably result in considerably more instructions, charts, schedules, etc., than are commonly employed today; but these should not be voluminous and complicated. The point is that the goal is strictly practical; to so plan, schedule, and control the work that every man is contributing to its accomplishment with minimum waste and interference. The charts and instructions must be clear, concise, and definite, and directed at understanding by the men actually doing the work. There is no need to restate all the manifold considerations that lie behind a set of instructions.

The highly technical nature of many prestressing operations makes it essential that prior planning be carried on in considerable detail. Most problems associated with prestressed concrete could have been prevented had proper planning been given before the time of the actual construction. Conversely, the economic results of prior planning have been very profitable and, in some cases, have actually resulted in halving of the labor costs in a particular prestressed concrete construction operation.

For many construction projects, the prestressing operations must be fitted into an overall critical path schedule. Prestressing is usually on the critical path; therefore, the selection of the method and procedures may be heavily influenced by the time allotted.

Unlike many other types of construction, with prestressed concrete the details of tendon layout, selection of prestressing system, mild steel details, etc., are often left up to the contractor, with the designer merely showing the final prestress and its profile, and setting forth criteria.

Thus, the constructor must understand enough about design considerations to help him select the most efficient and economical system. Such knowledge may often provide him a competitive edge in bidding.

Finally, in many aspects of construction, the contractor and engineer are functionally united. Industrial projects and commercial developments are frequently undertaken by a combination of the contractor and engineer, either joined within one firm or as a cooperative effort. The same is true of many overseas contracts where the owner and his engineer merely establish the basic requirements and criteria, leaving the final design to the contractor. Foundation projects in the United States are frequently bid on performance specifications only, or alternates are permitted. In many new areas of construction, such as nuclear reactor power plants, offshore structures, etc., the contractor must bid on a "design and construct" or "turnkey" basis. Therefore the basic principles of design are essential background knowledge for an enterprising construction engineer, even though he himself may not undertake the design task. He must understand the phenomena, the reasons behind the technical requirements, the advantages and the limitations of prestressing.

As in all other aspects of construction, but perhaps more intensively in prestressed concrete, design and construction must be integrated if the most effective results are to be obtained. The historical separation of the two disciplines, design and construction, is an artificial one, a division of labor in order to enable concentration of skills. The overall endeavor is to build a structure that will serve a specific purpose well. Designer and constructor are thus members of a single team with one goal.

I

MATERIALS AND TECHNIQUES FOR PRESTRESSED CONCRETE

1

Materials

1.1 Concrete

Concrete is a heterogeneous material composed of aggregates embedded in a matrix. Most commonly, the aggregates are natural sands and gravels, or crushed rock, and the matrix is Portland cement which has been hydrated by water.

Careful attention is invited to the above definition because, in this technological age, every possible variation is being explored and even commercially applied in an attempt to produce desired properties. Thus we have artificial aggregates, such as expanded shale, and pozzolanic and organic cements, being applied on a substantial scale in engineering construction. (Structural light weight concrete, produced from expanded slate, shale, and clay aggregates, is discussed in detail in subsection 3.1.2)

The remainder of this chapter will deal with conventional concrete, consisting of limestone or siliceous coarse aggregates (crushed rock or gravel), limestone or siliceous sand, Portland cement, water, and admixtures. This produces concrete weighing from 140 to 160 pounds per cubic foot ($2240kg/m^3$ to $2560kg/m^3$) and, when properly selected, batched, mixed, placed, and cured, capable of strengths up to 8000 psi ($570kg/cm^2$) in 28 days.

Concrete for prestressing is most commonly in the range of 3500 to 7000 psi ($250kg/cm^2$ to $500kg/cm^2$). Such concrete can be produced with reasonable economy in most parts of the world, provided proper care is taken in all phases of the concreting operation. The existence of an established concrete industry in metropolitan and technologically sophisticated centers must not be allowed to induce complacency or carelessness, as the uniform high qualities, rather easily obtainable in these centers, were attained only after many years of intensive efforts on the part of all segments of the industry and, even

11

then, there has been a substantial amount of trial-and-error which eliminated the unfit. So, when commencing operations in new areas, it is essential that fundamental considerations be resurrected and applied.

There are adequate and detailed specifications available for the selection and evaluation of concrete materials, such as those of the American Society for Testing Materials (ASTM). All that can be done in this chapter is to highlight certain qualities that can be particularly important or serious for prestressed concrete.

1.1.1   Aggregates

Since we are attempting to get high-strength concrete, the maximum size of coarse aggregate should be limited. For most applications, 3/4 inch is the optimum maximum size.

Coarse aggregates must not contain clay seams that produce excessive volume change, such as creep and shrinkage.

Both gravel and crushed rock are used successfully. For normal high-strength concrete for prestressed application, gravel will give better workability and compactability at low water/cement ratios. For extremely high-strength concrete, crushed rock of proper angularity is superior but requires very intensive vibration to achieve proper compaction.

For use in salt-water environments, aggregates must be sound and not subject to sulfate attack. (See ASTM C88.) An important iron-ore loading pier in South America, founded on prestressed piles, has suffered catastrophic deterioration apparently due to lack of soundness of the aggregates.

Aggregates must not be reactive to the alkali in the cement. Siliceous aggregates are more likely to be subject to alkali reaction than limestone, but a few rare cases have been reported with the latter. (See ASTM C 289 and C 227 for the standard tests for alkali reactivity.) A large prestressed bridge in Germany is reportedly disintegrating due to reactive aggregates.

Fine aggregates can be in the coarser ranges since, with the rich cement factors usually employed in prestressed concrete, perfect grading is not necessary and may be undesirable. Gap-grading, properly applied, can often reduce shrinkage and improve strength and modulus of elasticity.

Aggregates must be clean. Even a few percent of silt can make the dry mixes for prestressed concrete excessively sticky and difficult to place. Silt often gives a flash set. Silt reduces strength and increases shrinkage. Silt can usually be removed by rewashing, with very beneficial results.

Aggregates must not contain salt. Salt can be deposited on aggregates, particularly fine aggregates, from sea-water immersion, or even from salt fogs in desert countries like Kuwait. Even small percentages of salt reduce the corrosion-inhibiting value of the cement and may help initiate electrochemical corrosion. This is particularly dangerous with steam curing.

Aggregates must be of a proper temperature for incorporation in the mix. Since the aggregates are by far the largest component of the mix, it is often most effective and economical to cool the aggregate, as by water evaporation, in summer, or to heat it in winter. Water "soakers" running continuously over the aggregate piles will prevent dust and cool the aggregate by evaporation.

1.1.2 Cement

Almost all prestressed concrete employs Portland cement. ASTM C 150 designates five types. Type I is standard, Type II is moderate low-alkali, Type III is high early strength, Type IV is low-heat, and Type V is sulfate-resisting.

Most prestressed concrete employs Types I, II, or III, or a modification of these. The cement is usually selected on the basis of rapid early strength, minimum shrinkage, durability, and economy. Flash set is to be avoided.

Type I is suitable for most building work, but Type II is preferable for coastal and marine environments. Salt fogs can penetrate 50 miles or more inland, and Type II cement gives better durability to the concrete along with good corrosion-inhibiting properties for the steel.

A few Type III cements tend to develop flash sets or have excessive shrinkage under steam curing.

Recently, in an effort to obtain the optimum balance of properties, modified Type II cements have been developed. These generally are ground much finer than conventional Type II, for example, to a Blaine fineness of 4000 to 4200. They have been specifically developed to meet the needs of the prestressed industry and are often marketed as "prestress cements."

Type V cement (sulfate-resisting) is actually not as well suited for most prestressed applications as Type II. Type V is low in C_3A, which gives the concrete itself greater durability under sea water and sulfate attack but, unfortunately, reduces the corrosion protection for the steel.

1.1.3 Water

Until recently the standard requirement for water was merely that it be potable. However, water for use in prestressed work should be more definitely restricted in salt, silt, and organic contents. Suggested limitations are:

(a) No impurities that will cause a change in time of set greater than 25% nor a reduction in strength at 14 days age greater than 5% as compared with distilled water.

(b) Less than 650 parts per million of chloride ion (some authorities permit up to 1000 ppm).

(c) Less than 1300 parts per million of sulfate ion (some authorities limit this to 1000 ppm).

(d) Water shall be free from oil.

Water may be added to the mix in the form of ice, in order to reduce the ambient temperature of the fresh concrete mix, or in the form of steam when desired to raise the temperature.

1.1.4 Admixtures

These are very useful in prestressed concrete in permitting use of a lower water/cement ratio while maintaining workability. Certain admixtures also reduce shrinkage. Some are retarders at normal temperatures, but cause acceleration of strength gain under steam curing.

Many admixtures used in conventional concreting practice contain $CaCl_2$. This must be *absolutely prohibited* for prestressed work because there is substantial evidence that this can cause corrosion, particularly when steam curing is employed. Even with normal water curing, the $CaCl_2$ lowers the inhibiting powers of the cement and can cause corrosion.

For prestressed work, therefore, the admixture must not contain more than a trace of calcium chloride. Most suitable admixtures are organic by-products from the pulp industry.

Air entrainment is beneficial in improving freeze-thaw durability and in improving placeability of lightweight aggregate concrete. Air entrainment does slightly reduce the strength of rich mixes, such as those used in pre-stressed concrete. Therefore air entraining admixtures should be employed judiciously.

1.1.5 Storing of Aggregates and Cement

Much contamination and deterioration of aggregate quality can arise from improper storage. Stored on the ground or on a slab at grade, the aggregate can be contaminated by dirt, ice and undersized particles. Stored in bins, it is much better protected, except from snow and ice. In either event, considerable dust and chips can collect at the bottom. The only way to positively eliminate this last problem is re-screening above the batcher. The need for this depends on the character of the rock, the abrasion in transport and handling, and the quality of concrete desired.

Aggregates exposed to the summer sun can become overheated. They can be shaded by galvanized or aluminum corrugated roofing. Soaking or spraying of the aggregate pile will cool it very effectively through evaporation. Vacuum-cooling is one of the newest techniques.

Cement must be stored and used in such a manner that none of it is left to age excessively. Obviously, it must be completely protected from moisture; a tropical or a humid atmosphere may present difficulties in this regard and may require dehumidification of the storage shed in the case of sacked storage.

1.1.6 Batching, Mixing, and Transporting

Approved procedures are set forth in ACI 614. Accuracy of batching is essential for production of consistent high-quality concrete. Therefore, batching should be by weight and preferably by automatic rather than hand controls. A continuous correction must be made for water contained in the aggregates.

Mixing must be thorough, especially with low-slump mixes. The turbine mixers are especially adapted to this need. The more recently built ready-mix truck mixers have been improved so as to handle a fairly low-slump mix, but the blades must be kept in good condition. The horizontal turbine mixer is definitely preferable, therefore, for the highest quality mixes. Adequate mixing time improves the uniformity, strength, and impermeability of the concrete. Mixer blades must not rotate too fast.

Transporting may be successfully carried out in a number of ways. Mixes may be transported in a ready-mix truck, with the water added a few minutes before discharge. Mixes may be transported dry (dry-batched) and mixed at the point of use by either a mixer or paver. Wet mixes can be transported in ready-mix trucks or in hoppers (concrete buckets, "dumpcrete" bodies, rail-mounted hoppers, or scoop-loaders). The essential cautions here are to prevent segregation and premature set during transport. A low-water/cement ratio and a set-retarding admixture are helpful. Vibration or mixing during transport is of considerable value in preventing segregation and premature set.

1.1.7 Placing and Consolidation

This subject is set forth in detail in the ACI Manual of Concrete Practice. With the generally dry mixes employed in prestressed concrete, intensive vibration is necessary to insure complete filling of the voids, especially in congested areas, and to thoroughly consolidate it. Internal vibration is the most effective method for the great majority of cases, as it serves to ensure compaction around the tendons, embedded steel, anchorages, etc. Frequencies of 9000 rpm are most commonly employed.

External vibration can be used very effectively with thin products, particularly precast elements cast in heavy steel forms. Such vibration definitely helps placement and produces excellent surfaces on the vibrated face. Vibrators placed opposite each other tend to cancel out. It is usually best to stagger their location.

Frequently, a combination of internal and external vibration will prove most satisfactory.

In placing low-slump or no-slump concrete in forms, it is best to dump it on the advancing face of the concrete where it will get the full effect of the vibration. This will speed concreting and produce better consolidation.

External or form vibration places the forms themselves under high stresses and may even get into the fatigue range of the connections, etc. Therefore forms must be specially designed when form vibration is to be employed.

With dry mixes, there is a definite tendency for water and air pockets to form on vertical and overhang surfaces. Entrapped air and excess water try to escape from the mix under vibration, and are trapped under the overhang and, to some extent, along the side of a vertical surface.

There appears to be no practical technique available for completely eliminating these when dry mixes are used, but they can be minimized by the following steps:

(a) Selection of a type of form oil suited to the surface that reduces capillary attraction.

(b) Thorough internal vibration.

(c) External (form vibration) is especially valuable if continued after internal vibration is completed.

(d) Use of an admixture that prevents bleeding and promotes workability. (Air entertainment may frequently be beneficial—entrained air is not the same as entrapped air).

(e) Spading along the form sides, where accessible, following vibration.

Absorptive form liners will absorb air and excess water from the surface skin, but pockets may still be located a fraction of an inch below the surface.

Strangely enough, this problem is generally not so noticeable with wet mixes, as the pockets and voids are then distributed as pores in the mix. Of course this is generally not a satisfactory solution at all, as it results in low strengths and low durability.

Vacuum processing will remove entrapped air and excess water to the depth to which the vacuum is effective, but is generally not economically practicable.

Experience has shown that the previously cited techniques of placing and consolidation of low water/cement ratio mixes, will produce the strongest and most durable concrete, even though minor surface blemishes occur. For architectural or exposed surfaces, rubbing and sacking will produce the desired uniformity. Properly done, this finished surface is durable from both an architectural and structural point of view.

For concrete exposed in the splash zone or other region of attack, the "bleed" holes should be filled if they are deeper than a nominal amount, say 3/8 inch deep. Depth can be determined with a wire. Filling should be with a suitable grout or dry pack or epoxy mortar.

Concrete mixes will develop the best quality when the temperature of the mix at the time of placing is near 60°F. Ambient temperatures above 90°F will result in loss of strength.

Concrete can also be compacted to a greater density by the use of pressure. This is especially effective when combined with intensive vibration. Ramming is one means of compaction—for example, steel balls inside spun concrete pipe. Centrifugal spinning produces compaction. In the USSR, slabs are compacted by the use of mats and vacuum.

A newly developed process injects live steam into concrete during the mixing phase, raising its temperature to 65 to 75°C. The concrete is then placed immediately (within 10 minutes) and maintained at 60°C for three hours. At the end of this period, strengths up to 60% of the 28-day strength are reported. More investigations are still required as to the effect of these procedures on long-term properties.

1.1.8   Curing

Immediately after pouring, a fresh concrete surface exposed to hot sun or drying wind may lose so much water that it sets and shrinks, even while the body of concrete is still plastic. This can be prevented by one of the following means:

(a) Fog spray: especially adaptable to large, flat surfaces.

(b) Covering with wet burlap or polyethylene. Polyethylene is preferred by some as the moisture steams in the sun and improves curing.

(c) Covering and injecting low-temperature, low-pressure steam.

Curing of concrete should provide sufficient moisture to allow the completion of the chemical reactions which produce strong, durable concrete. Some of these reactions take place in a short time, others require a much longer period. The exact time required is highly temperature-dependent, with different reactions responding differently to the temperature. It is essential to continue curing sufficiently long to enable all the desired reactions to be completed, not just until the concrete has reached a minimum strength.

Concrete, especially dense concrete, will continue to cure internally due to the excess water of the mix. It is on the surface where moisture must be contained or supplied.

Containment may be by a membrane sealing compound, polyethylene sheet, or enclosure in steam hoods.

Moisture may be added by water ponding, water spraying, or water-soaked mats.

Curing is accelerated by heat. Practicable means of supplying heat include low- and high-pressure steam, radiant heat from hot water or hot oil pipes, and electric-resistance heating.

1.1.9 Steam Curing

Steam curing at atmospheric pressure is widely employed in precast pre-stressed concrete manufacture. A great deal of attention has been directed at the effect of this accelerated curing on long-term ultimate compressive strength, durability, shrinkage and creep, loss of prestress, etc. Other studies have been directed at the determination of the optimum cycle for the steam-curing process. Finally, the matter of cooling, release of prestress, and thermal effects must be considered, although here most of the data and recommended procedures have been developed by the precast plants themselves, since comparatively little laboratory research has been carried out on this phase.

As a general statement, it may be said that *properly applied, low-pressure steam curing improves the quality of the concrete product.*

The research laboratory studies referred to have measured the various properties of the steam-cured concrete in comparison with 73°F laboratory fog-chamber curing. First, what has been widely overlooked is the fact that concrete cured by water or fog spray, wet burlap, or sealing compound is not the same as that cured in the laboratory fog-chamber. Therefore many of the reports have been comparing apples and oranges. Steam curing is a superior method for *practicable* curing of concrete in the plant.

Second, with particular regard to such properties as long-term ultimate strength, comparisons between similar concrete cured in the fog-chamber and by steam curing are somewhat irrelevant if the steam-cured concrete is always substantially above the required value.

Third, steam curing has been one of the important techniques which has made it possible to attain economical production of prestressed concrete elements, permitting daily turnover of forms. It has also made it practicable to have a short cycle between manufacture and erection, thus eliminating in large part the need for stockpiles and inventory. Steam-cured prestressed piles have been successfully driven when only one day old.

In many cases steam curing reduces shrinkage.

The adoption of a proper cycle of steam curing is essential. The generally accepted optimum cycle is: (Fig. 1)

(*a*) A delay period of three to four hours until concrete has attained initial set. Concrete should be protected from drying out during this period.

(*b*) A heating period, with a temperature rise of 40° to 60°F (22° to 33°C) per hour to a maximum temperature of 145 to 160 F (63 to 70 C).

(*c*) A steaming period of six hours at 145° to 160°F.

(*d*) A cooling period (steam hoods still covering concrete). During this period, the exposed portions of the strand cool faster than the concrete, thus pulling on the concrete. Similarly, steel forms cool faster, as do the outer portions of the concrete. Thus tensile stresses may be set up in the concrete element. For this reason, with many products, especially massive ones where the inner core of concrete may retain its heat for a considerable period of time, it is necessary to release the prestress into the member during the cooling period, and then re-cover the units to permit a slower and thus more uniform rate of cooling.

(e) An exposure period (steam hoods removed from the concrete which is now exposed to the atmosphere). This may be a somewhat critical period insofar as surface shrinkage, crazing, and ultimate durability are concerned. The concrete is warm and moist. In winter the concrete surface may be subjected to a substantial temperature differential and to very drying winds. The inside core is still warm. A combination of thermal and drying shrinkage may cause crazing and hairline cracks. It is the writer's opinion that the rather mysterious appellation "thermal shock" derives primarily from the above sequence. In summer there may be hot drying winds which again cause drying shrinkage on the surface. Also, the inner heat of the concrete tends to accelerate evaporation of the water from the surface. Therefore in temperatures above freezing, it may be desirable with certain products to apply supplemental water cure immediately after exposure. The first few hours are the critical ones.

Prestressed concrete cylinder piles with their relatively thin walls are typical of a product for which this supplemental water cure is recommended.

For temperatures below freezing, two methods have been employed. The cooling period has been programmed over a longer period, so as to permit the concrete to attain a rather uniform low temperature before exposure to the atmosphere. This can, for example, mean merely storage inside at room temperature for two or three hours before exposure. The second method is to apply membrane sealing compounds after the conclusion of steam curing, in order to prevent rapid drying shrinkage of the surface.

Some plants apply membrane sealing compound before steam curing. Although the compound deteriorates somewhat during the steaming, it is apparently still sufficiently effective.

Steam curing has the following effects on precast products as compared with field water cure.

(a) Shrinkage may be reduced, depending on presence of gypsum in cement.

(b) Creep may be reduced if prestress is released into concrete when concrete strength is greater.

(c) Relationship between compressive strengths at various ages are shown in the following chart. (Fig. 2)

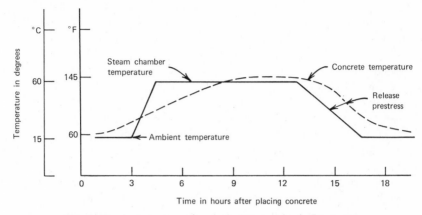

Fig. 1. Typical steam curing cycle (Atmospheric Pressure).

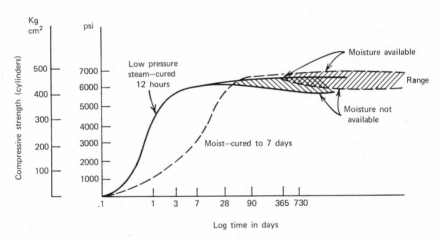

Fig. 2. Typical relationships between strengths at various ages and type of cure.

(*d*) There is some indication that the rapid rise in compressive strength by steam curing is not matched by the rise in tensile strength. At ages of a few days steam-cured concrete may not have the tensile strength that would be expected of a water-cured concrete of the same compressive strength, and thus may be more susceptive to drying shrinkage cracking. Hence the employment of supplemental water cure will improve the tensile strength of steam-cured concrete.

(*e*) With proper cycling and supplemental water cure, steam-cured concrete is approximately as durable as fog-chamber concrete. It is usually better than that of field water-cured concrete because of the difficulty or inability to truly maintain all surfaces in a moist condition by field water curing. This applies to both sea-water durability and freeze-thaw durability, with the supplemental water cure being particularly desirable for the latter.

(*f*) When the steam cycle is commenced, the metals inside the steam chamber respond much more rapidly to the heat than does the concrete. Steel forms expand while the concrete has low tensile strength; if there are changes in cross section, or fins, etc., the forms may crack the concrete. Internal ducts of metal should be plugged, otherwise they may expand under steam heat and cause longitudinal cracks in the concrete.

Satisfactory solutions to the form problem have been obtained by careful detailing, use of transverse elastic strips (neoprene), elimination of fins, and by raising the temperature slowly.

Other methods of accelerating the initial cure through heat have been applied. The beds or forms may be heated electrically or by pumping hot oil or hot water through them. Covering the surface with black nylon will cause radiant heat to be absorbed from the sun.*

Field-placed concrete may be similarly subjected to accelerated cure. With segmental construction (See Subsection 3.1.4), it may be particularly desirable to accelerate the rate of gain of strength of concreted joints. This may be done by steam jackets placed around the joint concrete or by electrically heated forms.

Heating the concrete internally by electrical resistance methods has been proposed for concreted joints, using embedded copper wires. The potential problem of electrolytic corrosion between copper and steel prestressing tendons must be considered because this matter has not yet been fully investigated to determine methods of positively insuring against such corrosion.

Concrete may also be heated internally by steam through piping or ducts. Steam should not be allowed to come in contact with tendons as it will cause rusting.

1.1.10   Hot Weather Concrete Practice

Refer to ACI 605. Coarse aggregates should be cooled by evaporation, letting soakers run over the stockpiles. If this is insufficient, the stockpiles can be shielded from the direct sun by aluminum roofing. Blowers or vacuum can be used to intensify the evaporation.

Ice can be utilized in lieu of mixing water.

In extremely hot climates, for example, desert, it may be preferable to pour at night or in the very early hours of the morning. This will help prevent flash set.

The forms should be cooled prior to placing concrete by shielding from the direct sun by means of aluminum roofing and/or by a fog spray. After placing concrete, cover promptly with wet burlap or polyethylene prior to start of steam cure.

Since finely ground cements develop greater heat of hydration, it may be preferable to use a standard or coarse grind of Type II cement (Blaine fineness of 2700 to 3200) in very hot weather.

*For Autoclaving, see Postscript, section iii.

1.1.11   Cold Weather Concreting Practice

Refer to ACI 306. Aggregates may be pre-heated by steam. The stockpiles should be kept free from ice or snow by providing coverage and draining.

The mixing water may be preheated to 140° F (60° C).

After placing and finishing, cover the concrete promptly. Place steam covers and apply low-pressure, low-temperature steam. Keep inside temperature about 65° F (20° C) until time to start the temperature rise in the steaming cycle.

After steam curing, protect concrete with blankets or tarpaulins or insulated forms, and bleed steam into them so as to keep them warm and moist.

Structural lightweight aggregates should not be fully saturated prior to mixing when they are to be used in winter concrete. Preheated lightweight aggregates hold the heat better for winter concreting than sand-and-gravel aggregates.

Type III cement and finely ground Type II cement develop more heat of hydration and thus may be preferable in cold weather.

1.1.12   Testing

Most design specifications are very strict as to the required cylinder strength at the time of release of prestress, and at 28 days. In addition, there may be strength requirements for removal from forms, for cessation of supplemental water curing, for transportation, and for erection or driving. It is important that attention be paid to the techniques of testing to eliminate radical and erroneous results.

Poor testing procedures usually result in lower indicated strengths than the true values.

Test cylinders should be produced from concrete samples taken during the middle of any flow (as down a chute), so as to eliminate the effect of segregation. They should then be well-rodded, or vibrated, using a small pencil vibrator, not a large vibrator. Cover cylinder tightly with a polyethylene bag to prevent surface evaporation. Burlap is not satisfactory; it acts as a wick to draw the water out. The cylinder should then be placed under the steam hoods or in the curing chamber at about the same height as the members themselves, so as to receive the same heat cycle. On removal from the steam hoods, the cylinder should be carefully transported (not bounced around in a truck) and immediately placed in a fog chamber at 73° F or in a tank of warm water at 73° F (23° C). It should be continuously stored until the day before the test, for example, 27 days in the case of a 28-day test.

Add lime (in a porous sack) to the water in the tank to prevent leaching of lime from the cylinder. The tank should have thermostat control to maintain temperature accurately.

On removal of cylinders, allow them to dry for one day and to cool. If testing a cylinder taken directly from steam hoods in order to determine strength for release, allow it to cool before testing.

Prepare a cylinder for testing as follows:

(*a*) Rub ends with a stone to remove any projections, as from date markings.

(*b*) Use new capping compound for highest results; keep cap as thin as possible. If capping job is not perfect, scrape it off and start again.

(*c*) Test at the specified rate of loading.

Warning: *With high-strength concretes, fracture may be explosive. Be sure to provide a wire screen protector.*

When testing cubes cut from a prestressed concrete member, cut cube between strands, polish ends and sides with a stone so they are perfectly plain and square. Throw away imperfect cubes. Test without capping.

1.2 Prestressing Tendons

1.2.1 General

The most prevalent means for inducing a compressive stress into concrete is by stretching a tendon and anchoring it to the concrete. This tendon may be located inside the concrete cross section, either directly embedded, or in a duct. Alternatively, it may be located external to the concrete cross section, either immediately adjacent, or at some finite distance away from it (e.g., a tensioned tie of a tied arch).

The tendon must have a very high tensile strength and an ability to sustain indefinitely a high state of stress, with little loss due to relaxation, corrosion, or fatigue. Cold-drawn steel wire and alloy steel wire and bars have these attributes and are the most common materials for tendons.

However, there are other materials (metals such as boron filaments and non-metals, such as fiberglass) which have potentially even better strength properties. Their practical use depends on technical aspects, such as chemical stability, and the economic aspects of cost of material and installation.

Cold-drawn steel wire is produced in diameters up to 0.276 inches (7 mm.). It has strengths ranging from 220,000 to 300,000 psi (17,000 to 21,000 kg/cm^2) and moduli of elasticity about 29,000,000 to 30,000,000 psi (2,000,000 to 2,100,000 kg/cm^2). However, the bond between smooth wires and concrete is low. This requires the use of large numbers of small wires in pretensioning, or a means of mechanical anchorage for post-tensioning.

Hot-rolled alloy-steel wires are extensively employed in Europe, especially in Germany. They are frequently rolled with a flattened or oval cross section, and may have a pattern or "profile" of indents rolled onto the section. This greatly improves the bond in pretensioning, enabling fewer larger wires to be employed.

By weaving several wires into a strand, a tendon of substantial capacity can be produced with excellent bonding properties. There is a slight reduction in modulus of elasticity (5 to 10%). Seven-wire strand is widely used in pretensioning and, to a lesser extent, in post-tensioning.* Three-wire strands are used in some lightly stressed, mass-produced pretensioned floor slabs. Nineteen-wire strands, similar to high-strength wire rope, are also available.

The wires used in stranding are usually round; thus there are interstices between the wires of a strand. A recent development has been to produce shaped wires in which the wires completely fill the cross section. This concentrates more force within a given cross-sectional area and is particularly valuable for highly stressed members.

Alloy-steel bars are extensively used for post-tensioning tendons. The bars generally have ultimate strengths ranging from 140,000 to 165,000 psi (10,000 to 12,000 kg/cm^2). End anchorages and splices are facilitated by either rolled threads or wedge-type grips. Bar diameters generally range up to 1-1/4 inch (3cm). They may be smooth or have corrugations rolled on.† While alloy-steel bars are less efficient in themselves than wire, they offer practical advantages in many installations.

By stress-relieving, the internal stresses of cold-drawn steel are eliminated and the elastic properties of the tendon are rendered more uniform. Initially the relaxation or creep is reduced; however, the ultimate relaxation may approach that of non-stress relieved steel.

Recently, cold-drawn steel strip has been produced for use in external winding applications. This strip has approximately the same properties as cold-drawn wire.

1.2.2 Stress-Relaxation

Tensioned steel tendons undergo a stress loss due to plastic flow. For tendons in common use, stress relaxation losses of 6 to 13% are to be expected.

In a process known as "stabilization" the steel wire is given a treatment combining temperature and stressing, which apparently reduces the long-term stress relaxation substantially.

Overstressing a tendon (to 75 or 80% ultimate) during the stressing operation and holding it for one or two minutes at normal temperatures (say 70° F=20°C) will induce performance similar to stress relieving.

Some European authorities prefer to limit the initial stress to 50% of ultimate, which practically eliminates stress relaxation. Others take the opposite approach, with initial stressing to 80%. Normal USA practice for initial stress is 70%.

*In nominal diameters from 1/4 inch to 0.7 inch (6.2mm. to 17.8mm.). See Postscript, section i.
†See Postscript, section ii.

1.2.3 Ductility

Ductility is an essential requirement for tendons in order to prevent brittle failure during installation and during service. Ductility is usually measured by bend tests and by elongation tests. Elongation tests must be conducted on substantial lengths of tendon to overcome the influence of the necked-down area. Most specifications require an elongation of 2% or more.

1.2.4 Corrosion

It is essential that tendons be protected from substantial corrosion. Corrosion may affect the ductility of the tendons or may simply reduce the cross section of tendon and thus reduce both the prestress and the ultimate strength. Corrosion may also reduce the fatigue strength.

Hydrogen embrittlement, while extremely serious, appears to be rare. Stress corrosion is likewise of rare occurrence. Atmospheric corrosion prior to installation is common but, fortunately, is rarely serious. Salt cell corrosion, and its related electrochemical types of corrosion (chlorides, etc.), are often a serious matter, and are primarily related to the concrete protection over the tendon.

All these forms of corrosion and means of protecting against them are discussed in detail in Chapter 4, Durability and Corrosion Protection.

The intent is to place a tendon in service and maintain it in service in substantially the same condition as when it left the manufacturer. This period extends during transport, storage at the site, handling, installing, stressing, and concreting or grouting. The following "rules" are recommended to give the best assurance of attaining the objective. Not all are currently followed as standard practice, but this does not necessarily mean that excessive corrosion will result. Corrosion is a matter of probability—by following the listed rules, the probability of corrosion will be reduced to a minimum:

Shipment. Tendons should be wrapped with waterproof paper, with vapor-phase-inhibiting crystals (VPI) inside. They should not be coiled very tightly; coils of very small radius produce micro-cracks in the steel surface, leading to stress corrosion.

Storage. Storage should be preferably enclosed, and heated to maintain a relative humidity of less than 20%. Never store in mud. Avoid open storage near refineries and industrial plants, especially those burning coal or oil and thus emitting sulfides into the air.

Handling. Tendons should not be dragged across the ground. They should not be subjected to sharp bends.

Installation. In pretensioning, strands should not be exposed more than 24 hours to normal atmosphere. In the vicinity of refineries or other plants emitting even minor amounts of sulfide gases ($H_2 S$), expose tendons in a stressed condition for the minimum time practicable.

For post-tensioning, be sure ducts are free of water by draining and by blowing out with compressed air. Avoid abrasion at point of entry into duct by providing, if necessary, a funnel entry piece. After installation, proceed with stressing within 48 hours, and grouting within 24 hours thereafter.

Where stressing is not possible within 48 hours, tendons should be thoroughly dusted with VPI powder during threading. (VPI powder is positively effective for only about 6 inches, so dusting must be thorough.) Seal ends of ducts with vapor-tight plastic covers (or tape).

Where stage stressing is required, it is better to stress some tendons to full value, and grout, leaving the remainder in an unstressed or low-stressed condition. These latter, of course, should be protected as noted above.

Concreting of Pretensioned Tendons. Select a plastic mix, rich in cement, low water/cement ratio, and consolidate thoroughly by vibration. Electro-chemical corrosion appears to require an oxygen-gradient (most usually produced by voids along the tendon) and proceeds most rapidly with permeable concrete.

Grouting of Post-Tensioned Tendons. Grout injection must be as complete as possible, with a mortar that is rich in cement and free from voids. Specific instructions are given in Chapter 5.

1.2.5   Fatigue

Since prestressing tendons undergo only a very small range of stress change while the live load is varied from zero to maximum, fatigue is generally not a problem with commonly used prestressing steels. Extensive tests by the American Association of Railroads and others have verified this.

However, there may be some reduction in bond between the tendon and the concrete due to cyclic loading. This will increase the transfer length and may be a problem with very short pretensioned members, such as railroad ties and overhanging cantilevers.

1.2.6   Bond

Bond (and transfer length) is a complex mechanism involving shrinkage of the concrete around the tendon, swelling of the tendon due to Poisson's effect, mechanical interlocking of deformations or strand twists, and adhesion. The adhesive effect is not yet completely understood but is, at least, highly dependent on minute variations in the surface. A highly polished surface will give much reduced bond. Similarly, any drawing lubricant left on the wires will reduce bond.

Lateral binding, such as spirals or stirrups, increases the bond.

The wire and strands generally available have a surface which is satisfactory but, from time to time, changes in manufacturing processes may give trouble in this regard. Once again, this is usually only critical in very short members and short cantilevered overhangs.

A slight coat of surface rust will improve bond significantly, but this is normally not to be encouraged. Attempts to purposely attain a rusted condition will inherently be non-uniform, leading to danger of pitting corrosion in some zones.

Bond tends to decrease somewhat with time or with cyclic loading. Any dirt, oil, or grease causes a significant decrease in bond.

Decreased bond (increased transfer length) may actually be of benefit in girders, piles, etc., by reducing the end-zone tensile stresses. Means of intentionally obtaining reduced bond are the following:

(a) Enclosure in a plastic sheath.

(b) Coating with wax or bitumastic and wrapping with paper.

(c) Greasing. A proprietary substance known as "Lubabond" is often used for this purpose.

1.2.7 Very Low Temperature

At very low temperatures prestressing steels in common use, such as cold-drawn wire, possess fully adequate properties. Strength and modulus of elasticity increase, ductility decreases.

1.2.8 Elevated Temperatures

As the temperature rises, the rate of stress relaxation increases substantially. Ductility increases. Above a critical temperature of 750 to 900° F (400 to 500° C) there is a substantial loss of stress.

1.2.9 Fiberglass Tendons

A considerable amount of research has been expended in the United States, Great Britain, and the USSR in an attempt to find practicable means of employing fiberglass tendons. Maintenance of chemical stability is the major problem. At present, this seems to be best provided by encasement of the fiberglass filaments in an inert matrix, thus making up a rod. The second problem is that of finding a practicable means of anchoage.

Fiberglass offers potential economies along with very high strength. However, it will require much development before practicable solutions are available for use.

1.3 Anchorages

Anchorages are mechanical devices used to transmit the tendon force to the concrete structure. They include the means of gripping and securing the tendon and the bearing plate or reinforced cone or other means by which the concrete reacts against the tendon's force.

Practically all anchorages are part of proprietary systems for post-tensioning. For competitive reasons, most European contractors found it necessary to have a system of their own, which explains the great number of devices available. Certain of the systems have now emerged as superior, usually because of more intensive and thorough engineering and testing.

Anchorages may grip the tendon by means of mechanical wedges. These may have serrations that dig into the tendon to grip it, or may be smooth, with friction furnishing the necessary grip.

Anchorages may also employ the wedging action of a swaged fitting, with molten zinc or cement mortar gripping the tendon through a combination of friction and adhesive bond.

Another group of anchorages depends on upset enlargements that bear directly on bearing plates. Buttonheads may be cold-formed on the ends of wires. Deformations may be hot rolled on bars. Upset threads may be rolled on bars.

Because of the tremendous forces involved, it is essential that the anchorage perform properly and safely. A failure may produce a serious or even fatal accident. Therefore only thoroughly developed and tested systems should be employed, and these should exactly follow the manufacturer's recommendations. Only anchorages approved by the responsible engineer should be used; serious slippages have occurred when minor changes in specifications, metal composition, or tolerances have been made.

Anchorages and the tools must be kept clean and free from corrosion during storage.

After installation, anchorages should be properly protected from corrosion or fire in accordance with the design engineer's details.

It is important that the anchorages are aligned axially with the tendons. Most anchorage systems are equipped with special chairs and jacks to accomplish this.

In some systems tendons may also be anchored at one end by looping around steel or concrete, or by simple embedment in concrete for a substantial length. These are the so-called "dead-end" anchorages and, with them, specified bending radii must be closely adhered to in order to prevent failure or rupture during tensioning.

After jacking the tendon, the anchorage must be anchored. This usually involves a small inward movement or "set" of the tendon, losing a slight amount of stress. This is usually of little importance, except on very short tendons such as those used for connections; on these, "set" may be serious. Certain types of anchorages provide for shimming or locking to overcome "set."

Before grouting, stressed tendon anchorages are under very high stress and, therefore, are vulnerable to accidental blows, accidental heat from a welding rod, etc. As such, they can become deadly missiles, shooting off with all the energy stored in the tendon.

1.3.1 Precautions for Anchorages

Use only anchorages of the selected system, in strict accordance with manufacturer's recommendations.

Store so as to keep clean and dry.

Use manufacturer's recommended practice and tools for installation.

Take necessary steps to insure axial alignment.

Protect workmen during tensioning; protect anchorage from accidental blows, heat, and electric arcing (as from welding).

Take special pains in applying protective cover (to protect against corrosion or fire) to anchorage. This is a highly stressed critical mechanism and deserves full protection.

1.4 Reinforcing Steel

This includes all unstressed steel, whether in the form of unstressed strands, mild-steel bars, alloy-steel bars, etc.

In a prestressed member the unstressed reinforcement usually reinforces against secondary stresses and shear stresses. It may also be used as additional primary steel to give greater ultimate capacity or to control behavior.

As a general statement, prestressed members will behave as intended only if the reinforcing steel is properly detailed and placed. Adequate reinforcing will serve to confine the member and, in effect, force it to function the way it was designed. Thus reinforcing makes up for much of the unknowns in design (e.g., shear, torsion) and in construction. Its proper installation, positioning, and securing becomes of fundamental importance.

Because, in prestressed members, the reinforcement frequently consists of closely- spaced small bars, and because the dry concrete mixes used require heavy vibration, the reinforcement is often dislocated by the vibrators. This can be prevented by proper tying.

Tack-welded cages have been extensively used in precast members, but the location of welds must be carefully detailed since a reduction in strength may be induced.

To prevent destructive corrosion, the specified cover must be maintained over the reinforcement. Plastic chairs and concrete "dobe" blocks should be used at sufficiently close spacing.

1.5 Ducts

"Ducts" is the term used to describe the conduit through which the post-tensioning tendons pass. In its broadest sense, the term "duct" may include a formed void, or a conduit of any material.

For reasons of economy, practicability and electro-chemical compatibility, the majority of ducts are of bright steel.

Flexible metal ducts are articulated, with non-watertight joints. They can be formed of very thin metal, easily shipped in coils, and are draped in the forms. However, their very flexibility is their greatest drawback; they have excessive "wobble" (erratic misalignment), thus requiring very frequent support by tying or other means. One of the "other" means is to insert a length of rigid electrical conduit in the flexible metal sleeve; during concreting this will hold the duct in an acceptable alignment and, after concreting, it is easily withdrawn.

Flexible metal conduit, being usually very thin, is subject to mechanical damage and to rusting. It must be protected by boxing during shipping and stored in a dry place till actual use.

Because it is non-watertight, blockage sometimes occurs when grouting an adjoining duct. When several ducts are in a vertical line, one above the other, and touching (e.g., at the point of maximum negative moment over a support), the act of stressing tendons may cut through or otherwise damage the conduit.

Rigid metal conduit, although more expensive (because of thicker gauge) and more difficult to ship and handle, is gaining popularity. It requires little tying or support, has a low friction factor (because of low "wobble") and is watertight (provided that joints are sealed).

To protect flexible and rigid conduit from corrosion and to reduce friction during stressing, coatings are sometimes employed. Galvanized coating has been widely and successfully used. A few warnings have been raised about the possibility of free hydrogen being liberated from the reaction of cement and zinc; however, such hydrogen embrittlement of tendons has not been duplicated in laboratory tests and the reported field instances are extremely scattered and not completely convincing.

At present in the United States, galvanizing is generally considered a beneficial means of reducing friction and providing corrosion protection. In France and Germany, current regulations warn against use of galvanized ducts with black wire tendons.

One alternative may be considered that will eliminate this doubt; lead coating is now being applied to some ducts in Europe and lead cannot generate hydrogen.

Plastic conduit has been used from time to time; it can be formulated so as to be inert in concrete. It has the proper rigidity and is easily joined (or spliced) in the field. However, with curved tendons, the tendons bite into the walls of the plastic conduit, and friction factors may be excessive. With low curvature or straight tendons, plastic conduit may be satisfactory.

Bond values (plastic to concrete and grout to plastic) are generally very low; thus there may be insufficient adhesion for fully bonded behavior. This places increased responsibility on the long-term behavior of the anchorage. The heat resistance of plastic is generally substantially less than that of the tendons so, in a fire, damage may occur to the sheath. In the course of many actual fire tests, this was not found to be of significance, nor did it lower the fire (heat) resistance of the member.

Where voids are formed in the concrete by removable ducts or forms, abrasion may take place while inserting and tensioning the tendon. Friction losses may be very high. If grease or a similar substance was used to permit removal of the duct form, bond may be reduced. Therefore formed voids are generally not used except for very short, straight ducts, as for connecting tendons.

Splices in ducts should be suitable to the material and the need for water-tightness. Wrapping with waterproof tape has proven very effective. Sleeves, plus taped joints, are useful in segmental construction. Plastic sleeves have similarly been used, but have been knocked askew during concreting.

For the highest integrity, as in pressure vessels or long-span bridges, screwed joints have been used. Care must be taken to prevent a kink in the duct profile due to the extra rigidity at the sleeve. Welded and brazed joints have also been used. With welding, care must be taken not to accidentally "burn" an adjacent tendon through grounding, and the welding technique must be suitable to the wall thickness of the conduit. Welding or brazing should never be used to splice a duct when the tendon is in that duct.

1.6 Embedded Fittings

Embedments are, of course, common to all concrete construction. With prestressed construction, certain precautions are necessary.

Volume Change. Prestressed members shorten in both elastic and plastic movement. This may change the dimensional relationship of critical embedments.

The high-strength, low water/cement ratio concrete generally requires more vigorous compaction from vibrators, etc., which may disturb inserts.

Steam curing with metal forms, produces an expansion of the form before that of the concrete, while the concrete is still weak. It may cause a loosening of inserts.

Corrosion. Electro-chemical compatibility must be assured where an electrolyte (eg. wet concrete) can electrically connect insert and prestressing tendon.

For this reason, copper and aluminum inserts are very suspect and can be used only where special insulating or other protective steps are taken.

Concern has often been voiced over the possibility of corrosion proceeding up an insert, particularly an insert of strand or steel bar, to a tendon. This fear is not borne out by experience, atmospheric corrosion generally being limited to 1/8 to 1/2 inch in depth. However, if the concrete around an insert is shattered or cracked in service, then corrosion will be accelerated.

Rust staining from inserts (and exposed strand ends) can be detrimental even in engineering structures. An epoxy sealant will prevent this. Alternatively, the inserts may be galvanized or cadmium-plated.

In such case it would probably be prudent to keep the galvanized portion separated from direct contact with the tendons (see remarks on galvanized ducts in preceding section).

Where high prestress crosses an opening, the hole should be formed in such a manner that the form for the hole will not be locked in by the distortion due to prestress.

Generally, tendons can just be spread around penetrations or openings or inserts without adverse effect on the stress pattern. As in all prestress work it is the concrete that is under stress; a plasto-elastic flow net is automatically formed, regardless of the exact location of the tendon.

1.7   Bearings

Prestressed concrete girders and beams are subject to considerable volume change in service. They continue to shorten under the influence of creep. Moisture differentials between top and bottom flanges may cause differential shrinkage. Temperature differentials may cause increase in camber. Service loads cause changes in camber (deflection), with accompanying rotation of the ends and change in length.

Although all these phenomena are to some extent true of conventional reinforced concrete, they are aggravated in prestressed concrete because of the generally thinner sections and the effect of continuously maintained prestress.

Bearings must allow for longitudinal movement, and rotation, while maintaining adequate vertical support.

Steel plate bearings, even stainless, are likely to "freeze" in service and become inoperative, due to minor corrosion. Lead plates have been used but are subject to plastic deformation in themselves under continued service.

Neoprene bearings are by far the most widely used type. They are now readily available with suitable durometer hardness, and of varying thicknesses to accommodate the anticipated total movement. For larger bearings a sandwich of neoprene and steel plates is frequently used. Neoprene permits rotation as well as movement.

Teflon provides an almost frictionless surface, hard and durable. Teflon bearings are being increasingly used on long-span girders and bridges.

Movement may also be provided for be designing appropriate flexibility into the columns or supports. Where dowels are used to secure the ends of girders, it is recommended that one end have an oversize hole and that it be filled with bitumastic, not grout.

Although it is the designer's responsibility to ensure that the connections and bearings provide for adequate movement, nevertheless, the construction engineer must be aware of the dangers of too rigid connections and the factors causing and influencing movement. When, for example, both ends of prestressed girders are welded to fixed plates, or embedded in cast-in-place concrete, distress will show up as cracks and spalling at the girder ends and in the seat. Usually the cast-in-place concrete seat shows the earliest and most severe distress. The constructor is usually charged with faulty construction practice and it is a long and costly matter to attempt to assign the blame to the design, even when poor design is clearly the cause. Probably more dissatisfaction has arisen over this matter of movement than any other aspect of prestressed concrete construction.

Therefore it is recommended that, as a matter of practice, careful check be made by the construction engineer of the connection and support details to be sure that there is adequate provision for movement of girders and beams. If provision appears inadequate, he should then undertake appropriate discussion with the design engineer.

2

Prestressing

A state of "prestress" or "pre-compression" may be introduced into concrete by internal or external means. External means include super-imposed weights and external jacking or thrusting. Internal means include pretensioning and post-tensioning. Composite means are those where internal expansion is resisted by external abutments or internal anchorages.

2.1 Pretensioning

Tendons are stressed and anchored against external abutments. The concrete is poured and cured so as to have substantial strength in compression and bond. The tendons are then released from the abutments and the "prestress" is transferred into the concrete member.

Prestress can be transferred only by elastic shortening of the concrete. Until the concrete has actually shortened, it is not prestressed.

In manufacture by the pretensioned method, the frictional weight of the concrete member, or binding in forms, may prevent longitudinal shortening until the member is lifted. During this period, shrinkage or thermal cracking may occur, unresisted by prestress, since the member has not yet shortened. Similarly, in picking from the forms, the initial lifting force may cause cracks, particularly at the point of pick, since the member has not yet shortened and become prestressed.

Pretensioning requires adequate transfer length to transfer the stress from the tendon to the concrete. Tendons for pretensioning are, therefore, designed to make the transfer in as short a distance as possible. Smooth wire is used only in smaller sizes, usually less than 3 mm (1/8 inch). Above this size, strands of 3, 7, or 19 wires are employed. The 7-wire strand is now almost universally applied in pretensioned manufacture in the United States, in diameters of 3/8 inch to 1/2 inch. Strand with a diameter of 0.6

inch has been extensively used in England and is recently available in the United States.

In Europe, deformed, "profiled" wires are used in 5-mm and 8-mm sizes. Cross sections are frequently oval, and the deformations provide mechanical bond at frequent intervals. Careful design of these ribs or protuberances has eliminated notch sensitivity and fatigue susceptibility.

Recent tests have indicated that the "adhesive" influence plays a very important role in transfer bond, and that this adhesion is largely determined by surface characteristics of the wire. Slightly rusted wire or strand has several times better bond than bright wire; however, the degree of rust is difficult to control and the dangers of pitting corrosion generally rule against intentional rusting.

However, recognition of the role of adhesion has led manufacturers to adope wire-drawing practices that are compatible with reasonable transfer bond characteristics.

Short transfer length is desirable when high moment must be resisted a short distance from the end; for example, cantilevered beams, railroad ties, and columns at rigid connections, for which, special efforts must be made to obtain the shortest transfer length. Such steps include:

1. Thorough consolidation of concrete at ends.
2. Higher compressive strength of concrete at release.
3. Reduce stress level in tendons by increasing steel cross-sectional area.
4. Increase surface area of tendons by using smaller wires or strands.
5. Gradual release, that is, by hydraulic de-jacking.
6. Adequate lateral binding, by spiral or stirrups, at ends of members.

On the other hand, for girders, piles, beams, and slabs—in fact, the great majority of all pretensioned concrete—a relatively long transfer length is desirable. The short transfer length causes high transverse tensile stresses in the end block and may even lead to cracks between strand groups, etc., in the end of the girder. With a slightly longer transfer length, these internal stresses are distributed over a longer zone.

To intentionally increase transfer length, the ends of tendons may be sheathed in plastic or coated with a grease. One such product, Lubabond, is especially designed to minimize early bond but not ultimate bond.

Release by Burning. When a longer transfer length is permissible, the tension may be released into the member by burning the strands. This burning should be performed for each strand by heating with a low-temperature flame (low oxygen) until the strand stretches, then by burning through. A careful sequence should be followed so as to make the release of prestress equalized over the cross section of the member. If the member is free to shorten, the last tendons may break before burning. Therefore this writer prefers to heat all the strands at one end with a low oxygen (yellow) flame before cutting any. "Shock" release automatically destroys the bond over a substantial distance (3 or 4 feet).

Because of the uncertainties associated with release by burning, most authorities now prefer or require gradual release, and the practice of release by burning is diminishing.

Pretensioned tendons, by the nature of their being stressed before concreting, must be straight between points of support. On the other hand, for many girders, etc., the specified tendon profile is a parabolic curve.

To achieve an approximation of such a curve, the strand may be stressed up over chairs and down under "hold-downs." These hold-downs can be attached to the soffit, or thrust down from above.

During stressing, a considerable longitudinal movement and force is developed. Later, on release, the longitudinal prestress must be transferred into the concrete before the hold-downs are released.

Provision, therefore, must be made for small movement and angular rotation of the hold-down devices.

The most accurate production method is that in which the strands are stressed to a pre-computed value in a straight profile, then pulled down (or up) to match the desired parabolic chord profile. This movement, of course, increases the stress in the tendons.

Provided calculations and practice are accurate, the final stress will equal that desired. This can be checked and adjusted by re-jacking the tendons longitudinally.

Some plants use the alternative method of stressing the strands in their deflected profile. Rollers or shoes are provided to minimize frictional loss. Variations of about 5% have been measured on strands deflected by this method. It should not be used for sharp profiles.

Since the vertical forces at hold-downs are high, it may be more economical and satisfactory to increase the number of hold-down points, thus distributing the vertical reaction along the bed.

Deflection of strands is a critical construction procedure. Should a hold-down fail, the stored energy in the strands is like a bow-string and the hold-down becomes a missile. Positive means must be taken so that this work can be done from the side; thus a worker does not have to expose any part of his body above the hold-down.

Remote hydraulic control is a preferred method for pulling down tensioned strands.

Again during concreting, a vibrator may hit a hold-down device. Hold-down positions should be carefully marked so the vibrator man does not expose himself. A protective device (steel bars) may be slipped in place above the point and fastened to the forms.

With deflected tendons, if the girder does not have its longitudinal prestress transferred into it, the release of the hold-downs will cause the girder to crack. The frictional force of the girder dead weight, plus the hold-down forces themselves, may prevent movement and transfer of prestress. Therefore longitudinal prestress should first be released, then the side forms broken clear to prevent binding. The hold-down devices should be fastened in such a way as to permit the necessary longitudinal movement and angular rotation due to shortening of the girder under release of prestress.

2.2 Post-Tensioning

Post-tensioning can be utilized for either precast or cast-in-place construction. The tendons are normally inserted after the concrete has hardened and cured. If inserted before casting, the possibility always exists for a grout leak into the conduit and a freezing there of the tendon. Corrosion is also an adverse factor where the tendon is placed before concreting, particularly when steam curing is employed. So the general rule is that the duct shall be formed, the concrete cast and hardened. Then the duct should be flushed with fresh water and blown out with compressed air. The tendon is then inserted, anchorages affixed, and the tendon stressed against the concrete.

When multiple tendons are involved, a balanced sequence of post-tensioning should be selected to prevent eccentric over-stress or cracking. Tendons may be required to be stressed in stages so as to compensate for losses due to elastic shortening and to prevent temporary over-stress. This stage stressing is quite adaptable to some anchorage systems, less adaptable to others. It is normally required only with tendons of very great force, especially where they are eccentric to the cross section.

While inserting the tendon, dusting on of a vapor-phase inhibiting powder (VPI), followed by sealing of the ends of the duct, will prevent corrosion. The minimum practicable time should ensue between stressing of tendons and grouting; many specifications limit this to 48 hours, but longer periods have been satisfactory when special precaustions (such as VPI) are employed.

Accurate alignment of the anchorages normal to the tendon is essential. The bearing plates must be accurately aligned and have complete bearing on the concrete beneath. The fixing of these anchor plates in the end forms so as to prevent dislocation during concreting is extremely important.

With some large, heavily stressed girders, the end blocks have been separately precast, in a horizontal position, with bearing plates, etc.; then these precast end blocks are set in the main girder forms and the ducts connected, the girder concreted, etc. This procedure assures highest quality concrete under the bearing plates and accuracy of their alignment.

Friction during post-tensioning can be verified by stressing first from one end, then from the other. Excessive friction can sometimes be minimized by moving the tendon at the start of the jacking operations; a few feet one direction, then back. Stressing from both ends is recommended for all double-curved tendons and for all single-curved tendons over 80 feet in length, and wherever sharp bends are involved. Where this is impracticable, calculations and tests must be made to determine friction losses and adequate compensation be provided.

Excessive friction can also be overcome by use of a water-soluble lubricating grease. This must, however, be thoroughly flushed from the duct after stressing. It should be employed only where the more conventional steps are inadequate.

With multiple tendons in a single duct that is curved in two planes, binding of one tendon on the other may occur. The effects of this can be minimized by making up parallel lay cables on the bench, and binding them with soft iron wire before insertion into the ducts.

In stressing, the theoretical and actual elongations should be checked against the jack gauge readings and calculated friction losses. Discrepancies of over 5% should be carefully investigated, the cause determined, and necessary corrective action instituted.

In anchoring tendons, there may be a seating loss or "set." With long tendons, this set (usually 1/8 inch) is of no importance, but with short tendons this loss may be serious. In the latter case the anchor may again be stressed against the concrete and shims inserted. Some types of anchorages are designed so as to automatically eliminate "set."

Stage stressing may be employed to keep stresses at proper levels during all stages of erection and superstructure construction. In this process, some tendons are stressed on erection, while others are left until a later stage of erection or until part or all of the deck is poured. Tendons which are to be stressed at different stages should be in separate ducts in order to minimize problems of friction and corrosion protection. In some instances a tendon may be partially stressed in one stage, then restressed to a higher level at a second stage. In such a case the anchorages shall be designed to facilitate this retensioning without damage to the tendon.

Warning: *A stressed and ungrouted tendon is very susceptible to corrosion.*

When multiple ducts are spaced closely on top of one another and the radius of curvature is small (as in negative moment areas), high tensile bursting stresses may occur during stressing. It is desirable therefore to have 2″ of cover over the duct and to stress and grout the lower ducts first. Confinement with mild steel helps, as does spacing the ducts as far apart vertically as practicable.

2.3   External Post-Tensioning, Wire Wound Systems

Post-tensioning tendons may also be external to the concrete cross section during the time of stressing. The most prevalent example is that of wire-wound concrete pipe and tanks. The concrete core is wrapped with wire under tension, then encased in a concrete mortar. The same principle has been applied to beams and other structural members, particularly ring beams and ring girders of domed roofs.

In the USSR, post-tensioning tendons are sometimes positioned in grooves in the sides of girders; the grooves are later filled with mortar. The use of grooves reduces the friction during stressing (as compared with ducts) and makes it easier to ensure full compaction of the mortar encasement.

Shrinkage, however, and lack of bond between mortar and concrete may present a durability problem. For this reason, application of a rich cement-sand flash coat over the wires, followed by a mortar application, is generally recommended.

The application of mortar is usually by rotating brush, in the case of pipe, and by pneumatic means (shotcrete, gunite) in the case of tanks. The difficulty with shotcrete is that varying degrees of porosity may occur, particularly if the operator allows rebound to be trapped. With a salt-fog or salt-spray atmosphere, electrolytic salt cells may form, and corrosion ensue. Therefore care must be taken to ensure the density of the coat and its water impermeability.

Use of a poured concrete coat, in lieu of shotcrete, may occasionally prove economically feasible and technically superior.

Placing of the wires under tension is generally accomplished by drawing through a die (the Preload method), by use of a differential-winding system, or by a braked drum system such as the Russian "turntable" and "DN-7" methods. These procedures are rapid and are probably the most economical and mechanized means for rapid placement of tensioning force.

The mortar cover for external tendons is not prestressed and therefore is theoretically subject to shrinkage cracks and tensile cracks in service. Actually some prestress does enter these coatings through the creep of the main concrete member. Favorable behavior of thin mortar coats is also assured by the continuous bond with the main concrete. When thick or large coatings are involved, consideration must be given to provision of mesh reinforcement or of other special protective means; for example, external tendons for large bridge girders may be enclosed in a watertight steel sheath, which is then pumped full of grout in the conventional manner.

For further discussion on external post-tensioning see "Unbonded External Post-Tensioning," in the following section, and Chapter 15, "Prestressed Concrete Tanks."

2.4   Unbonded Post-Tensioning

The tendons may be pre-wrapped in plastic or fiberglass sheaths, after coating with an appropriate bitumastic compound. The sheathed tendon is then placed in the forms, the concrete poured and hardened, and the tendon stressed. This is very adaptable to thin flat slabs, where ducts would take up an appreciable portion of the cross section. It facilitates the draping of the tendon into an up-and-down profile characteristic of flat slab moments, and thus is very economical. Many of the difficulties that have arisen with such systems are a result of this very ease of installation; it has led to carelessness. It is particularly important to insure adequate cover, through chairs or dobe blocks, and to insure thorough consolidation of the concrete under the anchorage bearing. The placement and support of the anchorage must be accurate and rigid enough to prevent displacement.

Obviously, the integrity of the sheath must be assured. Repairs, with bitumastic and fiberglass tape, must be completed before concreting. Excessive wobble or sharp bends must be eliminated.

There has been, and apparently will continue to be, a controversy over the merits and use of unbonded tendons. There are certain differences in behavior under static and dynamic loads which must be considered by the designer. However, the main concern of the construction engineer utilizing unbonded prestressed tendons must be to ensure their proper installation, their integrity during construction and durability thereafter, and the proper placement of the anchorages. Much of the criticism of unbonded tendons is really a criticism of careless and poor construction practice unrelated to the matter of "bonding" or "not bonding."

Unbonded tendons may also be used external to the concrete cross section as, for example, inside the box of a box girder. They transfer their vertical components of force at points of angle change through saddles, and their longitudinal force by means of cross-beams or slabs in the end block.

The saddles must, of course, be specially designed to minimize friction and to transmit the high forces generally involved. Some have been designed of stainless steel, covered with Teflon. The tendons must be protected against corrosion by encasement in grease. Frequently, both galvanizing and greasing are employed, together with a program of periodic inspection.

The suspender stays on such spans as the Lake Maracaibo Bridge, and the ties of arch roofs, etc., are really unbonded external tendons in behavior, even though they may be encased in concrete for rigidity and protection.

2.5   Flat Jacks

This generic term is used to describe all types of external jacks which prestress the member against abutments (or internal tendons). Flat jacks are thin neoprene bags which are inflated with air or water to exert a high force working over a very small distance. Several of these can be superimposed to gain greater movement. Since the prestress is an external, not an internal force, buckling is definitely a consideration and must be guarded against, particularly with thin slabs.

Use of water under pressure is the safest means of inflation (jacking). Then grout can be pumped in under high pressure to displace the water and to fix the jacking distance. A special valve is used that will permit the water to be ejected by the grout without loss of pressure.

Steel plates may be used to contain the flat jacks in batteries so that their expansion will be uniform and definite.

Flat jacks are extremely useful for de-centering of arches. By inserting them at several locations near the crown, the arch can be raised off the false-work and forced to act in true arch fashion. Flat jacks have been used to prestress pavements; they are very effective for a short time, but the losses due to creep and temperature are irrecoverable.

Flat jacks have also been used to raise extremely heavy weights, such as prestressed concrete nuclear reactor pressure vessels, from temporary supports, to balance reactions, which are then fixed by grouting.

A number of types of mechanical and hydraulic jacks have also been developed for utilization in the same manner.

3

Special Techniques

3.1 High-Strength Concrete

In the early days of prestressed concrete, it was quickly realized that high-strength concrete was essential to the technical and economical success of this new technique. The concrete must resist, during certain stages of its service life, a combined compressive stress due to prestress and dead-load stress. Higher strength produces a higher modulus of elasticity, reducing prestress loss. Higher strength often is concurrent with reduced creep and shrinkage. Therefore strengths of concrete in the range of 3500 to 5000 psi (250 to 350 kg/cm²) (cylinder) were utilized.

The advent of commerical precast pretensioned plant production required high strengths at early ages, often one day. It was quickly discovered that the techniques for producing high early strengths automatically led to concrete with 28-day strengths of 6000 to 8000 psi (400 to 500 kg/cm²). These strengths were achieved by using conventional concrete materials and practices, with emphasis on the following:

1. Use of coarse aggregate with a maximum size of 1 inch or 3/4 inch.
2. High cement factor (6.5 to 7.5 sacks/cy).
3. Low water/cement ratio (0.45).
4. Water-reducing and plasticizing admixture.
5. Thorough mixing.
6. Thorough vibration.
7. Excellent curing—low pressure steam and/or water curing.
8. In hot climates, precooling of aggregates.

Designers have been somewhat slow to realize the technical and economic advantages through the utilization of these higher strengths in design. Gradually, with experience, strengths of 6000 psi to 8000 psi are being specified for such uses as:

1. Prestressed concrete bridge girders. The higher strength permits longer spans or fewer girders per span.
2. Prestressed concrete piles and compression members. The load-carrying capacity is directly proportional to the strength.
3. Prestressed concrete railroad ties.

The modulus of elasticity of concrete increases approximately as the square root of the strength; thus deflection characteristics are improved. The tensile strength of concrete also increases approximately as the square root of the strength, thus improving behavior under overload or partial prestress conditions. Durability is generally proportional to strength in that the same factors that improve strength also improve durability by reducing porosity and permeability. An exception is air-entrainment, which reduces strength while improving freeze-thaw durability.

The actual experience with concrete strengths of 6000 to 7000 psi (400 to 500 kg/cm²) has been so satisfactory that extensive studies have been instituted of even higher-strength concrete. The Federation Internationale de la Precontrainte has had a commission investigating the techniques and evaluating the potential of concrete in strength ranges above 12,000 psi (cylinder) (1000 kg/cm² cube). The American Concrete Institute has been studying consolidation and other methods of achieving concrete strengths of 8000 to 10,000 psi. Much of what follows is taken from reports and studies of these groups.

Concrete can be reliably produced with strengths of 7000 to 9000 psi (and even 10,000 psi), using conventional materials and techniques, by adopting the following procedures:

1. Selection of aggregate for high strength.
2. Use of crushed rock instead of gravel.
3. Maximum size of coarse aggregate to be kept small, 3/8, 1/2, or 3/4 inch.
4. Gap grading, especially elimination of fines.
5. Cleanliness of aggregate, removal of silt and dust by rewashing and rescreening.
6. Cool aggregate by water soaking and evaporation, shielding from sun, etc.
7. Very rich cement factor—up to 9, 10, or 11 sacks per cubic yard.
8. Selection of cement for high strength, with attention to grind. This may often involve a compromise, since a fine grind may be desired for high early strength and a somewhat coarser grind for high late strength. There is, however, a trend to the use of reasonably fine grinds, such as 4000 Blaine.
9. Low water/cement ratios (0.3 to 0.4). To obtain a workable mix, water-reducing admixtures are necessary.

10. Heavy vibration to consolidate the harsh, dry mix that will usually result from application of previous steps.
11. Special curing, that is, continuous ponding or soaking in water, or steam curing followed by water curing.*

Careful attention to the above, using skilled and experienced personnel, has enabled top precast plant producers to consistently and reliably produce concrete to design (minimum) strengths of 8000 psi (550 kg/cm²). One of the most impressive of such projects was that for the prestressed girders for O'Hare Airport, near Chicago. On this project, the manufacturer paid particular attention to rewashing of the aggregate to remove dust and silt particles (which raise the water requirement and thus the water/cement ratio), and to the cooling of the aggregate piles by presoaking and evaporation, and to careful grading to obtain a maximum density. Another project involved the precast panels for the Fort Peck Dam, where a concrete strength of 8500 psi (cylinder) (600 kg/cm²) was specified. Actual strengths averaged 10,000 psi, with some values ranging to 12,000 psi.

Maximum aggregate size was 3/8 inch, a gap-graded mix was adopted, a water-reducing admixture employed, and the water/cement ratio was 0.35. Compaction was by intense table vibration.

Many studies, including those by the U.S. Waterways Experiment Station and Portland Cement Association, have shown that 10,000 to 11,000 psi is about the strength ceiling with conventional materials and techniques.

Currently, a 150-foot span pipe bridge in England has been designed for prestressed girders with a cube strength of 1000 kg/cm² (approx. 12,000 psi cylinder). 600 kg/cm² concrete was used on the Rotterdam Metro viaduct.

Measuring and testing techniques must be compatible with the strengths to be tested. Attention is directed to Chapter 1, Subsection 1.1.12, Testing. Correlation between test cylinders and actual concrete must be achieved; use of a testing hammer may prove valuable for such verification.

The studies of the F.I.P. Commission on High-Strength Concrete have recently been directed towards methods of achieving concrete with strengths above 1000 kg/cm² cube (12,000 psi cylinder). They have been directed toward increased cohesion, increased compaction, and triaxial stress.

1. Cohesion

For conventional concrete of 9000 to 10,000 psi, it is the adhesion between cement matrix and aggregate which fails. Use of cementitious aggregates, such as Portland cement clinker and aluminous cement clinker, would provide improved bond. In England there is commercial production of prestressed concrete beams of design strength of 12,000 psi (cube), using aluminous cement, fine aggregate, and a high-quality granite coarse aggregate. E is 9,000,000 to 10,000,000 psi (600,000 to 700,000 kg/cm²).

*Autoclaving is under development in East Germany and Japan. See Postscript, section iii.

Another approach is that of the silica-lime bond. Structures from ancient China, such as the Great Wall, have shown constant improvement in strength through the centuries. To obtain this reaction within a practicable time period, autoclave heating to 200°C or more is required. From a present-day practical viewpoint, it would seem that potential use would be limited to small-size products. One possibility, however, that is being studied is the production of high-strength segments which could be pre-stressed together to form heavily loaded and long-span beams.

Considerable research and development has been directed towards the use of epoxy-resin and polyester-resin binder. Problems are lack of fire-resist-ance, low E, and high cost. A variation is to precoat the aggregate with epoxy-resin, then incorporate it in a concrete mix having a conventional cement matrix. A major structure will soon be built in northern France to further develop this method.

2. *Compaction*

Present-day vibration frequencies compact the coarse aggregate only. Ultrasonic frequencies could be used to compact the sand, and even the cement particles. The practical problem has been to develop vibrators of such frequencies that have sufficient power. However, such vibrators do exist, and commercial availability appears entirely practicable. These will undoubtedly develop very high form pressures, necessitating more rigid forms and design of details to prevent fatigue failures of the forms. Frequen-cies of 5000 to 10,000 cps have been used in experimental work with great effectiveness in obtaining compaction and high strength. Commercially available vibrators have frequencies up to 3000 cps and rather high force.

Experimental work at Ohio State University has shown promising results when very high-frequency transponder vibrators are combined with conventional low-frequency, high-powered vibration.

Increased compaction and activation of the cement may be obtained by the electro-hydraulic impulse method. This latter, still in the development stage, consists of interrupted high-intensity electric discharges, which create a plasma sphere within the grout. Collapse of this sphere causes a break-down of the cement particles, formation of water films on the grain surface, and ionization of the grout. Gel formation is increased. The resulting grout is both stronger and more dense.

Certain chemical wetting agents (admixtures) also give promise of being effective in increasing the activation of the cement and the strength of the paste.

Revibration of the concrete mix has been found to be effective in achieving greater compaction and strength with some mixes. A retarder admixture will make revibration more practicable. Recent work in France has shown the benefit of pre-vibration of the concrete mix, followed by vibration and pressure after placement.

Compaction can also be increased by pressure. This method has been applied for many years in France for manufacture of prestressed poles. A combination of pressure and vibration appears most effective. Use of autoclaving or low-pressure steam curing appears to enhance the effects of pressure.

Pressure can be applied by the following:

(a) Spinning.

(b) Use of a strong, fully enclosed form, with an inflatable rubber tube which is expanded.

(c) In the USSR, a steel pad, with an inflatable rubber bag, is placed over a slab and locked to the soffit. Inflation, or vacuum, is employed to exert the compacting pressure.

(d) Ramming. This may be accomplished by shock dropping of the forms (as in the Schokbeton process) or by mechanical ramming. In the spinning process, a ramming ball may be used to increase compaction.

Strengths up to 20,000 psi (1400 kg/cm²) [and even to 35,000 psi (2500 kg/cm²)] have been achieved with pressures of four atmospheres combined with vibration. These are accompanied by values of E of 8,500,000 to 10,000,000 psi (600,000 to 700,000 kg/cm²) and specific gravities of 2.9 (180 pounds per cubic foot)(2900 kg/m³). Poisson's ratio was 0.28.

Finally, compaction can be improved by modification in the mix design. In some cases careful gradation has been adopted to ensure the maximum aggregate content, with the minimum requirement for cement paste. In what seems like a paradoxically opposite approach, high strengths have been achieved with mixes in which a small size of coarse aggregate (e.g., 3/8 inch) has been used with no fine aggregate whatsoever and a high cement content.

3. *Triaxial Prestress*

If binding, such as high-strength steel wire spirals, is used to restrain the concrete normal to the applied load, substantial increases in strength can be achieved [e.g., to an indicated strength of 50,000 psi (3500 kg/cm²)].

The problem is that the concrete cover over the binding spalls off long before ultimate strength is reached. Since E remains constant at the level of the concrete itself, initial strains are large. Preloading, however, will stiffen such a member, improving the apparent E.

Enclosure in a steel pipe shell has a similar effect and may be practicable in actual application (e.g., compression members, such as piles or columns). Concrete compacted inside steel cylinders has developed apparent strengths up to 100,000 psi (7000 kg/cm²).

Use of random fibers, such as finely divided wires and polypropylene and asbestos fibers, has been studied for many years.* Such research and development is still continuing. Tensile strengths are greatly improved; compressive failure is delayed. Strains remain large.

An interesting method involves the delaying of the progression of cracking by means of incorporation in the mix of 2 to 3% by weight of polystyrene "confetti" or "lenticules," 0.025 mm thick and 3 to 4 mm in diameter. These absorb no water and appear to act effectively as "crack stoppers."

4. *Polymerization*

Recent research has shown that substantial increases in strength and other properties can be achieved by impregnation of dried concrete with a monomer followed by treatment by irradiation or thermal processes. The monomers which have proven most effective are methyl methacrylate and styrene. Vinyl acetate and ethylene gas dissolved in sulfur dioxide have also been used. The concrete is usually thermally dried, and subjected to a vacuum of about 3 inches of mercury. Then the precast concrete element is soaked under a nitrogen blanket (excess pressure) of one-third atmosphere. This is followed by an irradiation treatment of Co^{60}, or thermal treatment at $75°C$, for about four hours, which converts the monomer to a polymer.

Such polymerization of conventional concrete of 5000 psi (350 kg/cm^2) improves the properties as follows:

Compressive strength, four times, to a maximum of 20,000 psi (1400 kg/cm^2)

Tensile strength, four times

Modulus of elasticity, 2 times

Freeze-thaw resistance, four times

Permeability—decreases almost to 0

Water absorption—decreases to about 1/20

Corrosion under sulfates, acids, and hot brines—decreases toward 0

Bond strength, 3 times

Creep, negative (slight expansion under continued load)

Hardness, 2 times

Such treatments appear to be economically feasible. The cost of impregnation and irradiation of 1-1/2 inch (4 cm) thick concrete pipe is estimated to run from $0.08 to $0.22 (U.S.) per square foot.

The lower values are for 1/8-inch (3 mm) penetration; full penetration increases the cost by approximately 40%.

The use of the Co^{60} irradiation source requires shielding of earth walls some 8 feet (2 meters) thick, or concrete block equivalent. Thermal treatment reduces the gain in strength and other properties by 30%.

Examination of polymerized concrete specimens shows substantially

*Concerning "Wirand," see Postscript, section vii.

increased bond between the matrix and aggregates, as well as essentially complete filing of the voids in the matrix.

Normal concrete (5000 psi) will absorb about 6% of monomer by weight.

Experiments continue on the incorporation of monomers in the mixing water; to date, results have been only about 50% as effective as with impregnation of dried concrete.

5. *Potential uses and advantages of high-strength concrete include the following:*

(*a*) High-strength concrete makes it possible to design long-span bridge girders of even lighter weight-to-strength ratio than steel. Precast segments would be factory-manufactured and assembled in place by post-tensioning (see Chapter 3, Section 3.1.4 Segmental Construction).

(*b*) Members subject to alternating or reversal of moments can best be made of high-strength concrete, prestressed axially to a high value.

(*c*) As suggested earlier in this section, compression elements such as piles and columns show a direct ratio of concrete strength to design load-bearing capacity.

(*d*) Trusses and space frames, in which pre-cast members are post-tensioned axially, are particularly appropriate for high-strength concrete. Trusses are widely used in Eastern Europe for bridges and the girders of industrial buildings.

(*e*) High-strength compression members may be used as "compression elements" in girders, columns, etc. Spirally-wrapped, triaxially prestressed, compression elements are used in the USSR by embedment in compression flanges of heavily loaded girders.

6. *Summary*

(*a*) Concrete of strengths of 10,000 psi (cylinder) (850 kg/cm² cube) are practicable with present techniques and materials.

(*b*) Concrete of strengths above 12,000 psi (cylinder) (1000 kg/cm² cube) are a practical possibility.

(*c*) Such concretes can be obtained by using special materials which are still within economic range.

(*d*) Alternatively, special manufacturing processes, while requiring high capital investment, may actually reduce the cost of the finished product when compared on an equivalent performance basis.

(*e*) Other means of obtaining high-strength concrete are being developed, showing great promise.

3.2   Prestressed Lightweight Concrete

Structural lightweight concrete is being employed very extensively in prestressed applications. The production of high-quality lightweight aggregates is well established in the United States, Canada, USSR, and Australia, and is being developed in Western Europe. The techniques of mix design and placement (vibration, finishing, etc.) have been developed to the point that lightweight concrete may be used with confidence for both highly technical construction, such as bridges, and mass production, such as roof and wall slabs. However, such high-quality structural lightweight concrete can be obtained only if these techniques and controls are carefully utilized by trained and skilled personnel. Failure to take this care may result in excessive creep and shrinkage, segregation, and nonuniform deflections.

Prestressed lightweight concrete is best considered as a new and unique material in its own right. Its particular properties can then be fully utilized and properly employed. Design and construction procedures can be adopted to give the desired results.

Structural lightweight aggregate concrete contains expanded clay, shale, or slate coarse aggregates. It may utilize expanded aggregates for fines, or natural sand, or a combination. The unit weight generally will run from 85 pounds per cubic foot to 120 pounds per cubic foot (1400 to 2000 kg/m³). The strength is quite dependent on the water/cement ratio and cement factor, but strengths of 4500 to 6500 psi (280 to 500 kg/cm²) are obtainable by following strict procedures. These strengths, and the associated properties, render it entirely suitable for prestressed applications.

The following list of properties is purposely general in its quantitative evaluation. Conventional concrete, made with sand and gravel (or crushed rock) aggregate, exhibits a broad spectrum of properties and, similarly, the properties of structural lightweight aggregate concrete can best be represented by a band, rather than a numerical figure or straight line.

Since structural lightweight aggregates are basically a manufactured product, it is to be expected that even higher quality and more closely controlled materials will become available in the future. This will undoubtedly make available higher strengths and reduced creep and shrinkage.

A parallel and recent development has been the exploitation of natural deposits of pumice. While the majority of natural aggregates are unsuitable for high-strength and prestressed applications, deposits do exist in Hawaii, Nevada, Japan, and the Caucasus which have very satisfactory properties and which are being used for prestressed panels, slabs, and even bridge girders.

Some interesting experiments are being conducted on the use of small glass balls as lightweight aggregate. However, before commercial use can be undertaken, matters such as chemical stability and aggregate-paste bond must be thoroughly evaluated.

3.2.1   Properties of Structural Lightweight Aggregate

Concrete.   Comparisons are between high-quality structural lightweight concrete, made from expanded clay, slate, or shale aggregates, and conventional (normal weight) concrete made from sand and gravel aggregates.

Unit Weight.   In air—60 to 80% of normal concrete. In water—50% of normal concrete.

Compression Strength.   With a 5 to 10% increase in cement content, strengths are about equal, up to a maximum of about 6000 psi (420 kg/cm^2) cylinder.

Shear (Diagonal Tension).   Ultimate strength 65% to 80%.

Tensile Splitting Strength.   Dry conditions, somewhat lower than normal concrete. Moist conditions, same.

Modulus of Rupture.   Dry conditions, somewhat lower than normal concrete. Moist conditions, same.

Bond.   Slightly less than normal concrete.

Transfer Length.   Same or slightly greater than normal concrete.

Modulus of Elasticity.   Approximately 50% to 60% of that of normal concrete.

Creep.   (Defined as the time-dependent deformation of concrete under sustained loading). Usually slightly greater than normal concrete of same strength (may run from 0 to 100% greater, depending on aggregates). The creep is proportional to the ratio of applied stress to the strength at time of loading. Thus, with a given prestress level, the creep may be reduced by having the concrete at a greater strength at the time of transfer of prestress.

Shrinkage.   For best lightweight aggregates, slightly greater than normal concrete. For some lightweight aggregates, 50 to 100% greater. Steam curing reduces shrinkage. Replacement of a portion of lightweight fines with natural sand reduces shrinkage.

Total Prestress Loss.   This loss is 110 to 115% when both are subjected to water cure; 120 to 125% when both are subjected to steam cure. (NOTE: Lightweight concrete subjected to steam cure has approximately the same total prestress loss as normal concrete subjected to water cure.)

Thermal Conductivity.   Forty percent of normal concrete.

Thermal Transmittance.   Fifty percent of normal concrete. This much greater thermal insulation has a decided effect on prestressing applications, such as the following:

(*a*) Greater camber when one side is exposed to sun.

(*b*) Better response to steam curing.

(*c*) Fire resistance.

Permeability.   Same, since structural lightweight aggregates are basically impermeable in themselves, i.e., their pores are sealed.

Water Absorption.   Runs 12 to 22% by volume as compared with 12% for normal concrete.

Abrasion Resistance.   Lower, where abrasion is local and concentrated. Lightweight coarse aggregate particles may be "plucked" out. Same for general wear, as in pavement.

Freeze-Thaw Durability.   Equal or superior, especially where de-icing salts are employed. Air-entrainment is recommended.

Marine Durability.   Believed generally equal; some authorities claim superior durability for lightweight concrete.

Atmospheric Durability.   Equal.

Corrosion Protection to Reinforcement.   Approximately equal. (*Note*: The cover is generally specified the same as for normal concrete. For lightweight aggregate concrete, the cover should always be at least 1-1/2 times the maximum size of lightweight coarse aggregate since carbonation does take place along the surfaces of externally exposed coarse aggregate particles.)

Fire Resistance.   Better by 20 to 50%.

Fatigue.   Equal or better, as shown by tests of railroad girders, etc., provided adequate shear reinforcement is specified.

Coefficient of Thermal Expansion.   With all lightweight coarse and fine aggregates, the coefficient is 80% of normal concrete. With lightweight coarse aggregate and replacement of 30% of fines by natural sand, then coefficient is 90% of normal concrete. While thermal expansion is lower than normal concrete, there have been no difficulties in service with reinforcement nor when lightweight aggregate concrete was used in composite action with normal concrete.

Poisson's Ratio.   From 0.17 to 0.21, with an average of 0.20.

Alkali-Aggregate Reaction.   Apparently nonexistent.

Dynamic Behavior Under Impact Stresses.   Stress wave velocity is about 20% less than normal concrete. Length of stress wave, same. Period of vibration, longer. Vibration damping, greater; probably due to the interface condition between paste and coarse aggregate. Shock and energy absorption, believed to be substantially greater than for normal concrete.

Ultimate Strength.   Research to date indicates greater ultimate strain under sustained loading; thus, probably greater long-term ultimate strength than normal concrete of same compressive strength by standard test.

3.2.   Utilization of Prestressed Lightweight Concrete

Prestressed Lightweight concrete has been utilized as a lightweight substitute for prestressed normal concrete in almost every type of application. The extent of such substitution has varied rather widely, depending on the immediate economic advantages obtained. Therefore, by far the largest use to date of prestressed lightweight concrete is for the roofs, walls, and floors of buildings.

The lower weight has also been the determining factor in the selection of lightweight concrete for prestressed bridge girders, particularly the suspended span of cantilever-suspended span bridges.

The reduced weight leads to a saving in the structural frame itself and particularly to a saving in foundations. In good soils, settlements can be reduced, or the size of footings reduced. In poor soils the number of piles may be reduced. In in-between soils it may be possible to use spread footings or a mat in lieu of piles. Savings in foundations have been cited as a major economic advantage for use of lightweight concrete.

The reduced weight also reduces the seismic loadings for design; this reduction in lateral force requirement can be of importance, not only in the structural frame, but also in the connection details and the foundations.

Reduced weight also facilitates transportation of prefabricated elements. In a number of instances studied in California, the saving in transport cost just about balanced out the increased cost of the lightweight aggregate when the transport distance was 100 miles; beyond that, the use of prestressed lightweight concrete offered a saving.

Moreover, since the maximum hauling weight is often limited, use of lightweight concrete may make it possible to haul a larger unit. Similarly, in erection, where crane capacities are limited, use of lightweight concrete permits larger single elements to be erected.

A rather dramatic extension of this latter advantage was realized in the reconstruction of a railroad trestle across San Francisco Bay. Because of restricted access on the tidal flats and the proximity of the deteriorated timber trestle, which had to be kept in full use, the weight of individual units was limited to about 50 tons. Use of lightweight concrete enabled the employment of prestressed cylinder piles, capable of taking lateral loads in bending. This, in turn, enabled direct structural framing, eliminating the need for cast-in-place pile caps. Thus the structural system was greatly simplified and a significant saving achieved in the time and cost of reconstruction.

The applications of prestressed lightweight concrete which offer the greatest ultimate or long-term benefit are those in which the unique properties of prestressed lightweight concrete are fully utilized, and it is selected not merely as a lightweight substitute but as a new material in its own right.

Foremost among these is the property of thermal insulation. Reduced heat transmission reduces the heat loss in cold climates and, conversely, reduces air-conditioning costs in hot climates. Condensation problems also are reduced.

The better fire resistance of prestressed lightweight concrete is primarily a result of this greater thermal insulation. This means that a reduced cover can be used for a specified fire rating. In turn, this reduces the dead weight even further.

Improved thermal insulation, combined with a slightly lower coefficient of thermal expansion reduces temperature stresses, especially where these are occasioned by short-term temperature cycles (e.g., daily). This is of direct importance where the major stresses are those due to temperature (e.g., pavements). This may also permit a reduction in the number of expansion joints, which are costly in both construction and maintenance. Future applications may well include furnaces, stacks, and nuclear reactor pressure vessels, where the major stresses are those due to temperature gradients.

The lower modulus of elasticity is a property of lightweight concrete that, although presenting problems in some applications, can be utilized advantageously in others.

Lower E increases the prestress loss. It leads to increased camber and increased deflections under load. Both of these can, and should be, recognized by the designer and properly provided for in the design.

Lower E permits greater energy absorption. An immediate application is to fender piles and dolphins, etc., where the lower E permits approximately twice the deflection under impact loading, before cracking. This increases the energy absorption and, at the same time, permits adjoining fender piles to come into action, thus achieving a distribution of the impact loading.

A new design concept, occasioned by the need for pile-supported pavements for aircraft runways, parking areas, and maintenance areas, is the distribution of concentrated loads by a flexible pavement slab, provided the piles shorten sufficiently under load. A significant reduction in the load which any one pile must carry is achieved by this "flexible" design concept. Use of prestressed lightweight concrete piles increases the "elastic" shortening under load. Future applications of this energy-absorbing capability may well include machinery frames and foundations.

Where the design stresses are induced by known strains (e.g., temperature stresses), then the lower E reduces the stress significantly.

Another property of great potential use is that of low submerged weight. Obviously this is directly attributable to low unit weight, but its field of application is quite different. Whereas, in air, prestressed lightweight concrete weights 75 to 80% of prestressed normal weight concrete, when submerged, the ratio is only 50%. This makes prestressed lightweight concrete an ideal material for barges, ships, drydocks, etc. Minimum wall thicknesses are generally established by the requirements of rigidity, safety against buckling, space for reinforcement and prestressing tendons, need for cover for durability, and ability to place the concrete. Thus, with an essentially fixed minimum wall thickness, the use of prestressed lightweight concrete offers a significant reduction in dead weight of the vessel. The drafts are

TABLE 3.1—Utilization of Prestressed Lightweight Concrete

Type of Application	Present Volume of Use	Form or Section Employed	Properties and Advantages Leading to Selection	Adverse Properties
Flooring	Substantial	Precast flat slabs. Precast cored slabs. Precast tee slabs, single or double tee. Precast joists with cast-in-place slab. Lift slabs. Cast-in-place composite deck slab on prestressed lightweight joists. Cast-in-place post-tensioned slab (or slab and beam) floor.	Lower dead weight. Reduced seismic loads. Reduced structure and foundation loads. Easier handling and erection. Better fire resistance. Lower transport costs.	Slightly greater camber and deflection. Reduced shear strength at bearings.
Roofing	Very large and widespread	Precast tee slabs, single and double tees. Precast and cast-in-place shells and folded plates. Cellular composite slabs. Precast slabs for cable-suspended roofs. Canopy roofs for grandstands and aircraft hangars. Roof trusses. Precast flat slabs, solid or cored.	Lower dead weight. Reduced seismic loads. Reduced structure and foundation loads. Better fire resistance. Better heat insulation. Reduced condensation. Lower thermal movements. Lower transport costs and erection costs.	Slightly greater camber and deflection. Reduced shear strength at bearings.
Walls for Buildings	Substantial	Solid panels. Tee panels. Sandwich panels. Cellular-composite panels. Panels with special architectural finish.	Prestressing of wall panels prevents cracking, makes thinner sections practicable, and facilitates handling and erection. Lower dead weight. Reduced seismic loads. Reduced structure and foundation loads. Better	

Application	Extent of Use	Typical Forms	Advantages	Disadvantages
Bridge Girders	Some, including some large and important structures.	Suspended span. I-girders with composite cast-in-place deck. Tee girders. Box beam girders. Hollow-core slabs. Segmental girder sections.	heat insulation. Better freeze-thaw durability. Lower thermal movements. Easier handling and erection. Better fire resistance. Less condensation. Reduced dead weight. Reduced seismic loads. Reduced structure loads. Reduced foundation loads. Easier handling and erection. For segmental girders, low modulus of elasticity permits better distribution of bearing pressures on adjoining faces when using "dry" joints. Better fire resistance (some installations in fire hazard areas).	Low modulus of elasticity. Greater camber. Reduced shear and diagonal tension strength.
Bridge Decks	Some use	Precast slabs. Cast-in-situ composite deck slab on prestressed lightweight I-girders.	Lower dead weight. Better resistance to freeze-thaw and de-icing salts.	Liability to local "plucking."
Foot Bridges	Considerable	I-girders. Tee-girders. Box beams.	Low dead weight of particular importance where dead weight is a major portion of total design load and where design load is seldom realized in service. Easier erection.	
Piling	Some	Bearing piles. Fender piles. Dolphin piles. Batter piles. Sheet piles.	Reduced submerged weight means greater capacity for design loads; also less bending	More subject to spalling from abrasion and local impact.

(Continued)

TABLE 3.1 – Utilization of Prestressed Lightweight Concrete (Continued)

Type of Application	Present Volume of Use	Form or Section Employed	Properties and Advantages Leading to Selection	Adverse Properties
			stress in batter piles. Reduced weight for transport and handling, especially important with very large piles. Better deflection and energy absorption. Better ability to deform without cracking during driving when obstructions are encountered.	
Craneway Girders and Industrial Building Elements	Some	I-girders. Rectangular beams. Box sections.	Lower dead weight. Reduced structure loads. Energy absorption. Fire resistance.	Greater camber. Lower diagonal tension. Greater deflection under live load.
Seismic-resistant and shock-absorbent structures	A few	Girders. Beams. Framed structures.	Energy absorption. Lower modulus of elasticity. Lower dead weight means lower seismic loads.	
Guard rails and fenders. Wales	A few	Rectangular beams.	Energy absorption. Greater deflection under impact. Fire resistance (as compared with timber).	Spalling under local impact and abrasion on edges.

Poles	Experimental & some production	Tapered. Constant section. Hollow core.	Reduced weight for transport and erection. Reduced seismic loads. Greater deflection under load.	Inserts require somewhat larger embedment or anchoring.
Railroad Sleepers (ties)	Experimental		Reduced weight for handling and installation. Better energy absorption. Greater deflection reduces "center binding."	
Floating Structures	Some	Precast elements, cast-in-situ.	Reduced weight where wall thickness is determined by cover over steel and by placement, water-tightness, and stiffness requirements. Greater durability anticipated. Better energy absorption. Reduced permeability anticipated. Submerged weight of lightweight aggregate concrete is only one-half that of normal concrete. Underwater tensile strength about same as for normal concrete.	

frequently reduced by 25% or more or, conversely, for a given maximum draft, the cargo capacity may be increased by 25 to 30%.

This lower submerged weight also has an application in extremely large and heavy piles and caissons, such as those for major bridge foundations. When these are driven through weak soils, such as mud, to bear in lower soils, such as sand, the dead load of the pile becomes a significant factor. Assuming a "fluid" density of the soil of 100 to 110 pounds per cubic foot, the use of lightweight concrete results in essentially no added load, whereas a normal weight pile has an effective dead weight of 40 to 50 pounds per cubic foot. In some soil conditions, this may permit a significant shortening of pile length.

Some of the most useful applications of lightweight concrete are in composite action with normal weight concrete. All possible combinations have been employed successfully. These combinations are the following:

(a) Prestressed lightweight concrete joists or girders with a cast-in-place normal weight composite slab.

(b) Prestressed lightweight concrete joists or girders with a cast-in-place lightweight concrete slab.

(c) Prestressed normal weight concrete joists or girders with a cast-in-place lightweight concrete slab.

All of these combinations, especially (c), require that the designer carefully consider the differential moduli of elasticity and shrinkage.

Composite construction has also been employed rather extensively in the USSR where cellular (aerated) concrete is joined with ribs or slabs or prestressed normal concrete. Cellular concrete, whose structure is produced by gas or foam, contains no coarse aggregates. Its unit weight ranges from 35 to 75 pounds per cubic foot (600 to 1200 kg/m^3). It is not suitable for direct prestressing; hence, the need to join it in composite action with ribs or slabs of conventional concrete or structural lightweight aggregate concrete. These ribs are highly stressed and provide the tensile element. The cellular concrete is poured around or over the ribs, then the entire unit is autoclaved. Elements of this type are utilized for roof and wall panels where the thermal insulating values are of greatest importance. The use of prestressed ribs or slabs gives greater strength and durability than is possible with mild steel reinforcement.

3.2.3   Manufacture and Construction

The techniques for manufacture and construction of prestressed lightweight concrete are essentially the same as for prestressed normal-weight concrete. The following specific recommendations are made:

(a) The substitution of natural sand to replace 30% or more of the lightweight fines will generally give more consistent results with only a small increase in unit weight. The mix design must be modified accordingly so that the volume of coarse aggregates in a cubic yard remains approximately the same. With sand replacement, the drying shrinkage is usually reduced and the tensile splitting strength in the dry condition is improved.

(b) Use of air-entrainment is always recommended for lightweight aggregate concrete to improve workability, reduce segregation, and increase freeze-thaw durability. Air-entrainment values of 4 to 6% are generally employed. For some applications where the lowest possible unit weight is required and medium strength (3500 to 4500 psi)(250 to 300 g/cm^2) is acceptable, air-entrainment up to 8% may be introduced.

(c) Use of water-reducing admixtures is beneficial in assuring adequate workability at low water/cement ratios and in achieving higher strengths at all ages.

(d) Careful selection and handling of lightweight aggregates to prevent crushing and contamination and to ensure uniform moisture content and uniform unit weight. (Note: Pre-saturation of aggregates will achieve uniformity but seriously reduces freeze-thaw durability. In warm climates pre-saturation is recommended by some but there is no unanimity of opinion on this.)

(e) Design of the mix to provide continuous control for variations in unit weight and moisture content of aggregates.

(f) Avoid excessive overvibration as this will cause the coarse aggregate particles to float to the surface.

(g) Curing by steam at atmospheric pressure (and at high pressures) generally improves properties by reducing drying, shrinkage, and creep. The optimum cycle for low-pressure steam curing is the following:

Delay (holding period)	3 to 5 hours
Rise in steam temperature	40° F (22° C) per hour
Maximum sustained temperature	140 to 160° F (60 C to 70° C)
Cure at maximum temperature	8 to 14 hours

Shorter delay periods, and higher maximum temperatures, may be employed to achieve a shorter production cycle, but generally reduce the ultimate strength, durability, and other properties.

Lightweight concrete gains heat more slowly and holds it longer due to its insulating properties. Lightweight aggregate, because of its water absorption, provides a measure of internal curing.

Steam curing at higher pressures (autoclave) is very beneficial in eliminating volume change but has so far not proven practicable for prestressed applications other than very small roof slabs. Electrothermal curing is practiced extensively in the USSR. The cycle is a two- to four-hour rise in temperature, followed by 12 hours of cure at 70 to 95° C. The higher temperatures are employed when the tendons are stretched directly against the forms.

(*h*) Supplemental moist curing, after steam curing, will reduce drying shrinkage.

(*i*) Due to the longer transmission length on release of pretensioning, slow release, by hydraulic jacks, is preferable to sudden release, as by burning.

(*j*) With post-tensioning, particular care must be taken to achieve dense concrete in the anchorage zone. Local reinforcement should be provided to prevent spalling and tensile cracking in the end block. In some special cases (e.g., long-span bridge girders) the end blocks have been precast in high-strength normal-weight concrete, then set in the forms and the lightweight concrete poured as the main body of the girder.

(*k*) Lightweight concrete is somewhat more apt than normal concrete to form air and water bubble pockets on vertical and return (overhang) faces. External vibration will minimize these.

(*l*) Surface or deck finishing should follow closely behind the placement, and care must be taken so as not to "drag" the coarse aggregate out of the mix.

(*m*) Avoid excessive soaking of surface during storage as this may cause drying shrinkage problems after erection.

(*n*) Protect edges and corners from impact as these are more subject to spalling.

(*o*) Protect thin elements (such as roof slabs) from differential exposure to the sun in prolonged storage as this may cause differential camber.

3.2.4   Architectural Treatments and Textures

Lightweight aggregate concrete is generally adaptable to a wide range of architectural finishes, paralleling those applied to normal concrete.

Light sand-blasting will give a bush-hammered effect to the surface.

White cement is often used with lightweight concrete wall panels, etc., although the basic brown or grey color of the lightweight aggregate particles may show through after sand-blasting. Use of white natural sand, such as quartz or dolomite, will generally give a very satisfactory appearance.

When colors are being applied to the matrix, use of white cement and white sand, at least in part, may help keep the color pure.

Acid-etching and exposed aggregate finishes are commonly employed. Where ceramics, marble, quartz, or other facing materials are employed in conjunction with lightweight aggregate concrete, and the composite section is prestressed, careful attention must be given to the difference in moduli of elasticity of the two materials. Then the strands can be positioned so as to minimize camber or sweep. Frequently this may require that the tendons be close to, or even in contact with, the facing material.

3.2.5   *Actual Performance of Prestressed Lightweight Concrete*

Actual fires have demonstrated that the better thermal insulation does provide greater fire protection for the same cover.

Experience from northern United States, the USSR, and Canada indicates better freeze-thaw resistance than normal concrete. Researchers in the USSR state this is due to the porous nature of the aggregates. Experience in New York has shown that where de-icing salts are applied to bridge decks, the lightweight concrete has shown markedly increased durability.

Experience with structural lightweight aggregate concrete (conventionally reinforced) in ships built in both World War I and II shows generally excellent marine durability despite the small cover. One especially well-documented case is the ship *Selma*, which has been beached since World War I at Galveston, in semitropical sea water. Despite only 3/8 inch (1 cm) cover of expanded shale aggregate, substantial portions of the ship remain in excellent condition.

Prestressed lightweight concrete piles have been in service in sea water since 1955. Inspection in 1968 showed no loss of durability.

Volume-change problems have occurred in a number of cases, principally with prestressed lightweight concrete floor and roof slabs. These are similar to the same problems that occur with prestressed normal concrete, but are aggravated because of lower E, lower dead load to live load ratio, greater shrinkage, and increased thermal insulation.

Because of lower E and lower dead load to live load ratio, design camber is often greater. (This can, of course, be compensated for in design.)

After installation, several factors may cause the camber to grow. The sun's heat on the top produces a thermal gradient, with the top fibers expanding more than the lower ones. The camber grows and the unit shortens. At night, friction restrains the ends; so the effect is cumulative, not reversible. This effect is very much more serious when the roof slab is covered with dark roofing which absorbs the sun's heat.

Similarly, the top flange may have a greater water content than the legs, thus less shrinkage, due to heating of the building and water ponding on the top. This results in an upward camber.

Use of neoprene bearing pads, good shear reinforcement of the end zone of the legs of the slabs, or joists, and proper curing time (to eliminate creep and shrinkage) before erection will eliminate most problems.

In some cases, it may be possible to spray-paint the roofing with a reflecting paint; this greatly reduces the thermal effect. Lastly, consideration must be given in the design to these problems, especially the dead load to live load ratio, and appropriate steps taken. These may include deeper sections, partial prestress, and, of course, proper detailing of bearings and adjacent zones.

Prestressed concrete piles up to 132 feet (40 meters) have been manufactured in lightweight concrete and successfully installed. Performance in driving was similar to normal concrete except for slightly greater head spalling.

Lightweight concrete has proven entirely satisfactory in resisting abrasion and wear. The most notable example is the roadway deck of the San Francisco-Oakland Bay Bridge. Decks of ships built of lightweight concrete have similarly withstood wear, vibration, and fatigue. However, lightweight concrete is sometimes subject to local "plucking" of coarse aggregate. Edges of curbs, rails, etc., are subject to greater spalling.

When concreting in winter, with heated aggregates, lightweight concrete has proven to retain its heat longer and more effectively.

Prestressed lightweight concrete appears destined to play an ever-increasing role in construction. Constant improvement in quality of aggregates, combined with the performance and behavioral advantages of prestressing, present opportunities for increasingly sophisticated design and new applications. The present differential in raw aggregate cost will probably narrow and disappear as close-in gravel deposits and quarries are depleted. Greater use will lead to a more widespread recognition of its many concomitant advantages.

The construction engineer has a major part to play in this development. By applying proper techniques and procedures, he can ensure consistent high quality. By his selection of prestressed lightweight concrete for those applications under his control, the direct and indirect economical and structural benefits will be made more obvious. It should be treated as a new material in its own right, possessing many advantages when wisely and properly employed.

3.3. Precasting

Precasting is one of the most important techniques by which prestressed concrete has been made practicable and economical. Precasting is so intimately bound up with prestressing that often within segments of the industry the two terms are used synonomously. From a factual and technical point of view this is, of course, not true; precast concrete can range from unreinforced concrete to conventionally reinforced (mild steel) concrete to prestressed concrete, and prestressing can be applied to cast-in-place concrete as well as prestressed concrete. Precasting, however, is the technique by which mass-produced elements of prestressed concrete are constructed. It offers all the advantages of prefabrication: quality, mechanized production, speed, and economy.

Precast prestressed concrete is applied to virtually all types of structural elements for every structural application. Among the widest applications of precast prestressed concrete are the following:

Buildings:	Floor slabs, roof slabs, wall slabs, piles
Bridges:	Piles, girders, deck slabs
Marine Structures:	Piles, deck slabs, sheet piles
Railroad Ties (Sleepers)	
Poles	
Pipes	

Precast members may be cast either at a permanent plant or a temporary job-site plant. They are cast and cured at the plant, then transported, erected, and joined into the completed structure. Joining may be accomplished by welding, bolting, or post-tensioning, or by cast-in-place concrete connections. Many efficient designs utilize precast and cast-in-place concrete in monolithic action.

Precast members may be pretensioned or post-tensioned. Post-tensioning can be utilized to make individual precast elements work together monolithically, or to increase the prestress values after erection (stage stressing).

There are two main uses of precasting. The first and most extensive use is in the mass-production of standardized elements.

The second is the casting of unique and complex units in the position and under conditions whereby the necessary quality and tolerance control and economy may be best achieved.

Standardization is an essential consideration for mass-production. Many building elements, bridge girders, piles, poles, railroad ties, pipes, etc., have been standardized by national associations, such as the Prestressed Concrete Institute, American Association of State Highway Officials, American Railway Engineering Association, American Concrete Pipe Association, etc.

3.3.1   Specific Advantages

Some of the specific advantages of precasting are the following:

1. *Control of Shrinkage*

Shrinkage may be reduced through lower water/cement ratio and through steam curing. By proper aging and drying before erection, the majority of shrinkage may be made to take place before erection.

2. *Reduction of Creep*

By proper curing and aging before erection, the strength and maturity of the element and its modulus of elasticity are all at higher levels at the time of loading, thus reducing the creep. Higher strength concrete of lower water/cement ratio may be utilized in manufacturing precast elements, thus producing concrete of more favorable creep characteristics.

3. *Control of Dead Load Deflection*

Precast concrete elements may be erected and adjusted to exact position on the theoretical profile, including an allowance for creep deflection, before final jointing and fixing. This is of particular importance on long-span bridges.

4. *Quality Control*

Precast concrete elements may be manufactured under the best conditions for forming, placement of reinforcement, placement and vibration of concrete, and curing. Uniformity and high quality may thus be obtained practically and economically.

Precasting makes it possible to assure greater durability through accuracy and uniformity of cover. Because it permits improved placement and vibration, while maintaining a low water/cement ratio, a more impermeable cover may be achieved. Precasting also facilitates a smooth surface finish.

Precasting also permits the application of special manufacturing techniques, such as combined vibration and pressure, vibro-stamping, extrusion, spinning, and steam curing, which, in general, cannot be applied to cast-in-place concrete. Similarly, precasting permits the use of pretensioned reinforcement.

Dimensional tolerance control is facilitated by precasting. Rigid forms, often of steel, prevent excessive deflection under vibration. Precast members may be checked for tolerance before shipment and erection. Prestressing steel, mild steel, and inserts may be accurately positioned and held during concreting.

5. *Timely Availability*

Many standardized, mass-produced elements can be furnished to a construction site on very short order. While some standardized elements are carried in inventory, most plants are able to produce to order at the desired rate and to meet even accelerated schedules; therefore, inventorying with its costs of rehandling, storage, and interest, is not a normal practice in the industry.

Precasting makes it practicable to utilize steam curing and thus to produce on a 1- or 2-day cycle, with full design strength generally achieved withing 7 to 14 days. (However, for some very long, thin members, such as long-span roof slabs, storage at the yard before shipment for a period of 30 to 60 days before erection may be desirable to minimize both shrinkage and creep.)

6. *Erection Over Existing Traffic and Minimization of Falsework*

Precast concrete girders may be erected without falsework over traffic, railroads, or waterways, etc. (See Fig. 3) Longer spans may be erected by the progressive cantilever scheme or cantilever-suspended span scheme.

Fig. 3. Night-Time erection of prestressed concrete girder over existing roadway.

Where falsework is required to temporarily support precast segments, the extent required may be reduced (although total loads remain the same as for cast-in-place construction). Since the time it must remain in place under any one area is reduced, maximum re-use of falsework is facilitated.

7. *Economy*

Precast concrete schemes offer substantial economies in construction. Site labor can be kept to a minimum. Erection may be performed during the most favorable seasons and conditions. By using higher strengths and thinner sections, etc., dead weight may be reduced, resulting in overall savings of concrete and steel.

Labor and equipment may be utilized with maximum efficiency. Forms may be reused a maximum number of times, and advantage taken of mechanization in manufacture.

8. *Suitability for Composite Construction*

Precast segments may frequently be combined with cast-in-place concrete to act in composite action, as a monolithic structure. Provision must be made for shear transfer. The precast units may serve as partial forms, and also as a support for forms for the cast-in-place concrete.

Consideration must be given to the various stages of loading to prevent overstressing of the precast elements while the cast-in-place concrete is still wet. The designer must consider the effect of differential shrinkage and of different moduli of elasticity.

3.3.2 Fatigue and Dynamic Loading

Consideration should be given to fatigue loading in all joint design, especially to welded reinforcing steel connections and to welded structural steel connections. The connection bars should be adequately embedded within the concrete to develop the full connection strength under dynamic loading wherever seismic forces or other dynamic forces may occur. Usually this will require an embedment length 200% of that required for static loading. Bars embedded in very thin flanges may tend to break out at right angles to their direct load stress due to eccentricities; these bars should be extended to attain sufficient embedment to hold even if local flange failure takes place under dynamic loading.

3.3.3 Manufacture

The manufacture of precast concrete girders, elements, and segments should follow applicable provisions of ACI Standards and Recommended Practices for Concrete Construction. Chapter 8 discusses the organization and layout for manufacture of precast pretensioned concrete, and lists special considerations applicable to all precast concrete construction.

3.4 Lifting and Erecting Precast Elements

Units should be picked and handled so as not to produce cracking. Buckling should be prevented by suitable stays or supports. Auxiliary trussing or support may be required for long, thin girders. Particular care should be taken at points of picking to avoid concentrated stresses and cracking; this may require careful design detailing and additional mild steel reinforcement. It should be noted that when precast units are removed from forms, there may be additional load due to friction, fins, and suction.

Picking and handling should be treated as dynamic loads.

Lifting loops for picking and handling must be designed with a safety factor of 6, that is, their ultimate capacity should be $3(DL + I)$ where $I = 100\% \ DL$. Embedment should be adequate to prevent pull-out bond failure under the above ultimate design load. Adequate concrete section and reinforcing should be available to develop the ultimate design load; be especially wary of thin flanges.

The angle which the sling or line makes with the lifting loop, at all positions during picking and handling, should be considered and provision made therefor. Either the angle may be fixed by using a picking beam, or the loop designed to be effective at all angles. Obviously, the increased force due to angular lead shall be used in design. The member should be checked for the axial component of load to insure against buckling. (See Fig. 4.)

Fig. 4. Lifting prestressed lightweight concrete double-tee unit for parking structure roof.

Consideration should also be given to sway or swing, i.e., bending of the picking loop sideways. This will cause sharp bending stresses in the picking loops and may cause local concrete crushing.

Bundled loops of strand may be employed provided they are equalized so all strands will work together. Their ends can be splayed out in the concrete. Smooth, mild-steel bars may be used, provided adequate embedment against pull-out is provided. The diameter must be such that localized failure cannot occur by bearing on the shackle pin.

Fabricated lifting inserts (e.g., fabricated plates) may be used, provided the following rules are applied:

(*a*) Their pull-out value is ensured by mechanical or positive fastening in the concrete.

(*b*) There are no welds transverse to the principal tension.

(*c*) Plates are thick enough in themselves to take bearing from shackle pins—no built-up washers or cheek plate reinforcement of the eye.

(*d*) Eyes are designed for shear, moment, and tension on the minimum section.

(*e*) Steel used is ductile (serious failures have occurred when hard-grade brittle steels were used).

For both mild-steel bars and mild-steel plates, the steel should have a Charpy Impact Value in the range of 15 foot-pounds at the temperature at which picks will be made. (Actual tests are expensive and are not normally required.) Most US domestically-produced steels (ASTM A-36) have adequate impact values at temperatures down to about $20°$ F (-5°C) but, at lower temperatures, a specially rated steel should be used. Prestressing strand and bars fortunately have good low-temperature impact strength. Particular care should be employed with unknown or foreign steels since, at low temperatures, they may fail in brittle fracture at 50% or less of their nominal strength.

Fabricated inserts may be prestressed by bars to the concrete, but care should be taken to provide adequate embedment and to use the above specified safety factors in design, including design for embedment.

Deformed reinforcing bars should not be used for picking loops as the deformations result in stress concentrations from the shackle pin. Also, reinforcing bars are often hard-grade or re-rolled rail steel, with low ductility and low impact strength at low temperature.

Transport.   Moderate-size precast concrete units can frequently be transported by truck and erected by crane. Rail may be used for long distance shipments and for over-length segments, but often requires supplemental transportation to the actual job site. Barge transportation is very economical and practicable for movement to water sites, and can be used to transport heavy and over-size units.

During transport, precast members are subjected to dead load stresses, dynamic (impact) stresses, and lateral instability. Ideally, a member being transported will be supported at the same points as in the final structure. However, this is not always practicable, and thus special design studies must be made for the transport condition. If there is excessive overhang, tensile stresses at the point of negative moment may be countered by additional mild reinforcing steel, external steel beams bolted on, or even temporary post-tensioning cap-cables.

When a deep, heavily stressed girder is rotated from the vertical, dead load stresses are reduced and the unit tends to buckle upward. This may be prevented by proper blocking and securing. Chains are preferable to wire rope because wire rope stretches.

Lateral instability is the cause of most accidents in transportation. Hog-rodding is often necessary on long, deep girders. Sometimes two or more such units can be transported together, side by side, and tied together to provide the necessary lateral strength.

Dynamic stresses are usually just considered as an increase in stress. However, for long, thin members supported at their ends only, a dangerous condition may arise due to bounce. The dead load of the member may be temporarily relieved by the bounce, and the prestress, no longer countered by dead load, may cause excessive tensile stress and even failure in reverse bending.

Care must be taken to prevent surface damage to precast concrete members during transportation. Spalling of edges may be prevented by proper attention to blocking and softeners, especially where tie-down chains or lifting slings pass around the member. Surfaces should be protected against staining and discoloration during transport; in some cases this may require covering.

When a precast girder is to be supported on more than one railroad car, a bolster or other means must be provided to permit rotation of the supports as the cars go around curves.

Erection Methods and Techniques.   These are determined by the span, height, and type of the structure, its location, the topography, the weight, size, and configuration of the precast elements, the method of jointing, and the erection equipment available. The demonstrated ingenuity of contractors in erecting precast members under extremely adverse conditions justifies confidence in specifying precast concrete construction. This ingenuity should be encouraged.(See Fig. 5)

The success of erection of large, heavy elements to close tolerances depends on the skill and care of the construction engineer. Thorough calculations of rigging, temporary support, and stress conditions are required for each stage of erection. Additional strengthening or support may be provided internally (by additional reinforcement or temporary stressing) or externally (by means of attached trusses or hog-rodding).

Lifting loops or similar devices must be provided for all precast units for handling and erection. (See Section 4, Lifting and Handling Precast Elements.)

Care must be taken to ensure lateral stability against twisting, buckling, and leaning, both during erection and after erection.

Consideration must be given to external forces on precast elements during erection, such as wind, current, waves, etc., as applicable. Erection should be suspended during unfavorable and unsafe conditions.

Temporary lateral support must be provided to prevent sidewise displacement, buckling, and torsion. With curved girders, guys and blocking will

Fig. 5. Lifting prestressed double-tee parking garage unit.

generally be necessary to prevent twisting and overturning. Cross-bracing to adjacent girders is often a practicable solution.

In seismic areas, it is necessary to insure sufficient tying in and bracing so that the structure *as a whole* is safe under earthquake. It is generally considered unnecessary to tie each member as it is erected against this unusual loading unless its dislodgment could have "stack-of-cards" or other catastrophic consequences.

Techniques pertaining to erection of specific types of structures are discussed in detail in Section II.

Welded Joints. Steel beams or shapes may be embedded in the precast segments, and extend from the ends. After assembly and erection bolting, etc., a pin may be driven, as in structural steel work, and a full welded or bolted connection made. Then the joint is encased in concrete.

The heat from the welding tends to produce edge spalling of adjacent concrete. For this reason, the weld should be kept as distant as practicable from the concrete surface, intermittent and low-heat procedures should be adopted, and the steel shape should be well anchored into the concrete to resist local distortion.

Continuous butt welds, with concentric joints, are preferred. Critical welds should be inspected by radiography. Welds should comply with the specifications of the American Welding Society, using certified welders and procedures. Steels used should be selected for weldability. In particular, avoid welding of hard-grade reinforcing steel. Some reinforcing bars sold as "intermediate grade" actually are high-carbon bars and are not adaptable to good field welding.

Special Concretes.   Structural lightweight concrete is frequently advantageously employed for precast prestressed construction. The lighter weight reduces costs and difficulties of transport and erection.

Very high strength concrete (fc = 8000 to 10,000 psi and above) can generally be achieved only through precasting. Its advantages in longer spans, higher capacities, and greater structural efficiency are extremely promising. However, the generally thinner, more highly stressed sections require even greater care in transport and erection.

3.4.1   Segmental Construction and Joints

Prestressing makes practicable the use of segmental construction in which precast concrete elements are joined together, so as to act in monolithic fashion. Segmental construction, in turn, permits higher quality and more economical constructions in prestressed concrete.

Segmental construction has been widely employed in buildings and bridges, particularly in Europe. Many of the most widely used floor systems involve precast segments. Some of the longest span bridges in Europe, notably the Oleron Viaduct in France, the Gladesville Bridge in Australia, the Oosterschelde Bridge in the Netherlands, and bridges across the Moscow, Dnieper, and Volga Rivers in the USSR, employ segmental construction. Segmental construction can be extended to two- and three-dimensional frames, as in roof trusses and space frames. It can be used to excellent advantage in cable-suspended construction, permitting dead-weight adjustments prior to jointing.

Although composite construction will be dealt with separately, it should be noted that a combination of segmental construction, employing precast elements, with cast-in-place concrete, all working monolithically, may be the most efficient solution in many cases.

Segmental construction obviously is a practicable means of utilizing very high-strength concrete. Precast elements, manufactured with special techniques under close control, can be joined together for efficient structural action in the largest and most complex structures.

Precast units can be made with very thin sections, cast in the most practicable position, so as to insure high quality and close tolerance. Dimensions and quality can be verified and corrections made, if necessary, prior to assembly into the structure.

Prestressing ducts and anchorage assemblies can be accurately located and fixed during concreting. Concrete can be properly placed and consolidated around congested reinforcing steel, as at the anchorage zones.

Segmental construction is especially applicable to winter construction, as on-site concreting and other work is kept to a minimum.

Precast segments for bridge construction usually comprise part of or the entire cross section of the bridge. They can be erected as segments, on falsework, or by progressive stressing back (cantilevering), or hung from suspenders; the joints are completed, and the segments stressed longitudinally.

Similarly, bridge pier shafts and columns may be constructed by erecting precast segments on top of each other, then stressing them vertically.

When a transverse cross-sectional segment would still be too large or heavy to erect, or when other structural requirements dictate, the segments may be longitudinal segments, stressed together transversely.

Segmental construction permits more rapid completion, since the segments can be manufactured while the site preparation work and foundations are being constructed. Proper techniques of erection, jointing, and stressing make it possible to attain rapid erection of the superstructure.

Segmental construction makes it possible to apply the structural and economical advantages of precasting to large and complex structures.

3.4.2 Joints

Joints must be detailed so as to meet the design requirements for transfer of bearing and shear. They must be practicable of construction and inspection under the actual conditions at the site. Since joints will generally be visible, it is essential that consideration be given to their appearance.

Joints in precast segments, which are subject to a reversal of stress at any section under the design criteria, should be detailed to provide adequate restraint against movement or be provided with flexible jointing material to prevent spalling of edges or fatigue.

Because proper jointing of segments is essential to the security and performance of the structure, a detailed description is presented of acceptable jointing techniques.

(a) Cast-in-Place Concrete Joints. Cast-in-place concrete joints are the most common method of joining precast segments into monolithic concrete. The length of joint is set sufficient to permit splicing of the mild reinforcing steel bars; thus, joint widths are usually in the range of 18 to 24 inches. The reinforcing steel bars extend from the segments and may be lapped the required number of diameters, or be spliced by an approved welded joint. Concentric welds should be employed, or eccentric welds checked for the effect of eccentricity. Welds should be checked for fatigue loadings, where applicable. All welds should conform to applicable AWS Specifications. Looped bars, with a vertical rod linking the overlapping loops, may permit some reduction in joint length.

Joint surfaces should be cleaned and roughened, as by sandblasting or by bush-hammering, and thoroughly pre-wetted or else an epoxy-bounding agent applied before concreting. Epoxy materials should be applied only to clean, dry surfaces, in strict accordance with manufacturers' recommendations.

High-strength concrete with 1/2- to 3/4-inch maximum coarse aggregate is then placed. Internal vibration is essential. Curing may be by water, but it is often accelerated by low-pressure steam in a temporary hood or jacket.

Ducts for prestressing tendons may be spliced and taped to prevent grout intrusion. However, because of the absolute importance of keeping these ducts open, it is usually desirable to insert an inflatable rubber tube all the way through the ducts and joints. If this is not practicable, a small wire line may be run the full length of the duct, and then run back and forth while the concrete in the joint is still green, in order to insure continuity of the duct.

(b) *Poured "Fine" Concrete Joints.* Poured "fine" concrete joints should be not less than 3-inches thick. No reinforcing steel splices are provided; but duct tubes must be spliced and taped and kept open as provided in (a) above.

Joint faces are prepared as in (a) and a "fine" concrete, using 3/8-inch maximum size coarse aggregate, is poured, vibrated internally, and properly cured.

(c) *Packed Mortar Joints.*   Packed mortar joints have been employed in the past, but are not generally recommended due to inherent problems of load concentrations. If used, they should be from 3/8- to 1-inch in width, and the opposing faces should be clean, smooth, and without high spots. The mortar should be of dry consistency and placed in stages by uniformly packing with a steel tool, working simultaneously from both sides of the joint. The extreme outer surface of the joint should employ a more workable mix so as to seal the surface.

(d) *Buttered Mortar Joints.*   Buttered mortar joints have been frequently used when one precast segment has been set on top of another; e.g., a hammerhead segment set on a pier. The edges may be sealed and protected against load concentrations and edge spalling by a perimeter strip of compressible material such as neoprene or asphalt-impregnated fiberboard. Then a thin coat of mortar is troweled onto the surface. The surfaces must have been properly prepared. An epoxy mortar will generally give best adhesion.

Buttered or mortar joints have not proven very satisfactory for vertical joints, such as joining segments of girders, due to uneven compaction and stress concentrations. If used, the grout should be nonshrink or employ expansive cement.

This type of joint has been used with great success in horizontal joints of vertical members (e.g., splicing columns and piles). Mild-steel reinforcing is used to dowel the segments together. Holes are formed in the bottom section, and filled with epoxy mortar, then the matching dowels are inserted while the mortar is still wet. The set and cure of epoxy mortar may be accelerated by heat. An interesting solution for this is practiced in Norway where the heat is provided by embedded electric resistance wiring (see "Pile Splices" in Chapter 11). Other types of mortar, such as expansive cement or nonshrink grout, have been found satisfactory where ample time is available for development of strength.

(e) *"Dry' (Exact Fit) Joints.* "Dry" (exact fit) joints are increasingly used for joining precast segments as they permit rapid progress, assure uniform properties of concrete at the joint, are unaffected by weather, such as freezing conditions, and are economical.

The methods of obtaining perfectly matching surfaces are described in Subsection 3.4.3(a), Manufacturing Techniques for "Dry" and Close Fit Joints.

Keys or dowels must be provided to insure that the matching faces are brought together in the correct alignment and position. Temporary bolts may be used to hold the segments in position until stressed.

With dry joints, consideration must be given to the problem of continuity of ducts across the joint. If all tendons can be grouted at the same time, leakage, if any, will not plug an adjoining duct. If all ducts across a joint cannot be grouted at the same time, the continuity of the ungrouted ducts should be insured by employing an inflatable tube, or by drawing a wire rope back and forth during and after grouting of the adjacent ducts.

The outer edges of dry joint faces should be chamfered slightly to prevent local edge stresses from causing spalling. Insofar as practicable, the joint location should be selected so that the effective stress across a dry-joint face is always compressive and as uniform as possible (including the effects of prestress and dead load stress). It must develop sufficient compression to insure adequate shear resistance under the most adverse loading e.g., maximum live load on one side of joint, no live load on the other).

(f) *Epoxy-Coated Dry Joints or "Glued" Joints.* Epoxy-coated dry joints or "glued" joints are formed as for "dry" joints. Usually the segment is fitted to ensure that the fit is "perfect," then separated and a relatively slow-setting epoxy is applied (a few mils in thickness). The jointing is then made and the segments stressed together wet (i.e., before the epoxy has set). Heat may then be used to accelerate set of the epoxy. Thicker joints, up to 1/4-inch thick, may be made by using viscous epoxy mortar containing silica (sand) filler, permitting application to vertical faces with minimum flow during the period of assembly. Epoxy materials should be used in accordance with manufacturers' recommendation. They should be applied only to dry, clean surfaces.

(g) *Shear Transfer.*   Roughened surfaces, formed by sandblasting, bush-hammering, chipping, or wire brushing, or a set-retarder plus jetting, etc., may be used to provide shear transfer from a precast joint face to a concreted joint and are usually sufficient for most cases. However, where shear forces are extremely high, shear keys or indentations may be formed in the faces of the segments; these should have dimensions two to three times the maximum size of coarse aggregate used in the joint concrete. Similar shear keys may be formed in matching faces of dry-joint segments.

Shear-keys should have sloping faces, not right-angled to the joint, so as to avoid edge concentrations.

With segmental construction, shear must be checked in the unreinforced concrete zone between the joint proper and the first shear (vertical) reinforcement. Although this zone may be only 2 inches (5 cm) wide or so, it may be critical for shear under some conditions of loading. This may make it necessary to provide steel across the joint in the form of diagonal bars, tendons or mild steel dowels.

(h) *Finish and Color.*   Exterior faces of girders made of segmental elements should either show the joint clearly or else attempts should be made to blend the joint with the segment.

Clear showing of the joint may be accentuated by chamfers. This treatment is particularly applicable to segments assembled with dry joints.

Blending requires a uniformity of finish and color. With cast-in-place concrete joints (a), use of the same forming material and curing method as was used for the precast segments themselves (e.g., steam curing) will tend to minimize color differences. With "fine" concrete joints, use of some white cement in the fine concrete mix will tend to achieve uniformity of color. Prior experimentation with a mock-up will help to develop the proper method of blending.

Finish uniformity may be obtained by rubbing the entire girder face.

(i) *Shrinkage in Joints.*   Shrinkage in the joints can occur with cast-in-place concrete or poured concrete joints. These shrinkage cracks at the faces of the joint may usually be prevented by keeping the joint concrete water-cured until stressing. Proper preparation of surfaces will help prevent shrinkage cracks, as will use of an epoxy-painted surface. Shrinkage-compensated expansive cement would appear to be useful for joints of these types.

3.4.3   Manufacture of Segments

In addition to the general procedures for manufacture of precast concrete, special techniques are often required for precast segments.

(a) *Manufacturing Techniques for "Dry" and Close Fit Joints.*   Dry joints require an essentially perfect fit. Permissible tolerances are of the order of 1/64-inch (0.4 mm).

Lightweight aggregate concrete may be slightly less sensitive to localized stress concentrations due to its lower modulus of elasticity; however, chamfers are essential to prevent edge spalling. (See Fig. 6)

Fig. 6. Precast lightweight girder segments are erected with "dry" perfect fit joints, then post-tensioned. Reconstruction of tunnel section, San Francisco-Oakland Bay Bridge.

Dry joints are most commonly produced by casting successive segments against each other and match-marking them.

The entire girder may be cast in segments in the casting yard and then moved, in segments, to the erection site for re-assembly.

Greater speed and economy may often be attained by moving each previously cast segment sideways to act as the end form for the adjoining segment; thus, a daily cycle can be attained.

Another method of obtaining dry joints is by grinding to a 1/64-inch tolerance. Reference and axial lines must be established. The wear of the grinding wheel must be compensated for. This method has not been as successful in practice as hoped for; despite the cost and effort, tolerances have usually exceeded the allowable.

Dry joint faces have also been obtained by casting the concrete against heavy rigid steel plates, 2-inches or more in thickness, which have been milled to 0.005-inch. If these plates must be removed for stripping, exact repositioning can be obtained by using close-threaded machine bolts. Particular care must be taken in mix design and curing to assure minimum shrinkage.

Means must be taken, through mix design and consolidation (or other special means, such as vacuum processing) to minimize water and air-bubble entrapment on joint faces, particularly if these joints are thin and highly stressed.

(*b*) *Schemes of Manufacture of Segments.*   For obtaining highest quality, closest tolerances, and greatest economy, segments may be manufactured in a position or attitude at an angle different from their final position.

Box girder segments have been manufactured as a series of flat slabs, then assembled in the yard into the box section, and the joints poured.

Similarly, precast subsegments may be set in forms in the yard and the remainder of the segment poured in a second step.

The selection of the proper attitude or position in which to cast the segment will be determined by the following:

1. Tolerances required.
2. Form setting and stripping.
3. Reinforcement or prestressing requirements.
4. Type of joint and method of forming.
5. Concrete placement.
6. Handling and storage in the yard.
7. Costs of manufacture.

(*c*) *Turning Segments.*   When precast elements are manufactured in positions normal or at an angle to their final position, they must be turned without damage.

Prestressed elements are very sensitive to positions other than their final position, and the temporary tensile stresses must be computed for all the positions in which they will be handled, turned, or stored. Additional mild-steel reinforcement, external steel beams bolted on, handling in pairs, and temporary additional prestressing, are techniques used to counter temporary excessive tensile stresses.

Crushing of the concrete edges in turning may be prevented by turning in the air, using rigging, or by resting the lower edge in sand, or by using wood-block softeners, or by turning in a turning frame (cradle). Very heavy units have been successfully tilted or turned in these structural-steel turning frames, which give full support to the unit during turning and prevent localized high-bearing stresses on the concrete edges.

(*d*) *Horizontal and Vertical Curves.*   With transverse segments, vertical and horizontal curvature may be achieved by varying the joint width progressively. For dry joints, the change in curvature can be accomplished in manufacture by adjusting the already-cast segment to its new angle before pouring the adjoining segment.

3.4.4   Assembly of Precast Segments

(*a*) *Alignment.*   Accurate alignment of precast segments must be obtained prior to final jointing and stressing.

With dry joints, alignment may be obtained by means of keys or dowels.

Precast segments can be jacked, wedged, or shimmed to proper alignment before joint concreting. Multiple hydraulic jacks, and flat jacks have been used for this purpose.

When erecting on falsework, final alignment should be made after all units are in place, so as to compensate for dead-load deflection and anticipated creep deflections. With successive cantilever construction, continuous observation and calculations must be made of the effect of creep and elastic deformation, and corrections made in aligning succeeding segments.

(*b*) *Temporary Vertical Support.*   Temporary vertical support for segments may be provided by blocking off of falsework trusses. With the cantilever method of erection, steps or seats may be provided. Bolts and steel frames, either underneath or above the new segment, may be used to support it.

Design of temporary vertical supports should take into account impact from setting, the possibility of eccentricity during initial setting, side sway or tipping in erection, and localized concentrated bearing.

When setting heavy units from a barge-mounted derrick, subject to wave or swell action, the precast segment may be cushioned by landing it on rubber bumpers, timbers, or on inflated rubber tires, then jacked down to final position.

(*c*) *Temporary Lateral Support.*   Temporary lateral support must be provided to prevent sidewise displacement, overturning, buckling, and torsion. Cross-bracing to adjacent segments is often practicable. Where segments are part of a curved girder, guys and blocking will probably be needed to prevent twisting and overturning.

(*d*) *Temporary Stressing.*   Temporary stressing, using post-tensioning tendons or bridge strands, is often an effective means for providing temporary support and for holding the segments in proper position during assembly. Tendons have been used inside the boxes of box girders, outside the webs of I-girders, and on or above the deck in cantilevered construction. Such temporary stressing may also be utilized to pull the segment into final position, and to counter temporary tensile stresses.

The ability to jack a tensioned or partially-tensioned external tendon up or down, and thus change its length or profile, has been utilized to make slight adjustments in position for assembly.

Temporary stressing may also be used to temporarily fix a hammerhead segment to a pier. The hammerhead section may be set on wedges or flat jacks, and tendons used to stress it vertically against the pier shaft. When the need for fixed connection is over, the tendons can be released and the wedges or flat jacks removed.

Precast deck sections and cable-suspended structures may be assembled in such a way that the main longitudinal and transverse stressing is achieved by tendons located in the joints. The precast sections are first assembled on falsework or, in some cases, on cables. The tendons are then placed in ducts through the zone of the joint, the joint is poured, and the entire assembly stressed. With cable-suspended roofs, the precast units are hung on the stretched cables in a carefully predetermined pattern, the stress is readjusted in stages, and finally, the joints are concreted.

3.4.5   Post-Tensioning—Special Provisions for Segmental Construction

(a) *Anchorages.*   Anchorage pockets may be provided in precast segments with a high degree of accuracy and assurance of quality. By precasting under most favorable plant conditions, very high strength, accurate alignment with the duct, and accurate bearing surface may be attained, even with the normal congestion of reinforcing steel in the anchorage area.

(b) *Ducts.*   Duct diameters for segmental construction should be slightly larger (say, 1/4-inch = 5 mm) than for other post-tensioned work to eliminate excessive friction at joints due to minor misalignment.

Duct splices may be made by using flexible metal duct, with well-taped joints, or by rubber or rigid metal sleeves. Ducts may be formed through joints by inflatable rubber tubes. Screwed pipe sleeves may be employed.

(c) *Tendon Splices.*   Where precast segments have embedded tendons, the continuity of stressing being provided by external jacking or flat jacks, the joint must have sufficient width or a pocket of sufficient size to permit placing of the coupler in its extended position. Where more than one tendon is involved, the initial tensions must be equalized; this may be done by calibrated ratchet jacks. In certain cases it may be necessary to stagger tendon splices to minimize loss of concrete section. Splice design should be checked for fatigue loading.

(d) *Corrosion Protection.*   With precast segments, the possibility of moisture entering through the joint must always be considered. Where tendons must remain in ducts but ungrouted across joints, special means should be taken to seal the duct or joint or both. Positive means must be taken to insure that the joints of precast segments are completely sealed against moisture penetration to the tendons.

(*e*) *External Jacking and Flat Jacks.*   External jacking and flat jacks are sometimes used to stress precast segments. With external jacking, pockets or brackets are provided in the segments, hydraulic rams are then placed symmetrically about the joint, and the segments are forced apart. The joint is then concreted, cured (usually by accelerated means), and the rams and brackets, etc., removed.

In some cases, flat jacks may be inserted between precast segments. After jacking to the required stress, they may be pumped full of grout.

3.5   Composite Construction

"Composite construction" in prestressed concrete construction generally connotes the combination of precast prestressed concrete beams or slabs with a cast-in-place concrete deck or top slab. The two act together structurally: the prestressed member serves as the tension flange, the cast-in-place section as the compression flange. This monolithic structural behavior is the basic premise.

3.5.2   Advantages

The advantages of composite systems are the following:

(*a*) The precast tension element may be prestressed with straight tendons or minimum deflection, since, in itself, a large eccentricity of prestress is not required. Manufacture is simplified and the precast element is more easily and safely transported and erected.

(*b*) The precast element is relatively light.

(*c*) Standard precast beams may be grouped close together or spread apart to accommodate heavy concentrated loads or openings.

(*d*) The cast-in-place top slab ties the structure together and gives a uniform unbroken surface which may be screeded for camber, super-elevation, and warped surfaces.

(*e*) The precast beams, with no shoring or a minimum of shoring, support the cast-in-place deck and its forms during construction.

(*f*) Since no prestress was induced in the compression flange, the composite section is, theoretically, more efficient structurally.

(*g*) Combinations of lightweight concrete with conventional sand-and-gravel concrete are facilitated. Thus, the lower, prestressed slab may be of structural lightweight concrete and the topping of sand-and-gravel concrete.

(*h*) Full or partial continuity is more easily achieved.

(*i*) Shear transfer is more easily provided.

Success with composite construction depends on insuring proper shear and contact at the interface.

In the case of precast prestressed slabs which are joined in building construction with a cast-in-place topping, suitable shear transfer may usually be obtained by bond alone, provided the upper surface of the precast member is roughened. Rough screeding, followed by a coarse broom, will generally prove most satisfactory. Then, just before pouring the top slab, the surface should be cleaned with a water jet.

Where shear loads are higher, as in bridge construction, then steel wire or mesh ties are used.

With precast beams and girders, steel ties in the form of mesh or bars are required. As before, the top surface should be rough.

Shear keys have been found in a number of tests to be of less benefit than formerly believed. The shear keys do not come into action until the surface bond has failed. Therefore, shear keys are generally not employed with composite slabs and then only at the ends of heavily loaded girders. When shear keys are used, they should be of greater dimension than the largest size of coarse aggregate.

One means of providing full or partial continuity over the support of two adjoining spans is to embed mild-steel bars in the topping slab. These bars should extend well beyond the point of inflection. Staggering the ends of the bars will also help prevent a transverse crack on each side of the support.

Many small bars are much better than a few large bars in preventing cracking. Precast prestressed tension elements are particularly useful in providing a crack-free surface over the joint [see Subsection 3.1.6, Internal Tension and Compression Elements (Stressed Concrete Bars)].

Because of the rotational effects of camber and live loads, shortening of the precast prestressed beams, and the possibility of positive moment at the support under unusual conditions of loading, it is usually desirable to tie the prestressed elements together at the bottom. This may usually be accomplished by letting the tendons extend through the joint, overlapping, or bending them around a transverse bar, if necessary, to develop bond. Alternatively, extended reinforcing bars may be joined by welding or hooked around a transverse bar. (See Fig. 7)

3.5.2 Shoring

Since the precast element (slab or beam) is prestressed for its final service and since, in the case of a beam, it may lack an adequate compression flange, it may be necessary to shore it at the center or third points until the

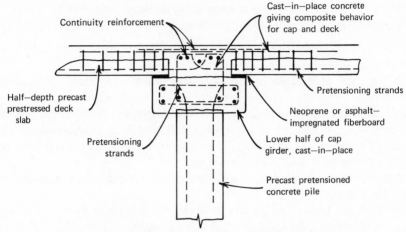

Continuity reinforcement

Cast—in—place concrete
giving composite behavior
for cap and deck

Half—depth precast
prestressed deck
slab

Pretensioning
strands

Pretensioning strands

Neoprene or asphalt—
impregnated fiberboard

Lower half of cap
girder, cast—in—place

Precast pretensioned
concrete pile

Note: As an alternative to bending up pretensioning strands
from deck slabs over cap, they may be simply lapped 30 diameters

Fig. 7. Composite construction: method of tying precast and cast-in-place concrete together for monolithic behavior.

top slab has been poured and attained strength. Shoring loads are usually light. Shores must be carefully placed and wedged or jacked so as not to break the beam in negative moment. Careful control is necessary.

3.5.3   Shrinkage of Top Slab

The drying shrinkage of the top slab reduces the load carrying capacity of the composite system. In effect, it applies a small counter prestress. Thus, every effort should be taken to minimize shrinkage. Possible means are:

(*a*) Low water-cement ratio.

(*b*) Selection of aggregates that give low drying shrinkage.

(*c*) Careful attention to curing: water cure is most effective.

(*d*) Use of shrinkage-compensating cements; a relatively new development with a rather spotty performance record to date. Most of the cases where this cement has not succeeded in reducing shrinkage have been traced to inadequate cure over a sufficient period.

(*e*) Thorough wetting of bottom slab before pouring.

3.5.4   Longitudinal Cracks

Longitudinal shrinkage (hair-line) cracks frequently form over the joint between precast slabs. This is due to the concentration of drying shrinkage; the rough surface of the precast slab prevents shrinkage cracks over the slab itself.

These can be minimized and/or prevented by taking steps to reduce shrinkage as noted above, and by proper detailing of transverse top steel. Many small bars or mesh are preferable to a few large bars. Mesh should be of sufficient steel area (gauge) to counter shrinkage stresses as well as provide the nominal transverse reinforcement required.

3.6 Internal Tension and Compression Elements
(Stressed Concrete Bars)

If a bar of small cross section or a thin slab is given a high degree of prestress and then incorporated in cast-in-place concrete, it will act as a tension bar, similar to reinforcing steel but with important differences. First, there is a comparatively large surface for bond, and the bond is both adhesive and crystalline in nature (secondary crystallization can take place across the interface under proper conditions). Secondly, due to creep in the highly stressed element, a degree of prestress is transferred to the adjacent cast-in-place concrete.

As a result of these and perhaps other phenomena, the full tensile value of the cast-in-place concrete (the full value under the stress-strain diagram) can be utilized effectively as the transverse reinforcement in prestressed pavements and floor slabs. They have proven extremely beneficial in crack control when used over supports to provide the negative reinforcement in continuous construction.

Since these bars normally contain only one central tendon, with that tendon encased in very dense concrete, they offer great durability.

Thin, highly prestressed slabs have been tested for use as facing panels on dams. In addition to their own inherent durability, they delay cracking in the adjoining cast-in-place concrete, thus enhancing the freeze-thaw durability and making it possible to use reduced cement contents for the cast-in-place concrete.

Similar highly stressed slabs have been used as the tensile face (and soffit form) for large prestressed girders in powerhouse construction in the USSR.

The ability to transfer prestress from the tensile element to the cast-in-place concrete, being associated with creep, is highly time-dependent. Ideally, prestressed tensile elements should be utlized within a short specific period after casting, say, one week.

Tensile elements, being concentrically stressed and fully bonded, are not subject to the same limits of prestress that we are accustomed to work with in normal prestressed construction. The higher the prestress, the more efficient the bar, subject to the limitation that creep rates are very high at these higher prestress levels. Values up to 40% f'_c would seem acceptable upper limits, pending further research.

Compression elements have been developed in the USSR and offer some very interesting possibilities. One type of compression element is a passive one; it takes the place of compressive reinforcement. Cylinders of high-strength concrete are wrapped with high-tensile wire under tension and are thus bi-axially compressed. Such passive compression elements have been incorporated in the compression flange of high-capacity girders.

Active compression elements have been proposed by Prof. V. V. Michailov for use in columns and the legs of towers. These would be wire-wrapped cylinders which, in addition, would be post-tensioned longitudinally with an unbonded tendon. They would be incorporated in the cast-in-place concrete of the column or tower-leg. After a sufficient amount of superimposed construction load had been applied, the post-tensioned anchor would be released. This would then expand like a jack, reducing the compressive stress in the cast-in-place concrete, and permitting deformation to be controlled. Its advantage lies in the fact that the triaxial prestress and the high quality of the factory-manufactured element allow the compressive element to take a very high temporary compressive stress. Since the compressive element is stressed to a maximum during manufacture, it has a similar relation between allowable stress in service and ultimate strength that a prestressed high-tensile wire has to its ultimate strength. Performance in fatigue is excellent because of the low range of stress variation.

Active compression elements could well be employed in reconstruction and underpinning projects in which access to the ends prevents use of normal post-tensioning procedures. As such, they would perform a function similar to flat jacks, with the important added benefit of exerting a continuing stressing force—not a fixed strain as in the case of flat jacks. The possibility of incorporating them in arch ribs, etc., is extremely intriguing.

3.7 Epoxies and Polyesters in Prestressed Concrete Construction

These exciting new materials are being adapted to prestressed construction in a number of ways:

1. For improving bond between the precast and the cast-in-place portions in composite construction.
2. For splicing piles and columns.
3. For connections between precast elements, both for primary (direct) load and for secondary forces such as seismic shear, load transfer, etc.
4. As a binder for ultra-high-strength concrete, or to coat aggregate particles to improve the cement paste-aggregate bond.
5. For coating prestressed concrete members that will be exposed to acid or other chemical attack.
6. For coating the contact faces in segmental construction.

7. As a wearing surface, epoxy-resin concrete may also be used in composite structural action with prestressed beams.
8. To coat tendons or to paint anchorages to protect them against corrosion.
9. As patches for repair of defects, such as cracks, plucking, pockets, etc.
10. For sealing joints between precast elements.

Epoxy resins have high tensile, shear, and compressive strength, generally good impact resistance, good wearing surface, and high durability. Their limiting characteristics are short pot life, high cost, and sensitivity to temperature.

The properties of epoxies vary widely, depending on the formulation, the relative amount of catalyst added, the method of placing (e.g., whether under pressure or not), and the curing. There are so many different epoxies available that it is wise to work directly with company technical representatives to insure selection of the one with the best properties for a particular application. Then it is extremely important to follow the directions exactly. Making one technician responsible for all the work helps to insure proper procedures. Most epoxy manufacturers offer technical service at the site.

Proper preparation of the surface is essential. Cleanliness is extremely important, for both the equipment and place of application.

Polyesters have not been as widely used in prestressed concrete construction applications because they lack some of the bonding and tensile strength characteristic of epoxies. However, polyesters are much less expensive and more easily applied wherever high compressive strength is required. They can, therefore, be used for applications requiring larger amounts of material, where the prime requirement is high compressive strength and accurate definition.

Sealants are formed of organic materials that ideally provide bond, watertightness, flexibility, and durability. They must be able to stretch and compress to accommodate volume changes in the precast units due to temperature and mosture differentials. They must be durable in water, heat, and sunlight; they must neither ignite, nor become brittle in cold weather. A number of proprietary materials are available and generally give excellent service if installed correctly. Accurate spacing of joint width is important, as is cleanliness and dryness of joint at time of applying sealant.

The entire subject of the use of epoxies and other organic materials in concrete construction is a complex and rapidly evolving matter. The ACI Manual of Concrete Practice, Part 3, contains two excellent chapters: "Guide for Use of Epoxy Compounds with Concrete" and "Guide for the Protection of Concrete Against Chemical Attack by Means of Coatings and other Resistant Materials."

3.8   Welding in Prestressed Construction

Welding is extensively used in joining precast elements in shear connections, and in other special work such as inserts. There are a few special precautions to take in prestressed concrete applications:

1. When welding connections between precast members, or shear keys, the effect of heat on the adjacent concrete must be carefully considered. Otherwise, the concrete will spall, and the anchorage of the connection plate be destroyed. Spalling is unsightly, usually requiring repairs. Burned and blackened concrete may also be unacceptable, particularly in exposed architectural applications.
   Heat may be reduced by the selection of low-heat rods, of small size, and by making a series of intermittent passes, allowing the heat to dissipate. Connections and inserts should be designed so as to minimize heat transfer. Inserts should be given adequate cover and be anchored well back.

2. Never permit accidental or intentional ground through a prestressing tendon. Never weld to a tendon under stress. Never weld to a tendon except in the special cases listed in 3 that follows. The point of ground will burn and destroy the high-strength properties of cold-drawn wire, usually causing a rupture if the wire is under stress.
   Care is especially necessary around pretensioning beds, with tendons stressed but not concreted.

3. Welding of projecting strands from a completed prestressed member is sometimes used for connections. It can safely be done provided there is no ground of the tendons elsewhere, as at the other end of the member. Also, the heat welding will destroy the high-strength properties so that the welded joint will have only the properties of mild steel. Use of a low-hydrogen rod is recommended.

4. Use of a welding torch to burn back the ends of strands at the ends of precast members can be safely done provided the tendon is not grounded, as at the other end or through inserts. However, to eliminate the possibility of such trouble, a small burning torch is preferable.

3.9   Power-Driven Tools in Prestressed Construction

These may be used to install fittings, studs, etc., as in all concrete construction with the following special precautions:

1. Prestressed concrete is usually higher strength with a higher modulus of elasticity; therefore, more likely to spall.

2. Never drive a power-driven stud in the area of prestressing strands where it might shear a tendon. Tendon locations should be marked on the member by an engineer prior to use of the power-driven tool.

The general safety precautions for power-driven tools must be enforced.

3.10   Cutting of Prestressed Concrete

Cutting of prestressed concrete with large jack-hammers may cause extensive spalling. The concrete is under stress and, in the zone near the point of cutting, there will be transverse tensile stresses, tending to burst the concrete.

Therefore, it is most desirable to make an initial cut with a concrete saw. This notch protects the concrete below and results in a relatively clean cut.

In lieu of or in addition to the saw cut, a heavy clamp exerting squeezing force on the concrete will counter bursting forces and help to control the cut zone.

Use of primacord or other explosives to cut prestressed concrete has not been widely practiced. However, primacord is used to cut relatively thin-walled, non-prestressed concrete pipe. With prestressed concrete, primacord is likely to cause greater spalling due to the transverse tensile stresses.

Drilling and coring of holes in prestressed concrete members is an excellent way of fitting piping and electrical leads through floor slabs. However, extreme care must be taken not to cut prestressing tendons. The permissible zones or areas for such coring must be approved by the structural design engineer and, when in proximity to tendons, the tendon location should be marked.

Lightweight concrete cuts and drills more easily, but also tends to spall to a greater extent.

3.11   Electro-Thermal Stressing

The use of electric resistance heating for prestressing is practiced rather extensively in the USSR and other Eastern European countries.

In one application, high-strength bars are heated (usually by electrical induction), then placed in forms and anchored to the forms. Concrete is poured, the bars contract, and a mild degree of prestress is introduced. The level of prestress is sufficient to overcome shrinkage cracking. However, precision is low because of the short length of tendon and the difficulty of accurately affixing the anchorages.

Electrothermal heating is also used with continuous-winding machines, in which wire is paid off under tension and laid in a predetermined pattern in the forms. The wire must be laid under stress around dowels, in a rather short radius bend. The considerable breakage of wires· has been a major problem. By electric-resistance heating of the wire, the ductility of the wire is increased and breakage reduced. Also, it may be laid under lower stress since a portion of the prestress will be supplied by the cooling contraction.

Such electrothermal heating has reportedly practically eliminated the problem of wire breakage in this application.

Electrothermal stressing has also been used experimentally in a rather unique manner. High-strength bars are coated with sulfur. They are placed in the forms, the concrete is poured and allowed to harden. Then the bars are heated by electro-resistance means. The sulfur melts, the bar expands. It is anchored against the ends of the concrete while still hot. Then, as it cools, the sulfur hardens and the bar contracts, thus prestressing the member.

This method, to the writer's knowledge, has not been used commercially. There are a number of possible problem areas, including the serious question of corrosion of the steel under high stress in contact with sulfur. It would seem that other more suitable materials may be found in the future, with which this concept may attain some practical value.

3.12   Chemical Prestressing

Several different expanding cements have been developed as a result of intensive research into gel chemistry. These have the property of first hardening, then expanding as secondary crystallization takes place. In general, the expansion is three-dimensional; thus, restraint must be applied to obtain a prestressing effect.

Chemical prestressing through expanding cement has a number of interesting possible applications. In slabs and pavements, expansion can counteract shrinkage and eliminate cracking. This, at present, is the most practicable application. If conventional stressing is applied on one axis, and the combination of weight and subgrade reaction provides the restraint on the second axis, then the expansion in the transverse axis will impart prestress.

Small-diameter pipes are produced in the USSR, using expanding cement. The expansion reacts against the spiral reinforcement. Longitudinal prestress is imparted by wires stressed in the conventional manner.*

Expanding cement requires continuous moisture to be present during the curing period in order to promote the expansion through secondary crystallization. This moist curing may be effected by total immersion (as of small pipe segments). Lightweight concrete aggregate, batched in a saturated condition, appears to offer ideal internal curing for expanding cement mixes, and experimental work along this line has been very promising.

Control of the degree of expansion is erratic at the present stage of development. Durability appears satisfactory, but still requires additional evaluation.

*For application to swimming pools, see Postscript, section vi.

Expanding cement concrete, and expanding grout may be utilized to fill connection pockets, such as the heads of hollow-core piles. Additional spiral should be installed in order to prevent excessive expansion. Expanding grout has been extensively used to set inserts and anchor bolts. The expansion is usually accomplished by the inclusion of iron filings or aluminum powder in the mix. The iron filings oxidize and expand. Aluminum powder reacts with the cement to produce hydrogen gas bubbles.

Expanding grout has also been extensively used in the grouting of post-tensioned ducts.

However, it is now believed that the use of aluminum powder may be dangerous in that mono-hydrogen molecules may occasionally be released. These can produce hydrogen embrittlement and sudden brittle fracture of the steel tendons. Although the probability is low, the consequences could be catastrophic; thus, the growing reluctance to use aluminum powder as an expanding agent in grouting post-tensioning ducts. Many recent regulations prohibit it.

There have also been some difficulties, particularly in Japan, with excessive expansion of grout in ducts causing longitudinal cracks in the concrete, along the duct walls.

Although not yet fully developed, use of shrinkage-compensating cement in grout for duct-filling would seem to offer considerable advantages. However, this should not be considered a substitute for such other shrinkage-reducing procedures as low water/cement ratio and shutting-off of grouting under pressure, but rather as a supplementary means to eliminate small shrinkage voids.

3.13   Prestressed Steel-Concrete Composite Beams (Preflex)

This unique Belgian system combines structural steel and prestressed concrete in composite action to produce a very efficient beam. It makes possible a very low depth to span ratio and reduced deflections.

The manufacturing process is very simple. Two steel beams are placed side-by-side, their ends secured together. They are jacked apart at the center. The outer flanges are then encased in concrete. (These will later be the tension flanges.) After the concrete hardens, the jacks are released and the beams straighten out. What results is a steel beam, working in composite action with a lower flange of pre-compressed concrete. The beam is then erected and the top flange encased in cast-in-place concrete, to act as the compressive flange.

This system obviously must give careful consideration to the losses in steel stress due to stress relaxation and creep and shrinkage in the concrete. Such losses are acceptable only because of the large steel area in relation to concrete area. It is essential that the lower flange encasement be bonded; stud bolts or shear lugs may be necessary. Also, adequate lateral ties (stirrups) are important, in order to hold the flange concrete.

Since this is a proprietary method, the procedures and techniques recommended by the parent organization should be followed in detail.

3.14   Continuous Wire Wrapping

Wrapping wire around concrete under tension is widely applied to prestressed concrete tanks. The wire is stressed either by passing through a die or by winding around braking drums. (See Chapter 15, Prestressed Concrete Tanks.)

Similarly, in prestressed concrete pipe manufacture, wire is wrapped under tension around either a steel liner or a concrete inner pipe. (See Chapter 18, Prestressed Concrete Pipe, Penstocks, and Aqueducts.)

Such methods have also been applied experimentally to beams, the wire having been wrapped under tension around the sides and ends of the beam.

Continuous laying of wire under stress is carried out by two basic methods in the USSR. In the first case, known as the turntable, the wire is wrapped around pegs or dowels that protrude up from the steel soffit. After concreting and accelerated curing, the dowels are hydraulically withdrawn.

In the second method, a machine travels down a longitudinal bed on rails. A feeding arm traverses back and forth, feeding wire through a freely rotating spindle. This wire also wraps around hydraulically-actuated dowels. (See Subsection 3.1.11, Electrothermal Stressing, for the use of electro-resistance heating to minimize wire breakage with this method.)

All of these methods lay out a considerable length of wire in rapid fashion. However, since most systems use only a single or double wire, the prestressing force in one pass is rather low and a large number of passes are needed.

Some newer machines, therefore, lay strand from off braked or differential-winding drums.

3.15   Jacking Apart

Concrete may be prestressed by jacking it apart at a central or interior location against internal or external restraint. External restraint may be in the form of abutments, similar to an arch rib's abutments. However, such abutments (and the member itself) are liable to creep, thus allowing a gradual loss of prestress. When using such external reactions, buckling must definitely be considered, as well as the probable failure mechanism, which may be explosive. For these reasons, this type of prestressing against external abutments is usually applied only to arch ribs. The initial need is for decentering, a temporary stage not affected by creep. In its long-term behavior, any creep of abutments or concrete is compensated by the arch-performance of the rib.

Jacking apart against internal restraint, on the other hand, is an entirely sound method, which can be effectively used in pavements, and in special cases where access to the ends is not available.

The tendons may be anchored at the ends, and encased in ducts, so as to be free during jacking. In this case, a post-tensioned construction results. The tendons must be joined at the central gap, equalized in length, and then the two segments jacked apart. Equalization of splices has been accomplished by using splice nuts, turned up by a torque wrench. Jacking tongs might similarly be employed, working against a fixed strut.

If the members are pretensioned, then the length of tendon to be stressed is only the short length at the gap. It is thus extremely difficult to ensure equalization of length of the several tendons, and to prevent excessive pre-stress loss in set.

However, it has been proposed to utilize this method in special cases by actually casting the full length (both segments) in one pretensioning line, with the gap formed; then fold one segment on top of the other and ship the two segments, still joined together by the tendons. After erection, the two halves are jacked apart. In this case, the tendons would automatically be of the right length (See Fig. 8).

Concreting of the gap should be with concrete of low shrinkage and accelerated strength gain. Low water/cement ratio, careful selection of aggregates and cements (including possibly shrinkage-compensating cement) will accomplish the first. Internal steam coils, or external steam jacket, or electric-resistance heating may be used to accelerate the gain of strength, along with use of a high early-strength cement.

Internal heating coils of copper are considered highly questionable from a corrosion viewpoint. Steel wires are much to be preferred.

3.16 Other Compressive Materials

Prestressing may be beneficially applied to other materials to produce a state of stress that will counter service-imposed stresses. The same principles apply as for prestressed concrete. Special consideration must be given to the modulus of elasticity, creep rates and duration, durability, etc.

Among the materials that have been prestressed with success are stone, brick, ceramics, steel, and wood.

Marble, granite, ceramics, and glass facings are often embedded in a thin, prestressed concrete slab, then erected as a large panel. The concrete slab and facing material work in composite action under prestress. Therefore it is

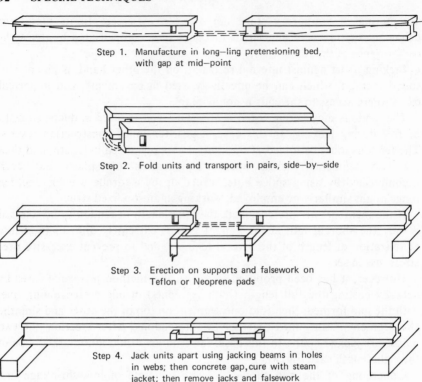

Step 1. Manufacture in long–ling pretensioning bed,
with gap at mid–point

Step 2. Fold units and transport in pairs, side–by–side

Step 3. Erection on supports and falsework on
Teflon or Neoprene pads

Step 4. Jack units apart using jacking beams in holes
in webs; then concrete gap, cure with steam
jacket; then remove jacks and falsework

Completed full–length girder

Fig. 8. Procedure for constructing long girders by using pretensioned segments.

necessary to evaluate the different moduli of elasticity of the facing materials
and the concrete, and locate the tendons so as to prevent excessive warping
or camber.

Native rock is sometimes prestressed to prevent its failure in tension and
shear. This is an extension of the rock-bolting principle. Because of the
higher stresses used, care must be taken to provide a proper bearing seat for
the anchorage, such as a concrete pad (precast or cast-in-place). Grout injection may be used in conjunction with prestressing to fill cracks or seams.
Usually the tendons must be re-stressed to compensate for early creep of
weaker material in fractured zones or seams.

Soil may be prestressed, or rather, anchors in soil may be prestressed.
Creep here is a major problem, being of a much greater order of magnitude

than with other materials. Therefore, it may be necessary to restress soil anchors several times.

Cast iron would seem to offer considerable opportunities for prestressing, using essentially segmental construction methods. Steel girders and trusses have been prestressed (e.g., for hangar roofs), using tendons to impart an upward camber, and making use of the efficiency of the high-strength steel tendon. Care must be taken to prevent buckling of the lower flange (chord) and to be sure that truss member connections are designed for the secondary stresses induced by axial shortening of the lower chord.

Finally, prestressing has been used very effectively in the restoration of ancient monuments, where existing segments are drilled, a tendon placed, and stressed, and the hole grouted. The temple columns of Baalbek, in Lebanon, are among the ruins restored in this manner.

3.17 Horizontally Curved Beams

Many structures can be most efficiently and aesthetically constructed with horizontal curvature. This especially applies to highway overpasses and bridges, and also to such structures as monorail girders and buildings. The tendency in the past was to form such horizontal curvature by a series of short chords; this applies to both precast and cast-in-place structures. However, such solutions were distinctly not elegant and presented deficiencies in aesthetics as well as in efficiency.

If the center of prestressing force is concentric with the center of gravity of the section, then the structural element will tend to retain its curvature; at every tranverse plane, the stress will be concentric.

Many cast-in-place prestressed concrete structures have been cast with horizontal curvature, some times of very short radius. Provided the ducts are accurately located, the subsequent post-tensioning will cause no subsequent deflection in curvature. However, such structures are sensitive to deviations in alignment; thus, ducts should either be rigid ducts, or else flexible metal ducts with an internal mandrel. Sometimes, especially with bars, the tendon itself may be used as the mandrel. Furthermore, accuracy of concrete section must be maintained in casting.

A number of important structures on horizontal and two-dimensional curvature have been constructed by using precast segments. This has permitted more accurate casting of thin sections and better control of duct position.

There has been a reluctance on the part of designers to recognize the potential for horizontal curvature of precast prestressed girders. In recent years, however, a significant use has been made of I-girders and box girders which were curved in plan. In manufacture the side forms have been made flexible, for example, by utilizing steel forms with vertical angle stiffening only. Pulling, jacking, or wedging them at a few critical points automatically forces the form to achieve a smooth curvature.

Most horizontally curved beams have been post-tensioned. However, pretensioning is possible if two beams are cast side by side, of opposite curvature, and the strands are deflected together (e.g., at quarter and mid-points). Then the two beams may be lifted and transported by a common lifting frame. Obviously each beam by itself is unstable, unless restrained horizontally. Thus, in erecting and immediately after erection, horizontal trussing or bracing must be employed.

A single girder may also have its strands deflected horizontally by jacking at side-pull points to a permanent girder. Before release, a horizontal restraining frame must be attached.

With such pretensioning, the tendon path is actually a series of short chords. Thus minor deflections of the beam will occur between side-pull points but can be kept within acceptable limits by calculation.

Side-pull devices may use a single steel rod or wire, which may be later cut back and patched as necessary.(See Fig. 9).

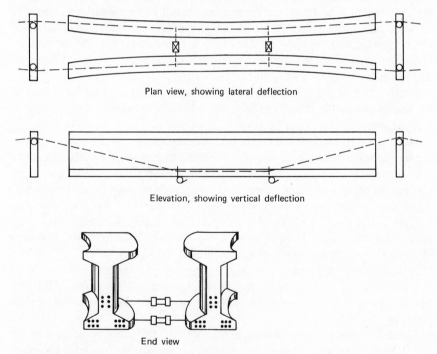

Plan view, showing lateral deflection

Elevation, showing vertical deflection

End view

Fig. 9. Manufacture of horizontally curved pretensioned concrete beams.

4

Durability and
Corrosion Protection

The designer and constructor-manufacturer share the responsibility for ensuring that the prestressed concrete structure will remain in essentially its same state despite the attack of the environmental conditions. This preservation of state is called durability.

Disruption may take several forms:

(*a*) Disintegration of the concrete.

(*b*) Chemical replacement in the concrete with a consequent loss of strength.

(*c*) Corrosion of prestressing tendons, causing loss of strength, fracture, or lower resistance to fatigue.

(*e*) Corrosion of the inserts and embedded fittings, connections, etc.

(*f*) Corrosion of anchorages.

The above forms may interact to intensify disruption. For example, corrosion of reinforcing bars produces products which swell and cause disintegration of the concrete cover.

There have been a great deal of research and many published reports on durability of concrete (by itself) and corrosion of steels (by themselves). It is essential in studying prestressed concrete structures that we consider the durability of the entire system, not the component parts. Otherwise, we might adopt measures to improve the durability of the concrete at the expense of increasing the corrosion of the steel.

Prestressed concrete is an extremely durable material and ranks high among all known structural materials for its resistance to the attack of natural environments. A recent survey of some 14,000 prestressed concrete

95

bridges in the USA showed that only 10 have shown any evidence of steel tendon corrosion. Freeze-thaw and salt-water immersion tests have demonstrated the inherent resistance of prestressed concrete. It is generally accepted that prestressed concrete piles are among the most durable piling for marine structures, even in tropical salt-spray environment.

Nevertheless, the maintenance of durability is achieved only by proper design and construction. The consequences of disintegration and corrosion are potentially catastrophic. The prestressing tendons have small cross-sectional areas under very high stress. Corrosion and disintegration are not random or sport occurrences. Rather, when disintegration and disruption do take place, it is usually due to some fundamental error or neglect; the cause is real and definite and often extends to the entire structure. Thus, except for some localized spot of impact or accident, if disintegration is found, a thorough investigation should be made of the entire structure.

Among the more common environmental attacks are the following:

I. Those causing or accelerating disintegration of or change in the concrete:
    1. Reactive aggregates.
    2. Unsound aggregates.
    3. Cement containing high percentages of alkalis or high C_3A.
    4. Freeze-thaw cycles.
    5. CO_2 in air or surrounding water.
    6. Erosion and abrasion from cavitation, ice, surf, moving sand.
    7. Marine organisms.
    8. Acids, sulphates, nitrates, or organic substances in mixing water or in surrounding water, as at discharge from chemical plants or in sewage structures.
II. Those causing or accelerating corrosion of steel:
    1. Salt or alkalis on aggregates.
    2. Chlorides in admixtures or water used for mixing and curing.
    3. Chlorides in water surrounding concrete (salt water), salt spray, salt fog.
    4. Oxygen.
    5. Sulphides combined with moisture on stressed tendons prior to encasement or protection.
    6. Stray electric currents.
    7. Alkalis in surrounding soils.
    8. High temperature.
    9. Embedded metals other than steel, particularly copper and aluminum.
    10. Inadequate thickness of concrete cover, or permeability of cover.
    11. Cracks.

12. Cement chemistry (e.g., too low C_3A).
13. De-icing salts, acids, or other aggressive chemicals.

Fortunately, the steps to be adopted to overcome these many forms of attack are complementary to each other. Most have been adopted as standard good practice. Occasionally, one has been neglected and, as pointed out above, the effects have been extremely serious.

4.1 Materials

All aggregates should be sound, non-reactive and abrasion resistant, and free from salt or alkalis. Particular care should be taken when working with aggregates from new sources, especially with siliceous rocks and desert areas.

Reactivity of aggregates and cement caused disintegration of an entire bridge in Europe. Reactivity and unsoundness of aggregates caused substantially every prestressed pile to disintegrate in a marine structure in South America.

Sands from deposits several miles from the shores of the Persian Gulf are heavily contaminated with salt from salt fog; and their use, unwashed, has led to serious corrosion in mild-steel reinforcing.

Aggregates should meet the requirements of ASTM C33 and, in addition, should be judged for their durability by a competent engineer, based on prior experience with the particular aggregates involved and tests. Tests are especially necessary when working with new aggregates. These tests, listed in ASTM C33, include tests for soundness (sodium-sulfate soundness test), alkali-aggregate reactivity, cement-aggregate reactivity, and for freeze-thaw durability.

Washing aggregates with fresh water will remove salt and dust from sand and aggregates.

The cement selected should have the proper chemistry. Portland cements, ASTM C-150 or C-175, are generally to be preferred.

Wherever chlorides will be present in the atmosphere (even up to 50 miles from the ocean in South Africa and California), low alkali content (less than 0.6% $Na_2O + K_2O$) is desirable. ASTM Type II, or Type III meeting the requirements of Type II, will give the best all-around durability. Type V, while giving the most sulphate resistance for the concrete, apparently decreases the corrosion-inhibiting power of the concrete cover for the steel due to low C_3A content.

Type I cement is entirely adequate for inland work not exposed to marine environment or atmosphere.

Special cements which were used in the past (not conforming to ASTM C-150 or C-175) often suffered magnesia replacement when immersed in sea water, with resultant loss in strength and impermeability.

Use of aluminous cement has been highly controversial. Too high a water/cement ratio may lead to swelling and breakdown of the resultant concrete. The present state of knowledge seems to indicate that aluminous cement may be safely used only if:

1. The water/cement ratio is kept below 0.4.
2. No steam curing.
3. The aluminous cement contains no sulfur or sulfides.
4. The environment is not continuously wet or submerged.
5. Aggregates contain no free alkali.
6. The sand used has a minimum of fines.
7. Proper curing (especially by water) for the first 48 hours.

Super-sulfated cement suffers severe loss of strength when steam cured.

The water used for washing aggregates, mixing, and curing shall not contain more than 650 parts per million of chlorides as Cl, nor more than 1300 ppm of sulfates as SO_4. The total amount of chlorides in sand and aggregates shall not exceed 0.02%. Admixtures shall not contain more than a trace of chloride ion. These limits on chlorides are of great importance in prestressed concrete, since the chloride reduces the pH of the concrete coating at the surface of the steel. Any chlorides or acids on the surface of concrete and especially in ducts should be thoroughly flushed clean prior to insertion of tendons.

4.2. Proportioning, Placing, and Consolidating the Mix

The mix should be proportioned to give the greatest density (least porosity). This is primarily a matter of low water/cement ratio, which should normally be kept below 0.45. Water-reducing admixtures are desirable for obtaining this low water/cement ratio, while maintaining adequate workability.

The cement factor should be chosen to give adequate alkalinity at the surface of the steel. Normally, this means not less than six sacks per cubic yard. An average value for prestressed concrete would be seven sacks per cubic yard.

The maximum size of coarse aggregate should be chosen, based on the size of the member and spacing of strands. Too large coarse aggregate generally reduces strength. For most structural members, 3/4-inch is the optimum maximum size.

The proportions of sand and coarse aggregate should be selected for workability and maximum consolidation of the resultant mix. Fine fractions are generally necessary in low cement factor mixes. With higher cement factors, the fines are less important and gap grading may be employed with beneficial results. The point to be kept in mind in such proportioning is that strength is of little value if the concrete is highly porous and permeable, so primary attention should be given to obtaining maximum consolidation.

To achieve consolidation and homogeneity with dry mixes, intensive internal vibration is required. External vibration may be beneficially used, in addition, as it achieves a dense, well-compacted surface cover.

4.3 Forms

The form surface should be selected to give a fine finish to the concrete. Fiberglass or steel forms, with suitable form oil, will produce a dense smooth surface. Plywood is satisfactory and, in addition, absorbs some of the water and air that moves to the surface under vibration.

4.4 Curing

Proper curing, by providing continuous moisture, will insure complete hydration of the cement particles. Furthermore, it will eliminate or minimize shrinkage cracks. The drying shrinkage will be delayed until the concrete has sufficient tensile strength to resist the strain without cracking.

Concrete which has been dried after curing generally exhibits better durability in sea water and splash zones than that which has been continuously wet.

4.5 Sufficient Cover

The concrete cover protects the steel by creating a passive condition of high pH at the surface of the steel. Too thin a cover allows carbonation to proceed, usually around the surface of the coarse aggregate particles. Carbonation lowers the pH.

Oxygen is necessary to the corrosion mechanism; a thicker cover minimizes the movement of oxygen to the steel surface. In sea water, chloride ion movement is also inhibited by thicker covers.

The cover should properly be related to the density and cement content. The exact relationships have not been thoroughly established, so arbitrary values are usually used as guides or standards. Thicker covers make it possible to achieve better compaction, fewer voids, and less permeability.

4.6 Prevention of Cracks

Cracks allow carbonation penetration and are also a route for oxygen to the surface of the steel. Cracks may further play a part in electrolytic cell formation in the concrete.

On the other hand, cracks are generally not as important a factor in the corrosion mechanism as commonly believed. There is considerable evidence, particularly the work by Dr. Abeles (see References), that cracks at the concrete surface narrower than 2 mm (0.01 inch) will not necessarily lead to serious corrosion. Subsequent research indicates this upper limit may be less than 1 mm (0.005 inch). In any event, hairline cracks are not the basic corrosion hazard that was once feared; they are certainly of far less importance than porosity.

4.7 Condition of Steel at Time of Concrete Encasement.

This is again a matter of considerable controversy, with some experts claiming the steel should be essentially bright and others permitting a light coat of rust. The concern of the former is that the degree of permissible rust is difficult to define, that substantial rusting may permit an oxygen-gradient at the surface, and lead to electrolytic corrosion. Those who permit a light coat of atmospheric rust define it by stating: "It should be capable of removal by wiping with a soft dry cloth." Some standard specifications, therefore, require that the prestressing steel be completely protected from rusting prior to actual placement in the forms, and that concrete encasement be accomplished within 24 to 48 hours thereafter.

Protection in transport and storage may be provided by wrapping in watertight paper and enclosing dehumidifying crystals, such as Shell Vapor-Phase-Inhibitor (VPI) powder. The tendons should never be coiled more tightly than manufacturer's recommendations permit.

German specifications require storage of tendons at the jobsite to be in watertight, heated enclosures, with the relative humidity kept below 20%.

Those specifications which do permit slight rusting are emphatic that there be no pitting.

The author prefers the more conservative provision, of having the steel bright at time of placement in the forms, for all main structural members. For less critical members, such as railroad ties, slight rusting is acceptable and acts to increase the bond.

4.8 Elimination of Voids at Steel Surface

Studies, such as those at the Denver Research Institute, have indicated that steel corrosion is associated with a void at the steel surface. In pretensioning, this can be prevented by mix design and thorough consolidation. In post-tensioning, grouting procedures should be adopted which will prevent or minimize these voids.

4.9 Air-Entrainment for Concrete Exposed to Freeze-Thaw Cycles and De-icing Salts

Concrete for prestressed structural elements that are to be exposed to freezing and thawing in a moist condition should contain entrained air. In

Western Norway, up to 8% entrained air is used with success in combatting the combination of freeze-thaw and marine environment.

4.10   Proper Grouting of Ducts in Post-tensioned Concrete

Proper grouting is essential for corrosion protection and prevention of bursting during freezing. (See Chapter 5, Section 5.1, Grouting, for details.)

4.11   Corrosion Phenomena

The corrosion of steel in concrete is a very complex phenomenon. Several forms of corrosion are of sufficient prevalence or seriousness to justify review in summary fashion.

Stress-corrosion cracking is an extremely rare phenomenon but, unfortunately, the occurrence of the word "stress" in both stress corrosion and prestress has lead to fear. Stress corrosion usually is associated with minute traces of chlorides or sulfides and, possibly, other negative ions, occuring in a humid atmosphere.

Heat-treated steel wire, as used extensively in Europe, and oil-quenched-and-tempered steel wire are believed to be more susceptible to stress corrosion than the cold drawn wire in common use in the United States.

A few isolated cases of serious corrosion, caused by a combination of stress and sulfide ions, have occurred when heat-treated wire, shipped in closely wound coils, was stored in contaminated mud. After installation, and indeed after grouting, the wires failed progressively. Similar cases have occurred when wires have been stressed and then left exposed to moist air in the vicinity of refinery or industrial plants. It is believed that the minute concentrations of H_2S in the moist air lead to corrosion and brittle fracture. Therefore tendons should preferably be protected in as short a time after stressing as possible. If they must remain stressed but ungrouted, then positive steps must be taken to prevent moisture entry. Tendons should never be coiled or bent more tightly than as originally shipped by the manufacturer.

These failures above are generally classified as one form of hydrogen embrittlement. Hydrogen embrittlement occurs when mono-molecular hydrogen ions are able to enter between the steel molecules. Hydrogen embrittlement is also theoretically possible when dissimilar metals, such as aluminum or zinc, are used in the vicinity of steel. In laboratory tests it has been very hard to produce such a reaction, but a few cases are believed to have occurred in actual practice. The author would prefer to ban aluminum ducts; however, galvanized ducts may, in his opinion, be used safely if sealed from moisture at all stages with VPI powder dusted on the tendons.

A new development of promise will provide lead coating for ducts; this has the lubricating and corrosion-resisting advantage of zinc but no risk of hydrogen liberation.

The F.I.P. Commission on Durability warns against the use of dissimilar materials, other than steel, in prestressed concrete, because of the possibility of electrolytic corrosion and hydrogen embrittlement. Aluminum and copper are particularly to be feared.

Salt cell electrolytic corrosion is the most common cause of corrosion for prestressing (or mild-steel-reinforcing) steel exposed to a marine environment. Complete immersion in salt water does not lead to salt cells. In the splash zone, however, the salt water splashes on the concrete, penetrates a short distance into permeable concrete, then the water evaporates. A salt-concentration cell will then be formed. Electrolytic corrosion may then take place with the steel, especially if a pore or void exists at the steel surface, or if chloride ions or carbonation have reduced the pH of the concrete cover.

Salt cells require both the dissolved salt in water or spray or fog, and subsequent evaporation. In both South Africa and California, salt cells have formed many miles from the sea as a result of salt fog.

The probability exists for this same condition to take place in concrete barges and tunnels if salt water enters from one face and evaporates from the other. However, as far as is known to the author, this has not occurred in practice, but this should not be taken as conclusive since most tubes, etc., have been waterproofed or otherwise sealed.

A case of serious corrosion of post-tensioned tendons is reported in which the precast architectural units were dipped in dilute acid for etching but apparently ducts were not flushed with water afterwards. Subsequent grouting was also not thorough, resulting in voids containing dilute acid. Failure of some tendons occurred several months after completion, and subsequent examination revealed serious corrosion on other tendons.

Electrolytic corrosion between dissimilar metals, such as copper or aluminum, embedded in the concrete, and prestressing steel, must be regarded as a distinct hazard if an electrolyte is present at any stage, especially salt water.

Some general observations concerning corrosion may be made:

1. In most pretensioning practice, a seven-wire strand is used. Concern has been expressed over the possible penetration of chlorides, etc., through the interstices between the center wire and outside wires. Numerous inspections of pretensioned elements in service have shown that this is not a problem, that the cement paste has indeed filled the interstices. However, this is not true of seven-wire strands grouted in post-tensioning; for this case, the anchorage or other means must serve to seal the interstices.

Extensive tests on large post-tensioned tendons show that grout penetrates strands more completely than tendons consisting of parallel wires, unless the wires are spaced by soft-iron wire wrapping.

2. Galvanizing* and epoxy-coatings give suitable protection to the steel, but only if the coatings are continuous. Extreme care must be taken to prevent abrasion of the coating in handling and insertion. Most specialists today would prefer not to employ these coatings except for special cases, and then only if integrity of the coating can be assured.

3. There are a number of chemicals which inhibit corrosion. Foremost among these is $NaNO_2$. This is added to concrete in the USSR for use in marine exposure. However, relatively large amounts are needed to be effective, and it is believed that the other measures listed earlier are more practicable.

4. An alternative solution to the use of VPI powder for corrosion inhibition during shipment, is coating of the wire with a sodium-silicate film. As yet, this has not been applied commercially, but appears to offer promise.

Water-soluble oils offer good corrosion protection to post-tensioning tendons prior to grouting, but they are difficult to remove completely so as not to impair bond.

5. The anchorages, in post-tensioned construction, must be protected from corrosion. The wires at the anchorages are under higher stress than anywhere else. In wedge-type anchorages bearing and bending stresses are added to direct tensile stress. In buttonhead anchorages, the head of the wire is cold-worked to the point of having microcracks.

This matter is especially important when high-capacity tendons are used. In such cases little help can be provided by the bond established by grouting. The anchorage must be maintained, and the terminals of the wires or tendon in the anchorage must be protected from the possibility of corrosion.

Epoxy concrete, of sufficient thickness, and well-keyed, appears to be the best means of providing such protection. Particular care should be taken to prevent leakage down the crack between the patch and the concrete; this may produce a worse condition than if the patch were left off. So the joint must be thoroughly bonded or sealed.

6. When the anchorage is seated in a pocket, and the encasement consists of filling the pocket flush with the ends, performance has been generally excellent, with no cracking, rust-staining, or other evidence of corrosion.

*See Postscript, section i.

Table 4.1 Techniques for Obtaining Maximum Durability
of Prestressed Concrete

Aggregates	Sound, nonreactive, abrasion resistant
Cement	Low alkali content (less than 0.6% Na_2O + K_2O) Moderate C_3A (for marine environments) High cement factor
Water	Fresh water Free from chlorides and sulphates
Concrete mix	Low water-cement ratio Clean aggregate Small size coarse aggregate Dense grading High-bond aggregate (where exposed to abrasion or cavitation) Limitation on chlorides from any source
Admixtures	Water-reducing Nonsegregating Air-entrainment (for freeze-thaw environments)
Forms	Smooth surfaced No re-entrant angles, projections, etc. Avoid sharp corners, edges
Placing	Thoroughly consolidated and compacted Minimize bleed holes
Cover	Adequate cover over mild steel Adequate cover over prestressing steel
Finish	Troweled
Construction joints	Well prepared Well bonded (by pre-soaking and grout or by epoxy)
Curing	Adequate water cure or steam cure Moisture available or sealed in during cooling period Drying after curing (for marine and freeze-thaw environments) Water free from chlorides or sulfates
Mild steel reinforcement	Free from pitting Well-distributed bars Avoid use of very large bars
Embedded metals	Avoid galvanic action, especially Cu and Al Ducts to be thoroughly clean and flushed with fresh water or fresh water containing inhibitor
Prestressing steel	Free from pitting and extensive surface rust Clean and dry, no salt Galvanized or plastic or epoxy-coated (for special cases only) (coating must not be abraded)

Kept free from corrosion until finally grouted or protected,
(a) VPI powder, (b) greases, (c) sealing, (d) limit time of exposure
High degree of prestress

Anchorages

"Flush-type," with the anchors themselves in pockets, are more
thoroughly protected
Epoxy concrete is best material

Grouting

Follow best grouting practice

Coatings

Bitumastic
Epoxy
Metallic sheathing   for special cases only (e.g., chemical)
Wood lagging

5

Protection of Tendons in Ducts

5.1 Grouting

Grouting with cement grout is the most widely used method of encasing tendons in the ducts of post-tensioning systems. The grout fulfills a number of purposes:

1. Encasement of the steel in an alkaline environment for corrosion protection.
2. Filling of the duct to prevent water from collecting and freezing.
3. Provision of bond between tendons and structural concrete.
4. Completion of the concrete cross-section.

Grouting should take place within 48 hours after placing the steel and within 24 hours after stressing. If tendons must remain longer than this, then, preferably, they should be left unstressed and should have special protection against corrosion. Dusting VPI powder on the tendons as they are inserted and then sealing the duct and vent tube ends with tape is one such means.

The technique of grouting is one of the most important aspects of the entire prestressing construction sequence. Since it is somewhat complex, and since it depends on proper and skilled performance by workmen, a number of rules and procedures have been developed to ensure proper performance.

The grout, ideally, will fill the entire cross section without voids. However, it should be free from substantial expansion pressures that might produce longitudinal cracks in the structural concrete. It should be dense and homogenous so as not to leave voids along the steel. (In present-day practice, the practical limit is "no voids larger than 1/8 inch diameter.") It must flow easily so as to fill the interstices between wires, and between wires and duct where the wires bear. The grout should not be susceptible to freezing and thawing disruption. It must have high compressive strength and high bond strength. It should attain reasonably high strength as soon as possible. It must have a fairly high cement content.

Materials, mixtures, and injection procedures should be selected to minimize bleeding of grout after injection. Bleeding is due to the fact that the water has a specific gravity only one-third that of cement; thus, sedimentation tends to take place and lenses of bleed water may be left in the ducts. This entrapped bleed water may freeze and rupture the member, or it may be reabsorbed, leaving air voids along the tendons, thus permitting an oxygen gradient to develop, with subsequent corrosion. Some specifications limit bleeding to 2% at 65° F (20° C) in 3 hours, with a maximum of 4%, and require that the separated water must be re-absorbed within 24 hours.

Expansion agents are frequently used in grout. They must not have too much expansive force or the duct may be ruptured. Total expansion should not exceed 10% (measured unconfined). During the test the testing tube should be covered to prevent evaporation. Use of aluminum powder, which evolves hydrogen gas, is *not* recommended because of the remote yet real possibility of the release of mono-molecular hydrogen and consequent hydrogen embrittlement of the steel. Admixtures are available commercially which liberate a relatively inert gas such as nitrogen.

Furthermore, the expansion is worse than useless if it takes place during mixing rather than after injection.

The best assurance of obtaining satisfactory grout and, therefore, the most important control is the water/cement ratio. This should not exceed 0.45. The grout must also be thoroughly mixed by machine to insure complete mixing and to obtain a workable grout for injection.

Neat-cement-and-water grout has proven satisfactory for ducts up to 3 to 5 inches in diameter.

In grouting large ducts a sand-and-cement grout is sometimes used. In these cases neat-cement grout should be injected first, followed by the sand-cement grout. After initial set (usually 1 1/2 to 2 hours), a second injection of neat-cement grout should be made.

Ducts. The duct size should be selected to give a ratio of steel tendon area to duct cross-sectional area of approximately 0.5. If too large a percentage of the duct is filled with steel, more voids will be formed. Tests have shown that fewer voids are found when strands are used as tendons than when wires are used, presumably because air bubbles can escape more easily around strand, and grout can more easily penetrate the spiral paths.

Proper grouting can be obtained only when the entire system is free of holes or joints which would allow grout to leak out of the ducts. This requires that a positive means of sealing be adopted for joints between precast members. The system must also be free of blockages or restrictions. These can be caused by collapsed or dented ducts, debris and foreign matter in the duct, and ice. Sudden changes in alignment and cross section of ducts should be avoided.

Leaks from one duct to another may occur at joints in precast members, unless properly sealed, or when two or more ducts are bundled next to each other and the ducts (or the joints in the ducts) are not grout-tight.

Ducts, particularly those of thin metal, are often rendered nontight by corrosion in transit or storage, by tearing and ripping in handling, or by tearing when placing adjoining reinforcing steel. Duct joints may be accidentally pulled apart when inserting a tendon prior to concreting. Ducts may be inadvertently holed by carpenters drilling holes for form ties or inserts. Ducts may also be holed by rough use of an internal vibrator.

Ducts can be inspected to insure integrity by visual inspection with the aid of a flashlight, by compressed air, or by water. Ducts are not required to be completely tight against air pressure, but use of air will help locate large leaks.

Ducts may be sealed or repaired by several wraps of waterproof tape. With holes or gaps larger than 1/4 inch, they should be sealed by a metal strip taped in place over the hole.

Grout ports and vents are required for the injection of grout and for the escape of air and water at all high points, changes in duct section, and at the terminus of any grouting length. These must also be sealed to the duct with tape. Vents or drains are also desirable at the low points to prevent water from collecting.

After concreting, but prior to insertion of the tendons, the ducts may be checked for blockages by passing through a rubber ball of smaller diameter on the end of a messenger wire or by the use of compressed air. After tendons are inserted, see-sawing back and forth will give an indication of any blockage (and will help remove minor restrictions). Compressed air admitted at one end should pass through with only nominal pressure build-up at the inlet. Then the outlet can be stopped and an air test made for tightness. If pressure falls slowly, then the duct is sufficiently tight; but, if it falls rapidly, a leak exists.

Materials.   The cement used should be ASTM Type I, II, or III. The cement must be free from any lumps, else serious difficulties are inevitable.

Water/cement ratio should be limited to 5 gallons per sack of cement (water/cement of 0.45 or less), using a retarding admixture as necessary.

The admixture, if used, may contain an expansive agent, giving 5 to 10% unrestrained expansion. The admixture must not contain chlorides as Cl in excess of 0.25 % by weight of the admixture.

Air entrainment in amounts of 4% to 8% should be achieved whenever there is danger of freezing and thawing.

"Intrusion Aid-R" and "Interplast-C" are two admixtures which have given satisfactory results. They are used in quantities of 0.5 to 0.7% by weight of cement and act as air-entraining agents as well as increasing flowability and causing expansion.

If sand is used (as in grouting large ducts), it should be fine sand and the mix should contain at least 10 sacks of cement per cubic yard. Use of sand is not recommended for normal size ducts (less than 5 inches). The mix should be proportioned by weight.

Grouting Equipment. This should be capable of producing grout of a uniform and, if possible, colloidal consistency. Mixing should be mechanically, not by hand. Water should be added first to the mixer, then cement (and sand, if any). Additives should be added during the latter part of the mixing time. Mixing time should normally be 2 to 4 minutes. After mixing, the grout should be kept continuously agitated or moving.

The grouting pump should be a positive displacement pump, producing at least 150 psi discharge pressure and have adequate seals to prevent introduction of oil, air, or other foreign substances into the grout and to prevent loss of grout or water. The pumping pressure should be controlled so as to prevent pressures in excess of 150 psi, so as to minimize leaks and prevent damage. A pressure gauge should be installed at the discharge of the pump.

The grout should be screened prior to introduction into the pump. Screens should have clear openings of about 0.07 inch (14 mesh). The grouting pump should be fed by gravity. A suction feed tends to suck air into the grout mix. Grout hoses, valves, and fittings should be watertight. Standby water-flushing equipment, powered by a separate power source, should be available, with a pressure of 250 psi.

A correct balance of materials, proportions of materials, method of adding to the mix, and mixing time will produce a grout which is

(*a*) easily pumped initially;
(*b*) will not take a false set during the planned grouting time;
(*c*) has good final strength.

The grout should not exceed 90°F in temperature. If necessary, the mixing water should be cooled. Blockages are more common in hot weather.

Injection by compressed air is usually *not* satisfactory and should not be used. There is danger of introducing a slug of air and of erratic, uneven grouting.

Procedure. The ducts should be flushed with water containing about 1% of lime. At this time, watertightness of the duct system can be checked against leaks. (Slow seepage is acceptable.) Then, immediately before grouting, the ducts should be blown free with oil-free compressed air.

Grout should then be injected, initially with a low pressure (about 40 psi), increasing it until grout runs out the vents. Then pressure can be slowly brought to the designed pressure. The optimum flow rate for grout in the normal-sized duct (3- to 5-inch diameter) is about 5 to 12 meters per minute (15 to 40 fpm). The slower rates are preferable in that they reduce the number of voids. Grouting must be continuous until the consistency of the grout emerging from the vents is the same as that being injected, without visible slugs of water or air. As grout flows out the nearest vent, it can be closed off, the grout then flows to the next vent, and so on.

All vents and the injection pipe must be closed off under pressure. Use of plastic or rubber nipples, with squeeze clips is one way to accomplish this. These nipples can be cut off after the grout has hardened. If the pressure is allowed to fall, serious voids due to entrapped air and bleed water will result. Injection pipes are preferably fitted with mechanical shut-off valves. Full injection pressure should be maintained for 30 seconds to a minute before closing off the injection pipe.

Tests have shown that holding the pressure on the grout tends to cause voids to disappear.

Recent tests on tendons with substantial curvature in the vertical plane, e.g., humped for negative moment, indicate an alternative procedure may produce superior results. A standpipe is provided at the high points in the tendon of sufficient capacity to hold all water displaced by sedimentation (up to 20% of the grout volume in extreme cases). The grouting then continues until grout of the same consistency as that injected appears at the vents, then the low ends are shut off but free expansion is permitted at the standpipe(s). Such procedure should also be followed for vertical or semi-vertical tendons.

For very large ducts (over 5 inches), the grout should be injected a second time after about two hours.

If grout pressure builds up above allowable, then injection may be transferred to a vent which has already been filled and capped, just as long as a one-way flow of grout is maintained.

If the grout pressure exceeds 150 psi and the one-way flow cannot be maintained, it indicates a blockage and the duct should immediately be flushed out with water. (See Fig. 10)

Grouting in Cold Weather. The freezing of fresh grout presents the danger of cracking of the structural walls and permanent damage to the structure. Therefore, if possible, postpone grouting until warmer weather (no danger of frost within 48 hours), keeping ducts sealed against accidental entry of water.

Before grouting after a period of cold or frosty weather, the ducts should be flushed with warm water (but not steam) to remove any ice. In temperatures below freezing the water must be blown clear with heated compressed air to prevent refreezing.

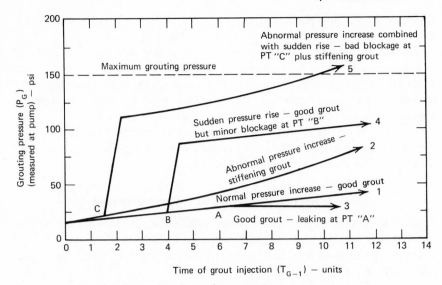

Fig. 10. Indicated performance of grouting operations.

Note.   Grouted duct length and grout injection time have
the following relationship:

$$T_{G\text{-}i} = \frac{V_v \times L}{C_p} \; ; \; \text{or} \; L = \frac{T_{G\text{-}i} \times C_p}{V_v}$$

where: $T_{G\text{-}i}$ = grout injection time (minutes)
    V_v = duct void volume (ft^3/ft)
    C_p = rate of grout pump (ft^3/minute)
    L = length of duct filled with grout (ft)

(Reproduced by courtesy of "Recommended Practice for Grouting Post-Tensioning Tendons, July 1967–Tentative," published by Prestressed Concrete Manufacturers' Association of California, Inc. and Western Concrete Reinforcing Steel Institute.)

If it is forecast that the temperature of the structure will not fall below 5°C (40°F) within the next 48 hours, grouting may be undertaken, using a grout containing 6 to 8% entrained air. If grouting must be carried out in cold weather, then means must be taken to keep grout (and adjoining structure) above 5°C (40°F) for 48 hours.

As noted earlier, do not use any admixtures containing calcium chloride.

After Final Injection.   Seal all openings and vents with cement grout or epoxy after final injection.

Records.   Records of grouting mixes, times of injection, quantity of grout used, and pressures should be maintained so as to be able to verify compliance with the specifications for that particular project. These records may show up radical problems such as leaks, blockages, etc.

Gamma radiography, using a Co^{60} source, is used extensively in England as a means of verifying the completeness of grouting of main tendons. When voids are shown to exist, the duct is drilled into from the girder side·and a secondary injection of grout is made. (Two holes are required—one for injection, one for venting.)

Grouting of post-tensioning tendons in nuclear reactors has been the subject of special studies and tests, and certain additional procedures have been developed. These are discussed in Chapter 18.

5.2   Corrosion Protection for Unbonded Tendons

Because all steel is subject to corrosion, prestressing tendons should be protected during storage, transit, construction, and after installation.

Unbonded tendons and their protective coating may be installed in one of the following ways:

(a) Tendons are coated against corrosion and for lubrication, sheathed so as to permit slippage during tensioning and for protection of the coating where required, then cast in the concrete. After hardening of the concrete, tendons are stressed.

(b) Tendons are coated against corrosion and placed externally to already-hardened concrete.

(c) Tendons are coated against corrosion and pulled into ducts.

(d) Tendons are pulled into ducts. The ducts are then filled with a suitable material, such as grease.

The coating material should:

(a) Remain free from cracks and not become brittle or fluid over the entire anticipated range of temperatures, at least $0°F$ (-20°C) to $160°F$ (70°C).

(b) Be chemically stable for the life of the structure.

(c) Be non-reactive with surrounding materials such as concrete, tendons, wrapping, or ducts.

(d) Be non-corrosive or corrosion-inhibiting.

(e) Be impervious to moisture.

The coating material should be continuous over the entire length of tendon to be protected, since the existence of a bare spot on a coated tendon is especially susceptible to electrolytic corrosion.

The minimum coating thickness depends on the particular coating material selected, but it should be adequate to insure full continuity and effectiveness with a sufficient allowance for variations in application. Where the coating material is applied before tendons are pulled into the ducts or casings, the coating material should be sufficiently tough to resist abrasion.

Coating material may be bitumastics, asphaltic mastics, greases, wax, epoxies, or plastics.

Galvanizing* or cadmium-plating should be by the hot-dip process and shall have the following minimum thickness:

For strand: Class A, as specified in Table 4 of ASTM Designation A-475;
For bars and wires: current commercial specifications call for between 0.6 and 0.9 ounce per square foot of surface.

Where tendons are exposed to a corrosive atmosphere, salt air, or high humidity, additional protection may be required.

For systems using friction grip anchorages, the coating may remain or be removed and cleared from that portion of the tendon to be gripped at the option of the tendon supplier. If the coating is removed, this portion of the tendon must be protected from corrosion after the anchorage has been installed.

Where the filling material is injected after the tendons are in place, the filling material should have these additional properties:

(*a*) No appreciable shrinkage or excessive volume increase.
(*b*) Suitable viscosity at ambient temperature or with moderate preheating to permit injection by pumping.

Wrapping and Ducts.   Wrapping must be continuous over the entire zone to be unbonded, and should prevent the intrusion of cement paste and the loss of coating material.

It should have sufficient tensile strength and water resistance to resist damage and deterioration during transit, storage at the job site, and during installation. Impregnated and reinforced paper, plastic, fiberglass, and metal have been used as wrapping material.

If the wrapping is damaged, the damaged section should be removed, and the tendon inspected. If the tendon is undamaged, it should be recoated with corrosion-protection material and the damaged portion of the wrapping replaced and sealed.

Ducts for unbonded tendons are similar to those for post-tensioned grouted tendons. They should be mortar-tight and non-reactive with concrete, tendons, or the filler material.

Protection of Anchorage Zones.   The anchorages of unbonded tendons must be adequately protected from corrosion and fire.

Except in special cases, anchorage zones should preferably be encased in concrete or grout free from any chlorides. Epoxy mortars have been widely used for this purpose.

Detailing of the concrete or grout encasement, design of the mix, and

*See Postscript, section i.

details of application are most important. Shrinkage cracks will permit moisture penetration.

Where concrete or grout encasement cannot be used, the tendon anchorage should be completely coated with a corrosion-resistant paint or grease equivalent to that applied to the tendons. A suitable enclosure should be placed where necessary to prevent the entrance of moisture or the deterioration or removal of this coating.

The anchorage encasement should provide fire resistance equal to that required for the structure.

Wire-wrapped tanks, pipes, and other externally stressed concrete elements require special protective steps to ensure against corrosion. Because the tendons are not within poured concrete encasement, they are more susceptible to corrosion possibilities. The cover is frequently of poured concrete or shotcrete. Neither of these are prestresed, except as a minor amount of prestress may be transmitted through creep.

Shotcrete (gunite), if carelessly applied, will have zones of high porosity: any lapse from the best workmanship may entrap some rebound.

Although the optimum proceudre for protecting exposed tendons is to coat each layer of wire with cement slurry, this is not always practicable. A flash-coat of shotcrete (cement and sand) will achieve satisfactory results. Then the external coating is placed by shotcrete or cast-in-place concrete. (See Chapter 15, Prestressed Concrete Tanks.)

Some new developments utilize plastic-encased tendons; these are highly corrosion-resistant provided the plastic coating is not broken or abraded. One commercially available product consists of a plastic-encased tendon inside a plastic sheath, with a friction-reducing compound between the two layers of plastic.

With external tendons on girders, which are later sealed with poured concrete, two steps may be taken to improve corrosion resistance. First, thorough coating with an epoxy paint after installation but before concreting. Second, painting or sealing the concrete cover with epoxy or bitumastic paint, especially at the joints between the cover concrete and girder concrete.

In prestressed concrete pressure vessel practice tendons are frequently left ungrouted. Attempts have been made to provide corrosion protection through dehumidification, as at Marcoule in France. Difficulties have occurred at points of discontinuity.

More recent practice in the United States is to encase the tendons in corrosion-inhibiting grease. In some French reactors, the tendons are first coated with an epoxy or corrosion-inhibiting wax, then encased in grout. (For further details, see Chapter 16, Prestressed Concrete Pressure Vessels.)

6

Manufacture of Precast Pretensioned Concrete

Precast pretensioned concrete elements are manufactured either in a permanent manufacturing plant or in a temporary job-site plant. The principles are the same: the permanent plant can economically incorporate more sophisticated machinery and equipment along with the flexibility to adjust to a variety of products, whereas the job-site plant is tailored to the specific needs of that particular project.

The manufacturing plant comprises the following minimum elements:

1. Concrete supply and delivery to forms.
2. Prestressing strand (or wire) storage on reels, with means for laying out tendons down the beds.
3. Reinforcing-steel storage and fabricating facility.
4. Beds upon which the tendons are stressed, forms placed, and concrete poured. These must be capable of taking the high compressive forces, as well as moment introduced by the height of tendons above the bed, and by deflection of tendons.
5. Stressing means (usually hydraulic jacking equipment and strand vises or grips).
6. Deflecting mechanisms (if tendons are to be deflected).
7. Forms.
8. Concrete placing and consolidation equipment.
9. Means for accelerated curing (usually a low-pressure steam system).
10. Lifting and handling equipment.

11. Storage area.
12. Transportation equipment; e.g., trucks, barges, etc.
13. Equipment for testing and inspection.
14. Facilities for maintenance and repair.
15. Utilities (water, power, fuel supply, compressed air).
16. Storage and fabrication of inserts, voids, picking loops, cushion blocks, etc.
17. Burning and welding equipment.
18. Shop engineering for shop and working drawings and computations.
19. Yard management and administration, cost records, accounting, purchasing, expediting, and estimating, sales engineering, and service engineering.

Obviously in special cases some of the above may be performed off-site, or by outside agencies.

While the individual items listed above are discussed in detail elsewhere, or are well-known standard practice, the coordination of all 19 items into a manufacturing system requires extremely careful planning and management.

Some guide-line principles are the following:

Communications must be provided, particularly from point of concrete placement in forms, to batch and mixing plant. Use of 2-way voice radio has been found useful. Communications are also needed, of course, to the vehicles or tugs, etc., involved in transport of the finished product.

Materials flow must be laid out so as to minimize distance of movement and present congestion. Materials delivery should have straight-line flow to point of use, insofar as practicable.

Beds must be designed to remain level and true despite repeated loading, and frequent wetting of ground. Use of short piles, or cast-in-place concrete posts in drilled holes, as a support for the beds, will often be found advantageous in this regard. Height of bed should be set at best working level, particularly where considerable hand work is required.

Proper storage must be provided for prestressing steel, mild reinforcing steel, and inserts, to keep them clean and dry.

Proper roads and drainage must be provided in all work areas and in storage area. It is extremely important to eliminate ruts or holes which might cause a crane or fork-lift truck to tip, possibly injuring a worker or damaging product or equipment.

Utilities should be brought to the work area at convenient outlets. Boxes or guards should be installed to keep outlets and receptacles dry and clean, and to protect them from accidental impact. Adequate lighting should be provided for night work.

Hydraulic jacks and strand vises must be properly maintained, clean and properly lubricated in accordance with manufacturer's recommendations.

Shop drawings and complete special instructions on inserts, strand tension, deflecting, etc., must be provided, in a readily readable form, at an accessible place at the work site (bed).

Tendons are usually pulled off the reels in groups by temporarily affixing a pulling nose, and pulling down the bed by a hoist. Once all tendons are laid out, properly anchored in their proper position, it is necessary to stretch them to assure equal length. If all tendons are to be jacked (stressed) as a group, then each strand should first be stressed to a nominal figure, say, 1000 lbs, and anchored.

If tendons are to be stressed one-at-a-time, then the single-strand jack will automatically equalize them. If tendons are deflected, the consequent change in longitudinal tendon stress should be verified to ensure its accord with design requirements.

After release of the stress into the members, the friction due to the dead weight of heavy girders, binding of hold-down devices, etc., and binding in the forms may prevent them from moving. Until a member actually shortens, it is *not* prestressed.

Therefore, removal of members from the bed should start at the free end (where the tendons are slack) and move progressively.

Similarly, in picking members from fixed forms, it may be necessary to overcome friction, suction, and dead weight. The concrete may even have a slight outward expansion against the forms due to the Poisson's ratio effect. Once the member is broken loose, the load drops to its dead-weight value only. This produces a dynamic "bounce," which may cause cracking due to impact. For this reason, when picking members from fixed forms (especially thin elements), it may be better to first break them loose with self-centering hydraulic jacks, reacting against the bed or forms. Refer to Subsection 3.3.4, "Lifting and Erecting Precast Elements."

Forms are generally designed for multiple re-use and should, therefore, be of steel, concrete, fiberglass, or heavy wood framing of equivalent strength. Attention should be paid to the laps and joints in the forms, as offsets and irregularities may produce incipient cracking. Jointing surfaces must be rigidly held in true alignment.

Forms must be cleaned immediately after removal of product. Particular attention must be paid to removal of grout from joints in working forms and from holes for affixing inserts. Proper taping of such joints and holes prior to pouring concrete will eliminate much cleaning.

Because of the generally thinner sections and higher concentrations of reinforcement, reinforcing steel must be secured by additional ties and carefully positioned in the forms. Tack welding should be avoided in high-tension areas. Prestressing tendons should never be welded.

Store elements on firm supports placed as nearly as practicable at points.

of final support or as specified. Protect thin slabs from excessive sun heat and rain. Strut inside of hollow units to prevent plastic deformation.

One of the main justifications for establishing a precast plant for the production of prestressed concrete is to get more efficient use of manpower. To achieve this requires a carefully developed work program. The rational way of achieving this is to prepare charts.

The first chart is a plan layout of the work area, showing each station of work and listing each operation at each station. Then, for each such operation, the number of men of a particular craft are listed along with the time required in the schedule. This chart is adjusted until the schedule fits the specified production cycle (usually 24 hours) and the desired work day. From this, the number of manhours of each craft and each man can be taken off. A second readjustment of the chart may then be necessary to properly utilize the full work day of each man.

Similarly, an evaluation can be made at this point of the use of staggered shifts—also of selective overtime—in order to minimize the total labor cost.

A second chart then lists for each man his assigned operation by location and function for each hour of the work day.

Charts such as the above require readjustment as experience is gained in actual production runs.

The above discussion deals with one form of plant manufacture: "long-line" pretensioning, which is widely employed in the United States, Canada, Australia, Sweden, and to a lesser extent elsewhere. It is characterized by the product being produced at a fixed location, with all the materials and production processes being brought to that location.

In the USSR and other Eastern European countries, a different approach is followed for many plants. In this case, the product is moved through the plant—that is, moving from one process to another. The product is produced on an individual pallet or soffit. This pallet moves from one station to the next, and at each station a process is performed, such as concreting, finishing, curing, etc.

If these several processes proceed in sequence, but are not interlocked, each is independent of the other as far as the production cycle is concerned. On the other hand, the under-carriage pallets may be connected (as by flexible connections or articulations) and the movement is continuous. A typical sequence, by successive stations, is as follows:

(a) Cleaning
(b) Oiling of forms
(c) Placement of tendons, reinforcement, and details (inserts)
(d) Stressing of tendons
(e) Placement of concrete mix

(*f*) Vibration and compaction of the mix
(*g*) Screeding and finishing surface
(*h*) Inspection
(*i*) Stripping of side forms
(*j*) Steam curing
(*k*) Removal of product for storage
(*l*) Shipment

In current practice in the USSR the conveyor line usually consists of individual pallets and the tendons are stressed against the frames on the pallets.

The author has proposed a system* whereby a continuous ribbon of concrete moves through the several processes. Tendons are held at one end on constant tension drums. At the other end (a complete cycle distant), the completed concrete is tensioned by means of rubber-tired or tracked tensioners similar to those used in the offshore pipe-laying industry. By installing this system on a slope, gravity may be used to assist the stressing operation.

Returning to the upper end, mesh is rolled out continuously onto the strands. A void-former may be installed at the concreting point, using stainless steel or Teflon-coated mandrels, with internal vibration and, perhaps, fore-and-aft jiggling vibrators to prevent bond. A rubber conveyor belt forms the soffit.

The concrete is fed on at a single point, vibrated, screeded, finished. The ribbon of concrete, containing mesh and strands, and riding the conveyor belt, enters a low-pressure steam curing tunnel. When it emerges from the other end, tensioning rollers, referred to above provide the tension. After moving off the rubber conveyor belt, a travelling saw cuts the precast slab (or other unit) to exact length. It then moves by roller conveyor to stacking pallets. The rubber conveyor belt returns underneath to the concreting station.

Despite the appeal of sophisticated production-line systems like that described above, currently the lowest labor consumption ratios (man-hours per unit of product) in the world are being achieved by the relatively simple "long-line" pretensioning system. This system has achieved its efficiency through the adoption of "advanced forms" incorporating the following improvements:

1. Use of multiple forms side-by-side, up to a workable width of 6 or 8 feet.
2. Multiple-handling (lifting equipment and gear capable of removing, storing, and loading several units simultaneously).

*U.S. Patent No. 3,055,073.

3. Use of fixed, flexible side forms, so that product may be raised from forms without moving them.
4. Where forms must be moved, use of hydraulically actuated rams or compressed air tubes to move them.
5. Synchronized deflection and stressing of strands by remotely controlled hydraulic system.
6. Combination of automatically controlled form vibrators with supplemental internal vibration.

In-plant quality control is important to the achievement of economy, schedules, erection, and performance. Definite inspection, checking, and testing procedures must be provided to insure that the product is correctly produced, not only for strength, but for dimensional accuracy (within prescribed tolerances), and accuracy of placement of inserts.

7

Cast-in-Place
Post-Tensioned
Concrete

Many of the most spectacular and important structures in prestressed concrete have been constructed in post-tensioned cast-in-place concrete. This includes many tall towers and long-span bridges of Europe. The use of cast-in-place concrete makes possible complicated shapes and curves and transitions. Where construction joints occur, only one face needs to be prepared and joined, as compared to the two faces of precast segmental construction.

Cast-in-place construction also minimizes the weights to be handled. The concrete may be transported to the forms by small bucket or pump and the forms may be handled by jacking or small hoists, as well as by cranes. Thus, it is adaptable to construction in lengths or heights beyond the practicable reach of large construction equipment.

Although the basic techniques of cast-in-place concrete construction are similar, whether reinforced by conventional mild steel or prestressed, there are a number of special considerations applicable to prestressed structures.

7.1 Forms

It is quite important in long-span construction to minimize the deflections that occur during concreting. Thus, the forms must be rigid and of sufficient section to reduce elastic deformations to a minimum.

Concrete to be prestressed must be high strength. This requirement automatically means drier and harsher concrete mixes (lower water/cement ratio) and this, in turn, requires more intensive vibration. Such vibration tends to cause local deformation of the forms, such as bulging, unless the forms have sufficient rigidity.

Form vibration is frequently used to obtain more complete consolidation of thin sections. This puts extremely high stresses in the forms, including fatigue stresses, requiring better form construction and securing.

The technique of prestressed concrete requires accurate location of the prestressing force relative to the center of gravity of the cross section. This means that the forms must be constructed with greater-than-normal accuracy.

After the concrete has hardened, forms are usually partially stripped; for example, the side forms are removed but the soffit forms are left in place. Shrinkage takes place. The remaining forms must be designed to permit this volume change to take place. Then at stressing, the member will shorten and may rotate (camber). The remaining forms must be designed and constructed so as to permit the shortening without undue restraint, and to accept the camber and also the shift in dead-load distribution of the member.

Finally, forms must be chosen to provide a dense, impermeable surface to the concrete, so that it will protect the prestressing tendons and mild steel from corrosion.

7.2 Mild Steel

Prestressing creates secondary stresses which require mild steel for containment. In addition, the thin sections, particularly webs, are subject to high shearing forces which are usually countered by mild steel, although occasionally post-tensioning also is used here.

The mild steel must be accurately fabricated, placed and held during concreting. Since accidental impact from the vibrators may dislodge the mild steel, it is often advisable to mark the location of bars on the forms prior to concreting, so they will be readily apparent to the vibrator operator.

Current design practice emphasizes the use of more small bars rather than a lesser number of large bars. This again requires greater care in placing and restricts the space for placement and vibration of concrete. This, in turn, may require the use of smaller coarse aggregate and smaller vibrators, and more frequent tying of the steel in position.

Because of the traditional separation of contractural relationships in the United States between the general contractor (performing the concrete construction) and the reinforcing steel-placing subcontractor (placing the steel), there has been a tendency for the general contractor to shrug off responsibility for the accuracy of cutting, bending, and placement of the reinforcing steel. In post-tensioned cast-in-place construction, it becomes of great importance that the steel be in full accordance with the design drawings. Therefore the general contractor must check and verify the accuracy of the reinforcing steel, whether placed by a subcontractor or by his own forces.

7.3 Post-Tensioning System

As for all post-tensioning, it is essential that ducts be truly located, free from gaps at splices in ducts, and free from blockages. Anchorages must be normal to the tensioning force and the concrete must be thoroughly consolidated under the bearing plates.

In particular, it is essential that the ducts not be pierced by accidental contact of the vibrators. This consideration may lead to the selection of heavier gauge metal for the ducts, or rigid ducts, or the insertion of an inflatable tube in the ducts during concreting. This latter solves both the problems of rupture and of joints or splices in the ducts. As with mild steel, duct locations may be marked on the forms as a guidance for the vibrator operator.

Frequently, anchorage assemblies are correctly set in the forms, only to be displaced during concreting. Therefore the template or form holding them must be rigid and sufficiently massive.

7.4 Concreting

To obviate or minimize many of the problems referred to above, a mix using smaller size (e.g., 3/4 inch) coarse aggregate will generally be found preferable. The mix should be workable; this requires adequate cement content and sand. Use of a water-reducing, retarding, and shrinkage-reducing admixture is beneficial and recommended.

Retarders are particularly effective in slow pours such as frequently required in prestressed cast-in-place structures. The rate of set can be controlled by varying the amount of retarder during the various stages of pour.

Aggregates should preferably be selected from sources which show minimum shrinkage and creep.

Chlorides must be absolutely prohibited because of risk of corrosion. This includes chlorides in the sand or aggregate, or in the water, or in admixtures. (See Chapter 4, Durability and Corrosion Protection.)

Shrinkage-compensating cements hold considerable promise. So far, however, they have been used (to any extent) only in cast-in-place post-tensioned roof slabs. The performance has been spotty, primarily because shrinkage-compensating cement is highly dependent upon receiving adequate moisture during the curing period.

High early-strength cements and finely ground cements are frequently employed to achieve early strength to enable prestressing. This is usually satisfactory, provided the cements do not contain chlorides and provided the sections are typical structural members. With large masses, their use may produce excessive shrinkage and excessive heat of hydration.

Steam curing is a very effective way of accelerating strength gain. The cast-in-place section is enclosed in a hood of rubberized nylon or polyethylene and steam is applied at atmospheric pressure. For smaller sections, metal hoods have been employed. In some complex sections containing well-distributed ducts, hot water has been circulated; however, this should be treated with a corrosion inhibitor to prevent corrosion of the duct.

In any event, cast-in-place concrete for prestressed construction requires more thorough curing than conventional cast-in-place concrete, because sections are thinner and more highly stressed. Therefore, adequate means of curing must be specified and enforced.

7.5   Stressing

In most cast-in-place construction, the tendons are inserted after concreting. Ducts should be flushed with water, then blown out with compressed air. The tendons may be inserted by pushing in, with a flexible nose shield fitted to the insertion end. Several tendons may be bound together, with a common nose piece. Alternatively, a messenger line may be sent through by attaching it to a rubber ball that is pushed through by compressed air, or to a wire leader similar to a "plumber's snake."

The writer prefers to insert a messenger line prior to concreting, since this may be used immediately after concreting to run back and forth to insure that the duct is free and unblocked. Then, later, the tendon may be attached and pulled through.

Stressing to design transfer stress should be performed only after the concrete has reached the specified strength for this prestress. Otherwise, excessive creep may result.

However, subject to the approval of the design engineer and provided it is compatible with the prestressing system adopted and the forms, it may often be found advantageous to impart a small degree of prestress shortly after concreting (e.g., 12 to 24 hours). This offsets shrinkage and prevents shrinkage cracks from developing. Later, when the concrete reaches the specified strength for prestressing, the tendons may be stressed to their full transfer value.

Such early stress values are usually very low (i.e., perhaps 10% of the final tendon stress) so their influence on corrosion-susceptibility is slight. However, whenever tendons are to be left in construction ungrouted for more than 24 hours, the dusting on of VPI powder and sealing off the ducts is recommended. (See Chapter 5, Protection of Tendons in Ducts.)

Generally speaking, tendons should never be left in place during steam curing. If this becomes essential in some cases, then extreme precautions must be taken to prevent corrosion, including use of VPI powder, sealing of ends of ducts, and possible injection of a water-soluble grease or water solution of an inhibitor.

7.6   Construction Joints

Much of the difficulty with post-tensioned cast-in-place concrete has arisen at construction joints. The concrete faces of construction joints must be properly prepared so as to present a roughened face with adequate shear transfer. One means is wet sandblasting, to expose the aggregate.

Another means is the painting or spraying on of a retardant on the forms for the construction joint, then jetting off after stripping.

A new tool is a water-jet which utilizes high pressure (1500 psi) water in a fine jet to cut away the laitance and cement paste.

Bush-hammering is an ancient technique that is still proving to be efficacious and economical.

Ducts must be made continuous across construction joints. With rigid or flexible metal ducts, the joints may be taped with waterproof tape. This is satisfactory, provided the ducts are somehow tied so they can't be accidentally pulled apart during subsequent operations, such as vibrating of the concrete. Rubber and plastic sleeves have been used, as have screwed connections.

Here again, the insertion of inflated duct-formers or of flexible rigid conduit (such as electrical conduit) in the ducts, will serve to preserve continuity across the joints.

7.7   Stripping

In all post-tensioned cast-in-place construction, after stressing, accurate measurements should be taken of camber, shortening, tendon forces, etc. Then these should be correlated and checked to be sure the structure is performing in accordance with the design calculations. Only after behavior has been fully verified is it safe to strip the supporting forms and falsework.

This is especially critical in thin members such as floor slabs where, if errors of placement, etc. have been made, the slab may not camber upward under prestress, but may actually be unstable. Thus a standard procedure should be instituted to verify performance prior to removal of supports.

8

Architectural Concrete

The use of concrete for architectural effects, with a wide variety of color, texture, and sculpture, has experienced a rapid growth in both application and technology. It has emerged as a wonderful medium for expression by artist, sculptor, and architect. Its moldability, permanence, and the flexibility with which it may be subjected to an endless variety of techniques, have been combined with concrete's performance as a structural and cladding material possessing strength, durability, weather and fire resistance, and economy.

The techniques for creating an architectural (or artistic) effect in concrete include those of integral color, sculptured or textured pattern, exposed aggregate, and embedded ornament. It is beyond the scope of this book to discuss these in detail, with their multitude of processes for obtaining a wide range of artistic effect. However, there is a strong trend to integrate this artistic function with the structural function, and architectural concrete techniques are being increasingly applied to prestressed concrete. This chapter will briefly examine some of the aspects of this combined or integrated service as it affects, and is affected by, prestressing. (See Fig. 11)

Prestressing enables the member to perform its structural function more efficiently, thus permitting thinner sections, longer spans, and overhangs. It can be used to control behavior, that is, prevent sag that might destroy the architectural effect. With sculptured members it gives increased freedom to the artist: prestressing can be used to counter adverse weight distributions.

For many panels the prestressing tendons may be placed at the center of the section, giving essentially uniform prestress. This increases the available cover over the steel, with reduced likelihood of rust-staining and spalling, of particular value in freeze-thaw and salt-air exposure.

Prestressing can be used to eliminate cracks. Not only are these unsightly in themselves, but they often lead to rust-staining.

Fig. 11. Pretensioned spandrel units utilize white cement and acid-etching.

Prestressing may be one- or two-dimensional. Techniques have been developed for two-dimensional pretensioning. One of these is the wire-winding system of the USSR, in which wire is wound in a pattern around dowels. The dowels are retracted after the concrete hardens, and the holes must be filled. The architect must be cognizant of these holes and, if possible, incorporate or hide them in his pattern as it is very difficult to accurately match color and texture in the patching.

Two-dimensional pretensioning has also been utilized in the USA, with the transverse tendons being anchored to a side frame along the bed. After curing, the transverse tendons must be released first, before the longitudinal strands are released. A cork, styrofoam, or rubber plug can be fitted at the edge, so that the strands may be cut back 1 inch or so from the edge, and the plug hole patched. Being on the edge, this patch is usually less exposed to view.

Patching seems to always turn out darker than the adjoining concrete, particularly when damp. Trial mixes, using varying proportions of white cement, and various curing techniques should be conducted well ahead of actual production patching in order to permit determination of that mix which most nearly matches. Such patches must also prevent rust staining. A tiny, carefully applied dab of epoxy on the end of the strand may help prevent rust-staining.

Precast plugs may be used, with a white epoxy bonding carefully applied to the hole (not to the plug) and the plug driven in. Care must be taken that excess epoxy does not drip or spread over the adjoining concrete.

Post-tensioned anchorages may be recessed and patched in the same manner as indicated above.

Tolerances are of great importance for architectural concrete; they must be realistic and practicable of manufacture, yet provide the required architectural appearance. "The Manual for Quality Control for Architectural Precast Concrete Products," published by the Prestressed Concrete Institute, lists recommended tolerances for warpage, thickness, squareness, location of anchors and inserts, block-outs and reinforcements, and joint widths. These may be summarized, in general, as follows:

1. Warpage: 1/8 inch per 6 feet length.
2. Thickness: minus 1/8 inch, plus 1/4 inch.
3. Squareness: 1/8 inch in 6 feet out of square as measured on the diagonal.
4. Anchors and inserts: 3/8 inch.
5. Blockouts and Reinforcement: plus or minus 1/4 or 1/2 inch.
6. Joint widths: specified widths are normally 3/8 inch to 5/8 inch. Acceptable limits are then 1/4 inch to 3/4 inch.

When special tolerance requirements are deemed desirable by the architect, they should be clearly designated.

If prestressed architectural panels are to be acid-etched, then the exposed ends of the tendons, etc. should be thoroughly flushed with fresh water after immersion. Ducts for post-tensioning tendons should be thoroughly flushed to remove all trace of acid.

Prior to acid etching, all exposed metal surfaces should be protected with acid-resistant coatings. In working with acid, safety precautions should be set, including protective clothing, breathing masks, and eye protection as required.

One of the early processes for exposed aggregate consisted of casting face downward, with the soffit painted with a retarder. After stripping, a water-jet was used to wash off the laitance and expose the aggregate. When a retarder is so applied, a *positive* means must be taken to prevent accidental spillage on the tendons. If the retarder is placed first, the tendons must be held up so they do not accidentally drag through the retarder.

There is a growing trend to cast architectural panels face-up, and to expose the aggregate after initial set by a water-and-air jet. In some cases, the concrete is cured and sandblasting or a hydraulic ram-jet is used to expose the aggregate. From a prestressing point of view, these processes are to be preferred as running less risk and giving less interference to normal production methods.

Honing and polishing are used to produce smooth, exposed aggregate surfaces. Honed surfaces are produced by grinding with carborundum particles bonded by resin, or by diamonds set in the cutting surface. Air voids must be filled before each of the first few grinding operations, using cement or a cement-sand mixture and allowing it to harden prior to the next grinding operation.

Sometimes a surface layer of different material is embedded in the concrete. This material may be glass, ceramic, marble or granite sheets, or just different stone aggregates. Prestressing serves an important function of locking these into the panel. These materials have a different modulus of elasticity, and thus an engineering analysis must be made as to the proper location for the tendons in order to prevent warping of the panel. In addition, it is very wise to run a test panel to verify the tendon pattern.

Steam curing may adversely affect certain matrix colors, particularly blues and greens. The dripping of condensate may also spoil a surface. For this reason, particular care should be taken in selecting and controlling the curing process. In some cases it may be preferable to employ conventional water cure, or hot water or hot oil circulating through pipes in the bed or forms if accelerated cure is desired.

Rust staining from exposed inserts, reinforcing dowels, tendon ends, etc., causes more trouble and costs than almost any other single item. Positive means must be taken to prevent this. Appropriate steps include the use of stainless steel or galvanized inserts and reinforcement, epoxy dabs on the ends of cut bars, plastic tubes placed over extending dowels and sealed to the concrete surface, and wax coatings, etc. The use of high-carbon steel lifting eyes should be avoided due to possibility of brittle failure under impact.

Similarly, grease, mud, and oil must be kept off panels during storage, transport, and erection. Many panels are wrapped in heavy kraft paper or polyethylene for this purpose.

Another source of difficulty with architectural panels is efflorescence in which lime leaches out over a period of time to form irregular white patterns on the surface. This is particularly objectionable on a colored panel. The ways of minimizing this are to use a lower cement content, to make the concrete more impermeable, and to seal the surface.

To keep the cement content low, 25% or so may be replaced by a pozzolan. Use of a very dry mix (low water/cement ratio) and heavy vibration will produce a more impermeable product. Sealing the surface is usually done with silicones; however, these coatings disappear in time and must be renewed.

Quality control of architectural precast concrete comprises the following elements:

1. Qualified personnel responsible for all stages of design, production, inspection, and installation.
2. Adequate testing and inspection of the various materials selected for use.
3. Clear and complete shop drawings.
4. Control of dimensions and tolerances, including adequate formwork.
5. Inspection of all embedded hardware.
6. Mix design, proportioning, and mixing of concrete.
7. Handling, placing, and consolidation of concrete.
8. Curing.
9. Finishing.
10. Handling, storing, transporting, and erection of elements.

Defects in architectural precast products which require correction in manufacturing and/or design procedures include the following:

1. Ragged or irregular edges.
2. Excessive air pockets on exposed surfaces.
3. Adjacent surfaces with different color or texture.
4. Construction joints or accidental cold joints readily visible.
5. Form joints visible, fins, honeycomb.
6. Rust or acid stains on surfaces.
7. Concentrations of aggregate or gaps, variations in aggregate appearance or surface.
8. Areas of back-up concrete showing through facing concrete.
9. Foreign material showing on face.
10. Cracks visible after wetting.
11. Visible repairs and patches.
12. Reinforcement shadow lines.

We have in concrete an ideal material for the incorporation of artistic effects into our structures. Prestressing enhances this property in many ways, but care and forethought are required in application if the maximum benefits are to be achieved.

9

Safety

Prestressing inherently involves the use of very high forces, with steel and concrete stressed to a high percentage of their ultimate load. Prestressed concrete structures frequently involve heavy masses and high lifts. During construction and erection, hydraulic forces, vibration, imbalance, and dynamic forces exist.

Safety of personnel and equipment can be achieved only by positive efforts, including planning, temporary additional bracing, the installation of safety guards and warnings, and the indoctrination of workmen through a continuing safety program.

"People who are constantly exposed to dangerous situations are prone to lose their conscious fear unless they are constantly reminded of it." Safety should never be subordinated to production expediency.

Tendons may be under tension to values of as much as 180,000 psi, which represents a tremendous stored energy. This can convert an anchorage into a deadly missile.

Safety precautions have been set forth in the "Manual for Quality Control," published by the Prestressed Concrete Institute (USA). Precautions are also set forth in the "Safety Precautions for Prestressing Operations," published by the Concrete Society (London).

Without in any way attempting to diminish the value of a thorough study and implementation of these references, the following abbreviated list of rules is set forth. Most of these have been extracted in principle from the references, with amplification or modification based on this author's experience.

9.1 Safety Measures for All Tensioning Operations

(*a*) The operation of tensioning has more potential for serious accidents than all phases of prestressed production combined. The following basic rules applicable to tensioning should be included in the safety requirements of all plants.

(*b*) Before tensioning, a visible and audible signal should be given and all personnel not required to perform the tensioning should leave the immediate area.

(*c*) Jacks should be held by means such that the jack will be prevented from flying longitudinally or laterally in case of tendon failure.

(*d*) Personnel should never be permitted to stand at either end of the member (or bed) directly in line with the tendon being tensioned.

(*e*) Personnel should not stand over tendons being tensioned to make elongation measurements. Such measurements should preferably be made by jigs or templates from the side or from behind shields.

(*f*) Eye protection should be provided for personnel engaged in wedging and anchoring operations as a protection from flying pieces of steel.

(*g*) Do not permit any welding near high-tensile prestressing steel; such material should not be used for grounding electrical equipment of any kind.

(*h*) Keep all equipment thoroughly clean and in a workmanlike condition. Badly maintained equipment always gives rise to trouble and, consequently, is dangerous.

(*i*) See that the wedges and the inside of barrels or cones of grips and anchorages are clean so that wedges are free to move inside the taper.

(*j*) Arrange for stressing to take place as soon as possible after the grips have been positioned.

(*k*) When assembling cables, check each individual wire for obvious flaws.

9.2 Additional Safety Measures for Pretensioning

(*a*) Avoid kinks or nicks in strand. Use care in handling strand to avoid damage. Do not tension strand that has been nicked.

(*b*) Prevent accidental heating of a tensioned strand. Keep all torches and welding equipment away from tensioned strand.

(*c*) Do not allow personnel to expose any part of their bodies above hold-down points of deflected strands. Block off directly above such hold-downs to prevent accidental dislodgement by vibrators during vibration of concrete. Hold-down inserts are potential cross-bow projectiles.

(*d*) Protective guards for personnel engaged in tensioning should be provided at both ends of the bed and should be of structural steel, concrete, or heavy timbers.

9.3 During Stressing

(*a*) Do not become casual because you have stressed hundreds of cables before. The forces you are handling are enormous, and carelessness may lead to loss of life.

(b) Regular examination of hydraulic hoses is essential, and oil in the pump reservoir must be regularly drained and filtered.

(c) Use only self-sealing couplings for hydraulic pressure pipes and take particular care that no bending stresses are applied to end connections.

(d) It is preferable to use only hydraulic equipment supplied with a by-pass valve which is pre-set to a maximum safe load before stressing. The maximum safe load should not be more than 90% of the minimum specified ultimate strength of the tendons.

(e) Never stand behind a jack during stressing operations.

(f) Do not strike the equipment with a hammer to adjust the alignment of the jack when the load is on.

9.4   Grouting

(a) A clear eye-shield should be worn by operator during grouting operations.

(b) Before grouting, check all ducts with compressed air to make sure that they are not blocked.

(c) It is preferable to use only threaded connectors between grout nozzles and grouting points. A sudden spurt of grout under pressure can cause severe injury, especially to the eyes.

(d) Do not peer into duct bleeders to see if grout is coming through. Grout may jam temporarily and, as pressure is applied, it may suddenly spurt from the bleeders, or the far end of the duct, causing serious injury.

9.5   Reports and Safety Meetings

(a) Proper reports should be made of all accidents and injuries, even those that are not of serious nature. Frequently, a serious accident is preceded by several close calls; proper attention to these may permit preventive steps to be instituted.

(b) A brief but rigorous safety meeting should be held once a week, and prior to starting any new operation or use of new equipment.

II

UTILIZATION OF
PRESTRESSED CONCRETE

10

Prestressed Concrete Buildings

Prestressed concrete has been widely and successfully applied to building construction of all types. Both precast pretensioned members and cast-in-place post-tensioned structures are extensively employed, sometimes in competition with one another, most effectively in combination with each other.

Nevertheless, a general building industry founded on prestressed concrete is currently at a crossroads: on the one hand, the potential growth appears inevitable and unlimited; on the other, problems of integration into the entire operating system, that is, the total building including all its services and interiors, etc., remain to be completely solved. A true building is far more than a structure (or "carcass," to borrow an East European word). It has an external environment, and an internal environment complete with landscaping, architectural facade, electrical and mechanical systems, air conditioning, fenestration, elevators, interior decoration, and furniture. In industrial buildings the total building includes the operating equipment, sources of power, means for material handling, and waste disposal.

Thus in the total picture the structural aspects represent 30% or less of the total cost. Strangely, in attempts at economy, 75% of the effort appears to be directed at reducing the cost of the structure. Perhaps this is a tribute to the dominant role of the basic structure in setting the architectural theme.

Prestressed concrete offers great advantages for incorporation in a total building. It is perhaps the "integrative" aspects of these, that is, structure plus other functions, that have made possible the present growth in use of prestressed concrete buildings. These advantages include the following:

137

Structural strength.
Structural rigidity.
Durability.
Moldability, into desired forms and shapes.
Fire resistance.
Architectural treatment of surfaces.
Sound insulation.
Heat insulation.
Economy.
Availability, through use of local materials and labor to a high degree.

Most of the above are also properties of conventionally reinforced concrete. Prestressing, however, makes the structural system more effective by enabling elimination of the technical sources of difficulty, e.g., cracks that spoil the architectural treatment. Prestressing greatly enhances the structural efficiency and economy permitting longer spans and thinner elements. (See Fig. 12) Above all, it gives to the architect-engineer a freedom for variation and an ability to control behavior under service conditions.

Fig. 12. Prestressed concrete double-tee roof elements complete bus terminal, San Jose, California.

Precast prestressed concrete has been selected in a number of Socialist countries (USSR and Eastern Europe) as the standard for building construction of all types. In free economies, prestressed concrete has been increasingly employed as architects have become familiar with its many advantages; as engineers have become accustomed to design; as manufacturers have achieved better and cheaper production; and as industry and professional groups have developed standards and details. Now there is a great impetus for "systems building" in the United States and Western Europe, and prestressed concrete is faced with both the challenge of competitive structural materials and the need to integrate its own capabilities into the systems concept.

Two basic alternatives are possible for the prestressed concrete industry. One course is the development of a system around a completely plant-produced prestressed concrete structure. The second alternative is the development of standard modular prestressed concrete structural elements which will fit into a number of systems. Although a discussion of the merits of these two choices lies outside the scope of this book, the author believes the second course may turn out to be the right one, because it is in accord with the basic philosophy of a system. The system concept must start with man and man's activities, develop a system to serve these, and then be put together with whatever materials and products fit best. Thus, within an overall system, different materials and products would be employed in different buildings, depending on their end use.

The first alternative puts the structure first; it is a prior determination by central authority. The decision may, for that matter, be right. However, times and needs change and such a system is inflexible and dictatorial.

The second alternative owes its ultimate allegiance to man and his environmental needs; thus can grow and change.

In a more immediately practicable vein, the second alternative means that efforts should be directed toward the widest possible utilization of those elements of the building which prestressed concrete does best, rather than an attempt to use prestressed concrete exclusively for all elements of a building.

Although prestressed concrete construction for buildings involves essentially the same considerations and practices as for all structures, a number of special points require emphasis or elaboration. These will be discussed under appropriate headings, limited to those aspects which are of peculiar importance to buildings.

The construction engineer is involved in design only to a limited extent. First, he must be able to furnish advice to the architect and engineer on what can be done. Because of his specialized knowledge of techniques relating to prestressed concrete construction, he supplies a very needed service to the architect-engineer.

Second, the construction engineer may be made contractually responsible for the working drawings; that is, the layout of tendons, anchorage details, etc. It is particularly important that he give careful attention to the mild steel and concrete details to ensure these are compatible with his prestressing details.

Third, the construction engineer is concerned with temporary stresses, stresses at release, stresses in picking, handling, and erection, and temporary conditions prior to final completion of the structure, such as the need for propping for a composite pour.

Fourth, although the responsibility for design rests with the design engineer, nevertheless the construction engineer is also vitally concerned that the structure be successful from the point of view of structural integrity and service behavior. Therefore he will want to look at the bearing and connection details, camber, creep, shrinkage, thermal movements, durability provisions, etc., and advise the design engineer of any deficiencies he encounters.

Information on new techniques and especially applications of prestressing to buildings are extensively available in the current technical literature of national and international societies. The International Federation of Prestressing (F.I.P.) has attempted to facilitate the dissemination of this information by establishing a Literature Exchange Service, in which the prestressing journals of some thirty countries are regularly exchanged. In addition, an Abstract Service is published intermittently by F.I.P. The Prestressed Concrete Institute (USA) regularly publishes a number of journals and pamphlets on techniques and applications, and procedures are set up for their dissemination to architects and engineers as well as directly to the construction engineer. It is important that he keep abreast of these national and worldwide developments, so as to be able to recommend the latest and best that is available in the art, and to encourage the architect and design engineer to make the fullest and most effective use of prestressed concrete in their buildings.

With regard to working drawings, the construction engineer must endeavor to translate the design requirements into the most practicable and economical details of accomplishment, in such a way that the completed element or structure fully complies with the design requirements; for example, the design may indicate only the center of gravity of prestressing and the effective prestress force. The working drawings will have to translate this into tendons having finite physical properties and dimensions. If the center of gravity of prestressing is a parabolic path, then, for pretensioning, an approximation by chords is required, with hold-down points suitably located.

The computation of prestress losses, from transfer stress to effective stress, must reflect the actual manufacturing and construction process used, as well as thorough knowledge of the properties of the particular aggregates and concrete mix to be employed.

With post-tensioning, anchorages and their bearing plates must be laid out in their physical dimensions. It is useful in the preparation of complex anchorage detail layouts to use full-scale drawings, so as to better appreciate the congestion of mild steel and anchorages, etc. at the end of the member. Tendons and reinforcing bars should be shown in full size rather than as dotted lines. This will permit consideration to be given as to how the concrete can be placed and consolidated.

The end zones of both pretensionsed and post-tensioned concrete members are subject to high transverse or bursting stresses. These stresses are also influenced by minor concrete details, such as chamfers. Provision of a grid of small bars (sometimes heavy wire mesh is used), as close to the end of a girder as possible, will help to confine and distribute the concentrated forces. Closely spaced stirrups and/or tightly spaced spiral is usually needed at the end of heavily stressed members. Recent tests have confirmed that closeness of spacing is much more effective than increase in the size of bars. Numerous small bars, closely spaced, are thus the best solution.

Additional mild-steel stirrups may also be required at hold-down points to resist the shear. This is also true wherever post-tensioned tendons make sharp bends. Practical considerations of concreting dictate the spacing of tendons and ducts. The general rules are that the clear spacing shall be one-and-one-half times the maximum size of coarse aggregate. In the overall section, provision must be made for the passage of the vibrator stinger. Thus prestressing tendons must either be spaced apart in the horizontal plane, or, in special cases, bundled.

In the vertical plane close contact between tendons is quite common. With post-tensioned ducts, however, in intimate vertical contact, careful consideration has to be given to prevent one tendon from squeezing into the adjacent duct during stressing. This depends on the size of duct and the material used for the duct. A full-scale layout of this critical cross section should be made. Usually, the best solution is to increase the thickness (and transverse strength) of the duct, so that it will span between the supporting shoulders of concrete (See Fig. 13).

As a last resort it may be necessary to stress and grout one duct before stressing the adjacent one. This is time-consuming and runs the risks of grout blockage due to leaks from one duct to the other. Therefore the author recommends the use of heavier duct material, or else the re-spacing of the ducts. The latter, of course, may increase the prestressing force required.

Temporary stresses are those imposed on the element or structure during the several phases of completion, until the structure is finally acting in its entirety. The obvious stresses are those caused by handling and transport.

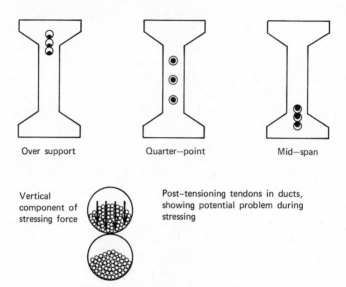

Over support   Quarter–point   Mid–span

Vertical component of stressing force

Post-tensioning tendons in ducts, showing potential problem during stressing

Fig. 13. Adjoining ducts require special consideration.

Less obvious are the effects of the time-dependent deformations that occur between first manufacture or construction and final completion.

Precast elements must be picked, stored, loaded, transported, and erected. Picking involves a very complex behavior of the member. If it is in fixed forms, with a side draft, as is so common in current manufacturing practice, there is the frictional resistance on the sides and the suction on the bottom to be overcome. Experience has shown a considerable reduction in picking forces when the side forms are flexible.

When rigid forms are used, for example, steel-lined concrete molds or polished concrete, suction may present a serious problem. The author, in his earlier years, attempted to overcome the side friction by the use of grease; this only insured that the suction effect would be greater. In some instances, injection of compressed air through the soffit has helped to free the unit.

Even with flexible forms, if these are multiple forms, the fact that adjacent forms are filled with the product at the time of pick may make them act as rigid forms.

Corner binding of closed channels is usually the most serious problem. Use of a loose corner insert in the forms, that comes out as the unit is picked, will prevent corner binding (See Fig. 14).

Perhaps the best method for overcoming side binding in rigid forms is to use self-centering jacks which raise the member a few inches vertically, breaking it loose.

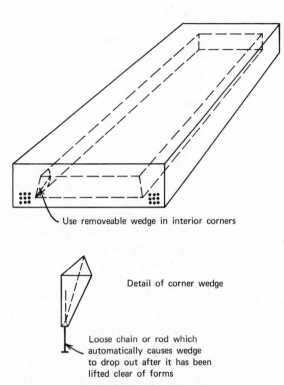

Use removeable wedge in interior corners

Detail of corner wedge

Loose chain or rod which
automatically causes wedge
to drop out after it has been
lifted clear of forms

Fig. 14. Method of preventing corner binding in precast prestressed slab manufacture.

A heavy member, such as a girder, will probably not move lengthwise when the prestress is released into it, because of friction. Thus at the moment of initial pick it is not prestressed. For this reason the picking slings may have to be slightly offset so as to pick one end slightly ahead of the other end.

Another cause of binding is fins of grout between side forms and soffit or between soffit and insert holes, etc. This can be prevented by forms manufactured to a tight fit or by scaling the joints with tape.

The effect of such picking loads, in addition to the dead weight of the member, combined with the dynamic effect of breaking loose, is to increase stresses by 50% to 100% (or more). This picking is, of course, done at a time when the concrete is relatively green. Although it has attained a specified minimum compressive strength, this is probably only 75% of its seven-day strength. The tensile strength at picking may be an even lesser proportion. Damage to the element can be prevented by using more picking points, properly equalized, and by adding mild-steel bars over the picking points (points of maximum negative moment).

During storage, the unit should be supported at essentially the same points as in its final condition. However, being at an early age, creep may be proceeding and provision should be made with long, heavily stressed members, to permit longitudinal shortening. In addition, a hand-tight support under the quarter-points (but not mid-point) of long girders or slabs may prevent undesirable permanent sag.

During transport, the members are again subjected to dynamic loading and the addition of temporary mild steel may be necessary to prevent overstress and cracking, particularly over the points of support. The up and down whip of a cantilever during transport may be a particularly serious condition, requiring addition of mild steel to prevent both cracking and bond failure.

In erecting buildings, the temporary stresses depend on the means of erection and the number of support changes that take place. Because the members do not have the design live load and perhaps not even all the dead load, the chance for primary failure in bending is slight. However, end-support conditions are not as accurately determined as in the final support; thus, concentration may occur in the bearing area, as well as negative moments and shear. Provision of adequate stirrups throughout the end zone, wherever temporary support may occur, plus possibly some longitudinal bars top and bottom, will minimize possible damage and failure.

Where beams or slabs are to be combined with cast-in-place concrete for composite action, temporary props or supports may be required to support the dead load of the fresh concrete without excessive stress or deflection. Means must be provided for removal of these, as by knock-out double wedges, or, better yet, screw jacks, so as to avoid upward deflection of the slab during removal. In some cases where props are not used, mild steel can be added at point of maximum positive moment to minimize crack width and increase ultimate strength should the dead load of the fresh concrete place the girder bottom flange in tension. A final alternative is to increase the value of prestress for this condition, provided this does not induce excessive camber or other undesirable effects.

The manufacture of precast prestressed concrete elements for buildings and the construction of prestressed building structures is differentiated by the relative thinness and sometimes fragility of sections used, and by the incorporation of architectural content.

Control of camber and warping becomes of great importance in long spans with slender beams and slabs. The main reasons for variation in camber are (a) variations in concrete thickness, (b) variations in strand position, especially deflection, (c) variation in water/cement ratio and curing (affecting the modulus of elasticity), and (d) variation in storage, and (e) variation in strength at time of stressing.

All of these are susceptible to control to ensure uniformity, but only if the problem is recognized beforehand. Forms and screeds can be designed so as to keep the tolerances in cross section to a minimum. Strand positioning is largely a matter of care and supervision. Water/cement ratios can be carefully controlled with proper allowance for moisture content of the aggregate. With lightweight aggregate in particular, more or less continuous adjustment may be needed to ensure uniform water/cement ratio at point of delivery. The curing cycle can be kept uniform by using automatic controls and recording charts and, particularly, by ensuring that the steam covers are tight enclosures to prevent loss of heat in a wind, etc. The temperature of the concrete mix should be kept approximately uniform, using ice or water-soaking in hot weather, and warm water or heated aggregates in extremely cold weather. Storage should be uniform, with the top slab in any pile covered to prevent earlier drying and excessive thermal response to the sun.

Recent experience with very long (over 110 feet) pretensioned building members has shown less camber than calculated. The probable explanation is increased shrinkage of the upper flange during the longer period required for concrete placement of these long girders. In some important cases, the remedy has been to add a single parabolic post-tensioning tendon, which is stressed or not, as needed, to achieve the desired camber. Limited investigation by the author indicates that greater efforts should be aimed at reduction of drying shrinkage in the top flange in mix design, placement technique, and curing procedure.

This phenomenon emphasizes the sensitivity of long thin members to such variations.

Many sections for buildings are so thin as to require especial care in handling. Lightweight aggregate concrete is particularly sensitive to chipping and spalling on corners and edges. Rough, careless handling can cause irreparable damage. In one factory, technically and architecturally perfect hollow-core floor slabs were observed being crudely handled by picking tongs and inserts gouged into the fresh concrete. These were later patched, with, of course, noticeable color variation and disturbance of the uniformity of stress distribution.

The growing use of precast prestressed concrete as architecturally exposed members means that ever greater care must be taken to prevent surface discoloration and imperfections. Discoloration may be caused by variations in water/cement ratio, variation in vibration duration, by impingement of steam during curing, and by variations in thickness or composition of form oil. Even more inexcusable is staining from grease or oil during storage or erection, etc. Steel forms seem to give slightly more discoloration than fiberglass or wood forms, probably due to the thermal conductivity of steel during steam cure.

Exposed architectural members may require extreme care, such as wrapping in polyethylene or heavy kraft paper during shipping.

The formation of "bug holes," that is, air and water pin holes, during manufacture has been discussed earlier. For architectural building members, where lower strengths may be permitted by specification, use of a slightly higher water/cement ratio may minimize these. Form vibration is extremely effective, even though the same pin holes may be buried below the surface. Absorbent forms, whether plywood, or special absorbent liners, also tend to keep the holes "submerged" so as not to present a visual problem.

In finishing, large bug holes should be patched first, using a mixture of white cement and regular cement. Then the entire surface can be rubbed. This will give a uniform surface appearance that will be quite lasting. Use of regular cement patches or, worse yet, epoxy-cement patches, will give a spotted appearance that worsens with time.

Architectural concrete can be ruined by rust. In one of the author's earlier experiences, stainless steel inserts were used to prevent rust staining, but projecting reinforcing bars, for later cast-in-place jointing, allowed rust stains to run down the panels. So any projecting steel or inserts, even if they are to be later encased, should be covered with plastic tubes and taped prior to shipment. In other cases, galvanized steel reinforcement has been employed in very thin sections to prevent later rust staining, only to have the ends of the bars burnt off and, thus ungalvanized, allow unsightly rust stains to run down the wall panels. Exposed ends of prestressing tendons can lead to similar discoloration. They should be cut back and touched with epoxy, and then patched with white cement. A trial series of patches should be made to determine the best mix of cement to produce even coloration.

One of the great advantages of bush-hammering and sand-blasting is that they erase and blend discolorations. When the structure permits sand-blasting in place, this can frequently prove the best and cheapest solution.

Coating of architectural units with silicones or pigmented linseed oil may prevent water-staining. Because such coatings do affect the coloration, they should be selected only after tests on samples are approved by the architect.

Leaching of lime from architectural units is particularly annoying. It shows up as a white efflorescence on the surface. Use of minimum cement content, low water/cement ratio, and possibly silicone waterproofing are steps to minimize this.

With open-web architectural units, and units with large openings, timber strutting or girting may facilitate handling, shipping, and erection and prevent damage.

With colored concretes, nonuniformity presents an intensified problem. The controls of mix and curing become of extreme importance. Judicious sand-blasting after manufacture may correct minor imperfections. Absolute cleanliness of mixers and transporting containers is essential as, of course, is absolute accuracy of batching, including the color addition. This should, wherever practicable, be added to a large quantity of cement or water, so as to attain a uniformity of mix. When colored panels are shipped and erected, it is frequently necessary to make shifts on the site, so as to match color variations as well as possible. Much erection time and cost can be saved by checking all panels in the plant with a color chart, and actually working out an erection sequence and pattern. It is an added effort, but much easier to perform in a drafting room than on the side of the building, and the architect will rightly insist on its performance. His criterion is properly uniformity of appearance, both texture and color.

With cast-in-place concrete construction, all of the previous suggestions are valid, but the techniques differ. With very thin sections, such as lift slabs, floor slabs, and shells, extreme care should be taken to ensure uniformity of thickness and accuracy of tendon positioning. The author has observed a thin shell in Europe in which, after one crew positioned the steel and tendons with care and accuracy, the concreting crew then walked on top of them!

There is a peculiar phenomena frequently observed with very thin concrete sections that the pattern of the reinforcing steel is shown by dark lines on the concrete surface. This is believed to be due to water content variations. The only suggestion the author has is to use denser concrete and smaller bars and, where practicable, greater cover.

With cast-in-place concrete, some patching is probably inevitable. The use of sand-blasting or bush-hammering would thus seem to be desirable if uniformity of appearance is to be achieved.

Some beautiful patterns have been specified and obtained, as at the Sydney Opera House, by casting against carefully patterned wood surfaces, and then forbidding patching. Since any patching spoils the entire effect, it is essential that placement and vibration techniques eliminate honeycomb and rock pockets. The use of a workable mix is essential for this. Wood surfaces should be field-checked to ensure reasonably uniform absorption characteristics. No form oil should be used, as it tends to absorb differentially and produce discoloration. After completion, such surfaces should be covered with paper or plywood to prevent disfiguration by workers (or engineers!) writing on them with marking crayon or other sources of discoloration and damage.

The type and location of picking inserts for exposed elements must be selected with care to prevent disfiguration of the element. In particular, they should be located and, if necessary, reinforced, to resist edge spalling adjacent to the insert. Inserts should be stainless, galvanized, or otherwise treated to prevent rust staining.

The erection methods and sequence will vary widely with the building to be erected. Some buildings permit easy crane erection and the elements are stable in themselves; others are difficult to reach, may require rehandling, and are not stable until a series of elements are erected and tied together.

For all structures an erection drawing should be prepared, showing the location of the crane, radius and weight of each pick, and point of delivery. Even in the simplest building, such a drawing will minimize crane moves, and guard against over-extending the crane reach. When delivery is by truck, the turning radius for the trucks should be laid out, with tolerances for backing, etc. In congested sites, this may be a major factor.

Such drawings may have to consider the three-dimensional effect, to insure that the crane boom will not encounter an obstruction. This can usually best be done by drawing sections in the vertical plane for each critical angle and position.

The need for insuring stability of the crane has been previously emphasized. It may require mats or rock pads, etc.

During erection, wind may cause difficulties in controlling the precast element, and may induce lateral and torsional stresses in the crane boom, to the point of endangering it. Therefore, guy lines are often necessary, particularly with panels, and may have to be activated by hand-jacking cable devices or air hoists. The lead and length of these guys varies during the lift, and the drawings must lay out the several positions.

For final positioning some form of close, positive control is necessary. Here ratchet or hand-jacking tools or "come-alongs" may be used effectively or, in some cases, screw jacks may be used to lower into place.

Stability must always be considered during erection of girders. If the girder tips sidewise, it may break, with serious danger to men and structure below. In general, all members should be picked so that the lifting eyes lie above the center of gravity of the member. The horizontal component due to the sling angle may be so great as to induce buckling; this must be particularly checked for thin I-girders.

When a girder is to be moved by support at its base (e.g., by rolling), lateral support must be provided at each vertical support point to prevent tipping.

After erection, each element must be stable against wind, accidental impact from a subsequent element being erected, and from other construction operations. The degree of securing must be carefully thought out with regard to consequences to the total structure and to personnel, and to the probability of occurrence. Certainly normal maximum wind forces for the season must always be considered. Seismic forces are usually only considered when the time of unsupported erection condition will be unduly prolonged (e.g., winter shut down) or when the results will be catastrophic. This, of course, was the case in a number of partially completed prestressed concrete buildings in Anchorage, Alaska, during the 1964 earthquake and led to several total building failures.

Accidental impact from subsequent erection is hard to appraise as far as force, but is, nevertheless, a very real hazard. Similarly, subsequent construction operations, e.g., raising structural steel or concreting, may produce lateral impact forces.

Of particular concern is the "stack-of-cards" type of failure, which has so often occurred with steel roof trusses. A pair of prestressed girders should be temporary trussing installed in the horizontal plane, then subsequent girders can be tied to the first pair.

During the erection and construction of very large shells, creep and shrinkage must be considered. Unbalanced snow and wind loads may be extremely troublesome, particularly the combined suction and direct wind force. Aerodynamic stability during construction must be investigated for large cable-supported roofs and some convex shells as well. Additional stability can sometimes be provided by properly located weights, e.g., concrete blocks or slabs, by strutting and tie cables or, in some cases, by shields or "spoilers" to minimize the suction effect. A snow-removal program can be instituted if necessary.

Temporary additional prestressing can sometimes be used effectively to ensure stability against wind. This stressing, for example, can be used to impose a downward load as well as a lateral tie.

Thermal effects during erection may assume serious proportions when large surfaces are involved. Diurnal movements of the order of 1 inch to 2 inches are not uncommon. Any temporary restraints, such as guys or scaffolding, must permit these thermal movements.

Scaffolding to support precast concrete buildings during erection is subject to the usual precautions against buckling and sidesway. The only difference with precast prestressed elements is that these units are usually longer, and are usually supported only at the ends. Thus, the scaffolding quantity is reduced, and some of the inherent space frame action of the scaffolding, whether so designed or not, is lost. Particular attention must therefore be paid to column strength with allowance for eccentricites in setting, etc., and lateral stability of the system against wind, etc.

Decentering of long span shells is always a "moment of truth." The shell must deflect, and uniform release is very hard to achieve. The possibility of collapse, while rare, must be considered. Therefore, screw jacks should be provided so that small retractions can be made uniformly, or in carefully determined sequence, without providing more than a very small gap in case uneven deflection or unusual sag occurs. The sequence of decentering must also be carefully engineered insofar as the stresses in the shell differ from the final condition.

Decentering by raising, whether by flat jacks or circumferential stressing, is attractive because the desired state of stress can be induced, the behavior checked progressively, and the shell raised slowly from the scaffolding.

10.1   Erection

Specific erection techniques for the erection of precast prestressed bridge girders are set forth in Chapter 13. Many of the methods and principles are applicable to the erection of buildings. In addition, there are several methods specifically applicable to building erection.

Sliding and Rolling.   Large structures in crowded urban areas frequently present a serious restriction of access for erection of interior members. One method of overcoming this is to erect members at one side of the structure, then roll or slide them laterally to final position. Extreme precautions must be taken to:

1.   Prevent tipping.
2.   Insure that both ends move together, under full control.
3.   Prevent lengthwise movement that would drop one end off its support.

These conditions may usually best be met by providing a structural steel cradle at each end, into which the girder end is set. This cradle, in turn, runs on a track, such as a steel channel, bolted to the cap-girder so that it is held in position. Movement is usually accomplished by air tugger or hydraulic winches, or long-stroke jacks may be used. Final positioning from the cradle to its permanent seat is by jacking down.

Tower Cranes.   Tower cranes are extremely versatile for erection of tall buildings. Since they usually have limited capacity at extreme reach, precast units may all be fed into a special bay or near side, from which they are hoisted and swung into final position.

Steel Erection Derricks.   These are frequently employed, especially where precast units are hung or combined with a structural steel frame. Since structural steel tolerances are often much greater than precast concrete tolerances and since the precast concrete is so rigid, provision should be made for accommodating dimensional variations. Double-slotted bolt holes and shims are most effective. Similarly, bolts with double-nuts may be used.

Helicopters.   Helicopters can be used very efficiently and economically for erection of precast units of small and moderate size, especially when the point of placement is difficult to reach, e.g., a tower. The helicopter lowers his hoist, slings are attached, the pilot raises his hoist. If trailing lines are attached to the unit, they must be individually handled by a competent man to prevent any possibility of snagging. Only those men directly connected with the operation should be within the proximity of the rotor blades.

The helicopter then lifts off to the placement position where the unit is lowered and affixed in position. Erection sequence may have to be altered, depending on the wind direction. Each unit should be immediately secured, as, for example, with erection bolts. The safety precautions for helicopter operation must be rigidly enforced, both for helicopter safety and safety of workmen. Communication between pilot, ground foreman, and top foreman is essential, both voice radio and visual signal.

10.2 Connections for Precast Prestressed Concrete Building Members

Whenever precast concrete elements are assembled into a structure, the connections and joints must be designed and executed so as to continuously perform over the life of the structure in the manner contemplated by the design. In other words, if the connection is an expansion joint, then freedom of movement must be assured regardless of the external environment, load history, and the effects of age. Hinged connections must continue to perform as hinges; fully continuous connections must maintain their integrity.

With prefabricated structures especially, the effects of variances in manufacture and in erection, of unequal foundation settlement, of differential thermal movements, etc., are concentrated at the joints.

Prestressing amplifies this problem due to creep and the generally thinner sections used, which permit greater rotation. Longer spans are more common with prestressing, thus again increasing the magnitude of the movements at the connections.

By far the greatest portion of difficulties and problems that occur with precast prestressed concrete building assemblies are at the connections. Yet this subject is treated very lightly, if at all, in most reference books on prestressed concrete. The reason is that the variety of connections and combinations and details is so great that classification and generalization are very difficulty.

Certain principles do emerge, however. The connections must be so designed and executed that they:

1. Transmit bearing, shear, moment, axial tension, and axial compression as required by the design.
2. Accommodate volume changes due to creep, shrinkage, and temperature without exceeding allowable stresses and permissible strains in the member, its support, and the connection assembly itself.
3. Accommodate all design loading combinations including superimposed live load, wind, and seismic loads within allowable stresses and permissible strains in the member, its supports, and the connection assembly.
4. Accept overloads, that is, ultimate load design with ductility, so that failure does not occur at the joints or connections before primary failure in the member. An exception is where a joint is specifically designed to break prior to primary failure of the member.
5. Perform their connecting function as designed, e.g., expansion, con-

tinuity, hinged, etc., continuously without marked change in performance despite the effects of age and anticipated environment.

6. Have adequate corrosion and fire protection. It should accommodate
   rotations and expansion due to fire temperature on the member while
   under fire exposure itself, for the prescribed fire duration.

7. Ensure adequate seating and performance despite the maximum permissible cumulative deviations in tolerance in manufacture and erection.

8. Insure watertightness, where appropriate, under maximum wind and
   volume-change conditions.

9. Provide mechanical means (stops) as necessary to prevent a member
   falling off its seat when the design limit is exceeded, as in a severe
   earthquake.

10. Be practicable and economical in attachment to the members and in
    erection.

The design and detailing of joints and connections is normally the responsibility of the design engineer. However, the constructor is often thrown into
a position of partial or entire responsibility when the specifications require
the constructor to "propose and submit details for approval" or where the
constructor has proposed a prefabricated alternative to a cast-in-place structure.

The constructor is always responsible for the careful and accurate execution of the jointing and connecting details.

When distress arises, such as cracking and spalling, the tendency is to
immediately charge the constructor with failure to make the connections
properly; it then is up to him to prove faulty design. Thus it behooves the
constructor to examine joint and connection details carefully and call attention to any inadequacies he discovers in the design details.

The materials used in connections generally consist of fabricated structural steel assemblies (plates, angles, and bars), reinforcing steel (of varying
grades and yield strengths), prestressing tendons (wire, strands, and especially bars), epoxies (and similar organic compounds), cement grout, concrete, and neoprene. Occasionally, other materials such as stainless steel and
bronze plates and Teflon are employed. Such materials and their structural
details must be designed in accordance with applicable codes and good
design practice. The stress distribution within connection details is usually
very complex, requiring careful analysis under the whole range of criteria.

Entirely too frequently, errors are made in the embedment or securing of
the connection into the concrete. This is usually a zone of high shear and
also of transverse tensile bursting stress, due to prestress. Minor eccentricity
or misalignment may produce stress concentrations. Under dynamic or
shock loading, bond may fail. Finally, this is usually a congested area where
it is difficult to place and consolidate the concrete.

Most of these difficulties can be overcome by increasing the depth of embedment and by liberal use of binding ties of reinforcement (stirrups or spiral).

In the member itself, shear and diagonal tension are usually at their maximum at the ends adjacent to the joints or connections. Adequate stirrups are essential to bind the concrete. It must be recognized that even with binding, there is always a zone (i.e., the cover) outside and beyond the last stirrup, which is unreinforced.

Welding of connections may cause spalling of concrete due to heat. This may be minimized by an intermittent welding procedure, careful selection of electrodes, and provision of adequate length of connection steel, both that extending beyond the face of the concrete and that embedded in the concrete.

The many connection schemes that have been developed and successfully used fall generally within the following categories:

1. Steel plates and angles, welded to reinforcing bars or anchors.
2. Bolts through sleeves in the concrete or through embedded steel plates. Double nuts may be used to facilitate adjustment.
3. Dowels entered into corresponding holes in the adjoining member and secured with low-shrink grout or epoxy. Polyesters have been used, with their set accelerated with an embedded wire through which a current is passed.
4. Socket connection, with the precast member set in a socket and fixed with cement grout or concrete.
5. Cast-in-place concrete joints, enclosing reinforcing bars projecting from the precast members, which are lapped or welded.
6. Post-tensioned bars or other tendons.
7. Fabricated steel hangers.
8. "Glued" epoxy joints.

Earlier in this section, the importance of permitting movement and rotation was emphasized. Neoprene bearing pads have proven widely applicable for this purpose. Some excellent connection details utilize neoprene or sponge rubber inserts within the concrete member itself to relieve stress concentrations and permit a restricted degree of translation and rotation. Seats for the support of precast members must be adequate to ensure full bearing without excessive edge concentration, even if members are cumulatively out of tolerance by their maximum permissible deviation. Edges should preferably be chamferred or rounded to prevent spalling.

When post-tensioning is extended by means of couplers at the joints, the sleeve for each coupler must permit the required longitudinal movement of the tendon without jamming. It must permit grout to flow around it without blockage. Since sleeves for couplers take out so much of the gross concrete

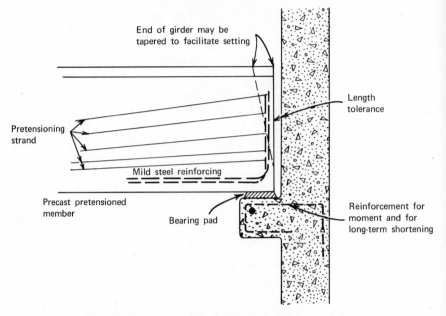

Fig. 15. Typical bearing detail: slab or girder on wall.

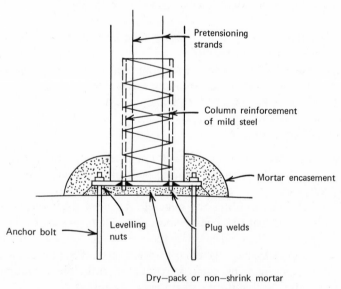

Fig. 16. Typical column base detail.

area, a check should be made to be sure that the remaining net concrete area can sustain the initial prestressing compression without crushing. This is a

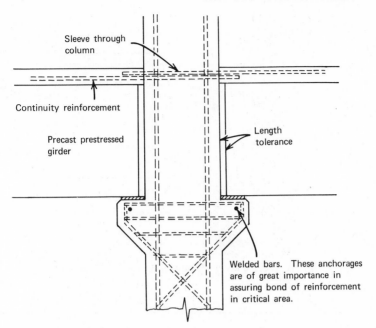

Sleeve through column

Continuity reinforcement

Precast prestressed girder

Length tolerance

Welded bars. These anchorages are of great importance in assuring bond of reinforcement in critical area.

Fig. 17. Girder to column connection.

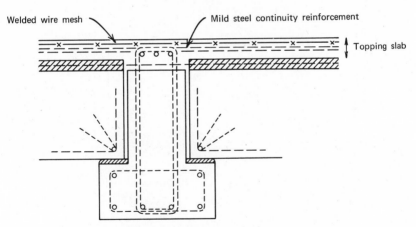

Welded wire mesh

Mild steel continuity reinforcement

Topping slab

Fig. 18. Double-tee to inverted-tee connection.

temporary condition and a factor of safety of 1.5 on ultimate should be sufficient (i.e., a safety factor of 1.5 on $.85 f'_c$.)

Connection details have received a great deal of study by the Prestressed Concrete Institute and their Connection Committee, who have issued reports on the subject from time to time. Additional information is found in

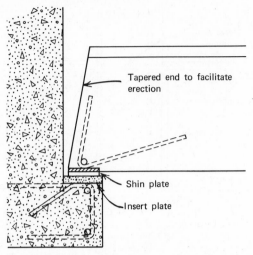

Tapered end to facilitate erection

Shin plate

Insert plate

Fig. 19. Welded connection: beam to girder or column.

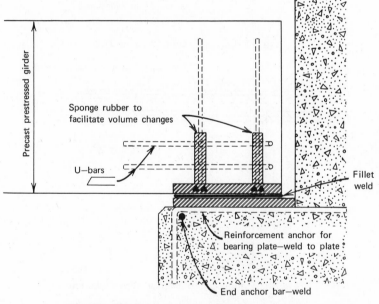

Precast prestressed girder

Sponge rubber to facilitate volume changes

U—bars

Fillet weld

Reinforcement anchor for bearing plate—weld to plate

End anchor bar—weld

Fig. 20. Welded connection: girder to column.

the literature of the precast concrete building industry (not necessarily prestressed). (See Figs. 15-21) Joints are further discussed in this book in Section 3.4.2 (Segmental Construction); 3.7 (Epoxies and Polyesters); and 3.8 (Welding).

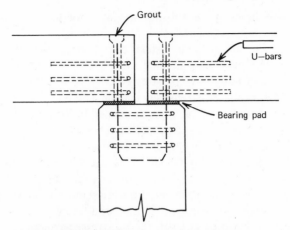

Fig. 21. Doweled simple spans: top of column.

10.3 Industrial Buildings

The utilization of prestressed concrete for industrial buildings has been very extensive in Eastern Europe, especially in the USSR, where it predominates. Typical elements consist of double columns, prestressed roof trusses, wall slabs, roof slabs, and prestressed crane girders. Wall and roof slabs are designed to give thermal insulation as well as structural strength. Three types of slabs are used: sandwich panels, all-lightweight concrete panels, and composite panels. The composite panels consist of a thin slab of prestressed normal or structural lightweight aggregate concrete, combined with a thick slab of aerated (cellular) concrete (See Fig. 22).

It is believed that the prestressed structural lightweight aggregate panels will emerge as the predominant roof slab and wall panel for this framed system.

The roof trusses are generally prestressed in the lower chord only, mild steel being used for the web and upper chord. Deep web girders are also

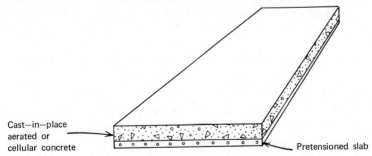

Fig. 22. Composite construction for industrial buildings. Originally developed by Dr. Jacob Feld in 1948 and utilized in Israel. It is occasionally used in U.S.S.R.

used, with large holes in the webs for weight reduction and passage of utilities.

Throughout Europe, most industrial construction has followed a similar framed approach, although thin shell roofs are more common than trusses. These thin shells usually have prestressed edge beams for the longer spans. North-light shell roofs are extensively used in The Netherlands, for example.

An interesting roof shell has been developed in Western Europe. This is a hyperbolic paraboloid, permitting use of straight pretensioning strands to produce a thin shell roof element for medium spans.

In the United States and Canada, the use of prestressed concrete for industrial buildings has lagged, for a variety of reasons. Tradition, the ready availability of steel, the trend for incorporating more services into the building structure (such as lighting and air conditioning), and the lack of a mature industrial building concept for prestressed concrete have been partially responsible. Earliest inroads have been made by roof slabs, where the long spans, fire resistance, and excellent appearance have led to their selection. Some long-span barrel vault shells, with prestressed edge beams have been used. The single tee, however, has been selected most frequently for long spans, with the double tee for shorter spans (70 feet and below).

Recently, storage buildings for a variety of bulk products such as potash and coke, have been constructed of prestressed concrete. These have mostly been A-frame buildings. One type uses prestressed inclined girders, supporting horizontal slabs on the slope. The most promising type, however, uses very long (150 feet or more) single or double tees to form the A-frame, thus supplying roof slab and beam in the same element. This trend is significant because it has also been employed for vertical wall panels, with single or double tees serving both the supporting and the cladding functions.

The emerging pattern in the United States and Canada, therefore, is expected to follow this pattern. Large elements, such as single or double tees, of maximum width, will be used to form both the structure and the cladding. Connections will be welded or, preferably, post-tensioned. Joints will be grouted, with plain or epoxy grout, or filled with a water proof expansion material, such as Thiokol, to permit differential thermal movements.

For special types of industrial buildings requiring large volumes of air circulation, hollow-box girders may combine the function of a fireproof, sound resistant, duct with that of long-span structural support.

Provision must be made for volume changes due to thermal movement, shrinkage and creep, and differential settlement of supports. Shrinkage may be somewhat more complicated than usual if the inside is maintained at high humidity, as, for example, is done with coke storage. Creep will be of a relatively high magnitude due to the long spans and highly-stressed members, and to the increasing use of structural lightweight aggregate concrete, with its somewhat greater creep factor.

Differential settlement of supports is a problem in bulk storage, where the high loadings cause consolidation and possibly lateral movement of the soil. Ties beneath the building floor may be more harmful than beneficial since any settlement of the floor tends to pull the base of the walls inward. Pile supports are usually preferable.

10.4 Commercial Buildings

Prestressed concrete has enjoyed its greatest growth in this field, because it has enabled structural and architectural functions to be economically combined. Shopping centers, for example, need moderately long spans to permit flexibility and alteration of partitions. External walls must be attractive as well as functional, leading to the use of precast panels or double-tees, either as plain concrete or with exposed aggregate and, occasionally, color.

The predominant structural system has been the column, inverted tee girder, precast hollow core floor slab or double-tee floor slabs, and double or single tee roof slabs.

Fire resistance is a major factor in the selection of prestresssed concrete, and performance to date in actual fires, as in tests, has been excellent.

Proper connection details are essential to ensure provision for volume change such as shrinkage and creep. Failure to allow for this has produced minor distress conditions at the supports in a number of the earlier commercial buildings. A multi-bay layout tends to make volume changes additive.

Provision must also be made to ensure waterproof joints in the roof despite thermal and other volume change movements. Leaks cannot be tolerated in shopping centers.

In erecting precast prestressed roof members, it is frequently necessary to match the sections to ensure approximately equal camber, and then to tie adjoining members together at the third or quarter points by bolted or welded connections.

Prestressed concrete has enjoyed some of its widest popularity in parking garages, because of its long-span capabilities combined with fire-proofing. Rectangular layouts have utilized the double-tee most extensively, supported on inverted tee girders. Circular layouts have utilized pie-shaped members; for example, a double-tee at the outer end tapering to a single tee at the inner end.

One popular concept utilizes pretensioned single tees which are post-tensioned after erection to provide a full-moment connection to the columns. A cast-in-place floor slab then serves in composite action. It, in turn, may be post-tensioned transversely with bonded or unbonded tendons, thus tying the whole structure together. (See Fig. 23)

Fig. 23. Parking garage utilizes precast pretensioned girders, made continuous by post-tensioning and composite with cast-in-place slab.

It is increasingly common to utilize the roofs of garages for either parking or playgrounds. This frequently calls for the roof to be black asphalt. Heat absorption causes accentuated thermal movements of the long-span roof slabs, and this must be taken into account in the connection details.

Columns and girders for prestressed concrete buildings are sometimes precast, sometimes cast-in-place. No clear cut economic preference has yet emerged. The design for either method must provide for lateral support and integrity, and for volume-changes of the slabs.

Post-tensioned cast-in-place construction is also being increasingly employed for garage construction, often using structural lightweight concrete. The advantage of long spans and thin slabs is maintained. Joint problems are eliminated, but shrinkage and plastic flow may be accentuated. Shrinkage-compensating cement is thus used for a number of such structures. The relative economies of the two systems (precast or cast-in-place) will continue to vary in individual areas due to the interplay of many economic and competitive factors. What is clear is the ever-wider employment of prestressed concrete in one form or another.

10.5 Housing

Prestressed concrete has been employed only to a limited extent in apartment housing construction, primarily because of the ready availability of low-cost alternatives. It is believed that this situation is in the midst of a radical change as a result of a number of concurrent factors:

(a) Greater emphasis on fire resistance.

(b) Greater emphasis on sound (acoustic) insulation.

(c) Increasing costs of alternative materials and methods of construction.

(d) The mass market developing for prestressed concrete will result in more efficient manufacture and comparatively lower costs.

Systems building, as discussed earlier, may take one of two directions as far as prestressed concrete is concerned. However, regardless of the outcome, the emphasis on the systems concept will stimulate the more careful evaluation of materials and the greater use of prestressed concrete.

The hollow-core floor slab is being widely promoted for housing construction. Experience to date indicates that the voids are of only marginal benefit for utility runs. The voids, therefore, are primarily a means of reducing weight, and reducing concrete and steel quantities. Greater emphasis on acoustic insulation is causing an increased interest in some parts of Western Europe in solid slabs.

Use of a cast-in-place topping permits shear ties to resist lateral forces of wind and earthquake. This can act in composite action with the prestressed slab, and provide the needed thickness for heat transmission during fire, for acoustic insulation, and greater rigidity. However, the use of a cast-in-place topping is more expensive than if full-depth slabs are employed. Development work continues on economical methods of jointing full-depth slabs to insure adequate shear transfer. Welded shear connectors often cause black stains on the concrete, and in actual earthquake, some have failed either in the weld (brittle fracture) or by the anchors pulling out of the concrete (bond failure). Therefore welded shear connectors require great care in detailing. Cast-in-place joints, with overlapping reinforcing bars, are structurally satisfactory if of sufficient width. Unfortunately, all of the transverse shrinkage is concentrated at the joints, and cracks will appear, unless special means are taken, such as use of shrinkage-compensating cement. (This also applies to composite topping described above.)

Grouted joints, with transverse post-tensioning, much used in the earlier bridges, have not been widely applied to buildings but would seem worthy of development, especially with full thickness slabs (See Fig. 24).

Lift-slab construction continues to be widely used for apartment housing construction. It depends on prestressing to keep the floor slabs level under normal (service) loading. Many lift-slabs in the United States are post-tensioned with unbonded tendons. In other countries, bonded tendons are generally used, placed in ducts, tensioned, and then grouted. Arguments continue, rather vociferously, as to the merits of the two systems: the low labor requirement and speed of the unbonded system versus the better behavior at ultimate (failure) condition of the bonded system. A major step in reconcili-

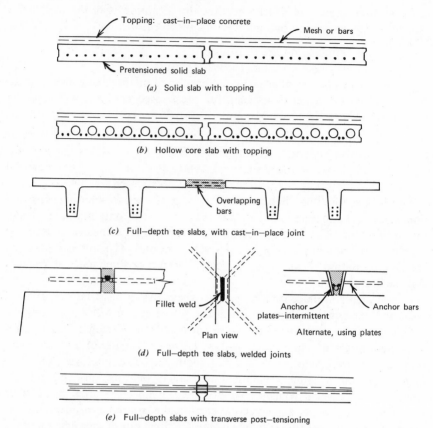

Fig. 24. Methods of connecting precast prestressed floor slabs.

ation has been the issuance of the "Tentative Recommendations for Members Prestressed with Unbonded Tendons" by ACI—ASCE Committee 423, which prescribes additional mild steel and other measures to ensure proper ultimate behavior of the unbonded system.

Lift slabs can be produced by pretensioning on site, using the basement walls as abutments. This is the contribution of a system developed in England. Greater development work on this approach is indicated because of its economy and potential technical advantages.

10.6   School Buildings

Prestressed concrete is selected very frequently for school buildings, because of its long clear spans, especially for auditoriums, gymnasiums, and cafeteria spaces. It is fireproof and has excellent acoustic properties.

In seismic areas connection details and shear connectors must be given special attention to insure their ductile behavior under dynamic loading.

Long, slender roof spans must be designed with attention to sag under rain and snow loading, so that they will always drain and not allow ponding to occur. Because of the long spans, particular care must be taken to provide lateral staying or guying during erection.

10.7   Office Buildings

Prestressed concrete is increasingly employed because of its long clear spans, permitting maximum flexibility in partitioning, and its fire resistance. Properly designed, it is suitable for the higher floor loadings that the computers and office machines impose.

At the present time prestressed concrete is most extensively employed in the form of floor slabs, such as hollow-core slabs and wall panels, the latter of architectural concrete, often with exposed aggregate or special coloring, and including the window openings. Proper detailing is essential to ensure waterproof joints despite thermal changes (e.g., the sun's daily movement) and wind. Connection details must be sufficiently strong, including the embedment of anchors into the concrete, to ensure adequate securing in wind and earthquake.

10.8   Churches and Civic Structures

These structures are often designed to convey an architectural theme as well as to provide utility of function. They must be permanent, durable, of clean, and attractive finish. Prestressed concrete is ideal for this use; the concrete can be molded to almost any shape and the prestress varied to suit the varying structural requirements.

One of the foremost of such monumental structures, the Sydney Opera House, uses precast segments to form inclined arch ribs. These are then prestressed to create a stable structural system and a favorable state of stress. Indeed, without prestressing, this form could probably not have been achieved and maintained.

A recent trend has been the use of standard single and double tee elements in a variety of combinations to achieve striking architectural effects consistent with the dignity and spirit of the theme. The stems of the tees may be turned in or out and architectural treatments applied to one or both faces.

Some of the finest of such structures have integrated precast and cast-in-place concrete, with both pretensioning and post-tensioning, thus showing a maturity of engineering design and construction skill.

10.9   Systems (Technical Aspects)

The general philosophy of systems building has been discussed earlier. This section is concerned only with the technical aspects of design, manufacture, and construction of components for incorporation in systems.

The concept of systems implies the integration of many functions. The integration of architectural and structural function is inherent in prestressed concrete. Insulation of various types may be applied by spray or gluing on of sheets. Holes may be formed and inserts placed for mechanical and electrical systems; in a truly integrated system, most of the runs will themselves be embedded in the concrete. Concrete girders may be hollow to serve as air ducts or vents. Window and door frames may be placed and the windows glazed and doors hung.

All of this will be successful only as each and every operation is integrated into the manufacturing process. A basic process must be established. Either the product moves to each operation station in turn, or the operational processes move in succession to the product. In either case, adequate time and space must be provided for the operation. Thorough consideration to material flow and work simplification must be given to each operation, even those like mechanical and electrical, which have not historically been concerned with these aspects.

By such a program it then becomes economically feasible to employ a highly skilled artisan (plumber or electrician) at the final fitting station. This practice may lead toward a solution of the problem that has plagued many previous efforts at systems building and prefabrication: the resistance of labor unions and building code officials.

Inserts must be rigidly held in forms so as to prevent their displacement during placing and vibration of the concrete. They may require covers or other protection to protect them from damage during handling of the product.

Certain processes, such as sprayed-on insulation, etc., require enclosures to protect both the product from contamination and the adjoining workmen from hazards. Prefabrication should lend itself to better safety practices in this regard.

Completed products must be protected by lagging or paper from damage and staining during storage, transit, and erection.

Code marking of all units is an essential ingredient of system prefabrication. Products must be tied in with a working drawing, a definite production sequence, storage space, shipping schedule, and erection sequence. Such a program has been computerized in at least one plant in the eastern United States, which utilized computer-drawn strand patterns as well as printed numerical instructions, coded to both product and forms, as well as to storage, shipping, and erection.

Whether such sophisticated programming will be useful in all aspects of systems building or not is irrelevant. The basic principles of carefully organized production are applicable. They do make possible that acme of mass production theory: the multiple use of standardized elements, with alterations and modifications so incorporated into the production process that the full benefits of standardization are retained.

10.10   Shells

Prestressing has provided an excellent means of countering the tensile stresses in the edge beams of shells. It permits a degree of control and assurance that would otherwise be unobtainable. Edge beams, therefore, are frequently post-tensioned, with the sequence of stressing carefully dovetailed with the construction of the roof.

The edge beams of domes are usually post-tensioned with internal tendons covering a 90° to 135° arc between buttresses. On some notable domes, the edge beam was stressed by wrapping wire under tension, in a manner similar to a tank. With other shapes, simple internal post-tensioning tendons are employed. Where such tendons lap at corners, adequate mild steel must be provided. Precast edge beams have been used on occasion, especially when they were separate from the shell, and then post-tensioned (e.g., as tie-beams).

With cable-supported roofs, the compression ring may be constructed, the cables strung and stressed to their specified length. Then forms may be hung and the roof concreted in a predetermined pattern. This procedure produces a varying state of final stress in each tendon. If precast slabs are hung from the tendons, then the tendons may be stressed in several stages as the dead weight increases, with final bonding with mortar in the joints between the slabs.

Precast pretensioned hyperbolic paraboloid roof shells are produced under one European system in which the pretensioned wires or strands are stretched on the generatrices.

10.11   Foundation Mats

Foundation mats may be subject to high bending stresses. They are usually very heavily reinforced both top and bottom. Post-tensioning has been used with great advantage in specific cases, where supports and/or columns are widely spaced, or where unequal loadings are applied to a continuous mat. Provision must be made for movement of the mat despite the friction of the ground. This is a problem similar to that of pavements, except the dead weight, and hence the friction, is usually much greater. Sand layers, between polyethylene sheets, are one solution. For a proposed multistory building in Hong Kong a 1/8-inch asphalt coating will be applied over a 3-inch thick concrete blinding layer. With post-tensioned mats, adequate room must be provided at the edges and jacking pockets formed.

The post-tensioning must be performed in stages, as the building loads are imposed on the foundation mat. For the Hong Kong building, four stages of stressing will be employed, together with prestressed soil anchors at the edges of the mat.

Tension elements (highly stressed concrete bars) (see Section 3.6) would appear to be particularly well adapted to foundation mats in lieu of large diameter mild reinforcing steel. They have inherent durability, excellent bonding, and can be furnished in long lengths. Splicing is facilitated by the lapping of extended strands or by overlapping the tension elements themselves. By grouping these, problems of temporary support could be minimized. For one proposed project, highly stressed pile sections (10" x 10" x 150' long) are being considered.

10.12   Summary

Buildings represent perhaps the largest potential area for the application of prestressed concrete. The nature of buildings varies widely. As pointed out in the Indian Symposium on Human Habitation, "habitations" includes all the structures which house human services and which serve human needs. To be truly useful, they must serve the whole man. Prestressed concrete has a major role to play in buildings. It will realize its full potential only when the technical knowledge of the prestressing engineer is integrated with the principles of plant layout, material handling, and production engineering, and these, in turn, are utilized by the architect to serve the "whole man."

11

Prestressed
Concrete Piling

11.1 Introduction

Prestressed concrete piles are widely used for marine structures and building foundations throughout the world. Actually, the growth in their use and the constant extension of the fields of application have been phenomenal. The original impetus was based on their durability in adverse environments, especially sea water, their strength as a beam for handling, and their strength as an unsupported column. Further advantages emerged quickly, especially their strength in bending and in tension. Wider use gave rise to lower costs of manufacture, which made prestressed piles an economical competitor for foundation piles on land. Their ability to withstand hard driving, and their good soil-pile interaction behavior has led to rapidly increasing use in this application. More recently their use has been extended to fender piles, sheet piles, and soldier piles.

Prestressed piles have been utilized successfully in a wide variety of environmental conditions, from sub-arctic to tropical and desert. Properly designed, manufactured, and installed, they are extremely durable under both freeze-thaw and salt-spray attack.

Prestressed piles have been used as end-bearing piles on rock; as friction piles driven (or jetted) into sands, silts, and clays; and as piles set and grouted in predrilled holes.

Prestressed piles have been as small as 10 inches (25 cm) in diameter [with lengths up to 120 feet (36 m)] and as large as 13 feet (4 m) in diameter. The longest piles made and driven in a single piece, up to the date of this writing, are 36 inches (90 cm) diameter cylinder piles, 260 feet (80 m) in length, in

Lake Maracaibo. Prestressed piles have been installed in increments, spliced during driving in lengths of 200 feet (60 m) and more.

Prestressed piles offer these advantages:

Durability.
Crack-free under handling and driving.
High load carrying capacity.
High moment capacity.
Excellent combined load-moment capacity.
Ability to take uplift (tension).
Ease of handling, transporting, and driving.
Economy.
Ability to take hard driving and to penetrate hard strata and debris.
High column strength.
Readily spliced and connected.

Prestressed concrete piling can be driven with an underwater hammer below the surface of the water, or with a follower below the surface of the ground, thus permitting piles to be installed prior to excavation and dewatering.

High-capacity prestressed concrete piles are particularly advantageous for deep foundations with heavy loads in weak soils. They can be successfully driven through riprap, debris, and old fills, and even through hard strata such as coral, and can penetrate into soft or partially-decomposed rock.

The use of prestressed concrete has enabled the designer to utilize, in an economical and practicable way, the pile as an integral structural member of the overall structure.

11.2   Design

The design of prestressed concrete piling for direct bearing (short column) loads has gradually evolved from an arbitrary, highly empirical percentage of concrete cylinder compressive strength (e.g., allowable capacity = P = $Kf'_c A_c$) to a more rational evaluation of the actual capacity, including the effect of prestress at time of failure. To this a suitable factor of safety is applied. For example, a current code formula reads $P = (0.33f'_c - 0.27f_e) A_c$

Current codes, however, do not yet take into account the difference in behavior of the concrete inside and outside the spiral. The constants used do reflect an allowance for safety factors to cover such matters as variation of actual strength from cylinder strength, accidental eccentricity, etc.

It should be always recognized that the structural system includes both the pile and the soil, and that almost always a failure under actual load test occurs in the soil, not in the pile.

Where piles have an unsupported length, their behavior as a column must be investigated, so as to prevent failure through buckling.

When prestressed piles must resist both moment and direct load, the interaction requires careful analysis. The presence of direct load up to about 30% of the ultimate short-column compressive strength, causes an increase in moment-carrying capacity. As the direct load increases above about 30%, the moment-carrying capacity is reduced. (See Fig. 25)

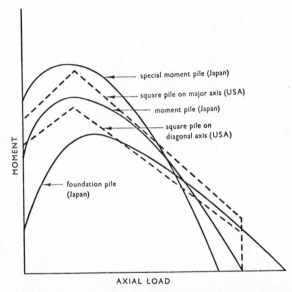

Fig. 25. Typical moment-interaction curve for prestressed piles.

At the pile head, where combined moment and direct load are often critical, the favorable effect of the transfer length (prestress varying from zero to full value over the transfer length) may be taken into account. Additional mild steel may often be effectively utilized in areas of high combined moment and load, such as the pile head.

The construction engineer is usually responsible for insuring that the pile is manufactured, delivered, and installed in its design condition, that is, without cracking, damage, or permanent deformation from overstress. He, therefore, must investigate stresses during these phases. (See Fig. 26)

For the handling of prestressed piles, the value of f'_c should be the ultimate compressive strength of the concrete at the age in question.

Impact must be considered and may impose additional stresses of 50% or more of the static stress. During handling and transporting, tension up to 40 to 50% of the modulus of rupture may be allowed to resist impact. It must

Fig. 26. Handling 110 foot long 10 inch pretensioned concrete piling.

be noted that at early ages, the modulus of rupture (tensile strength) of concrete is somewhat lower in relation to its compressive strength than at later ages.

Many pretensioned piles are manufactured in fixed forms with tapered sides. At the time of lifting of these piles from the forms, a great many conditions must be considered:

1. The pile may not have been able to fully shorten when the prestress was released due to friction on sides and soffit.
2. Lifting will have to overcome friction on the sides.
3. Suction.
4. Binding at any irregularities, such as dents in the forms.
5. Dynamic loads from the crane.

6. The concrete is still young (green) and fully saturated, except at the surface where tensile stresses may already exist due to drying and cooling.

To prevent overstressing during this condition, pick-up points may have to be spaced more closely, additional mild steel (or short pieces of unstressed strand) may be needed at the picking points, and it may be desirable to raise one end a few inches ahead of the other.

Under seismic and wind loads a combination of axial load and moment may occur at the head of piles serving as a foundation for chimneys and stacks. This condition can best be provided for by means of auxiliary mild-steel bars embedded in the pile head. Another solution is post-tensioning of the pile to the pile cap or footing by means of short tendons such as bars, or embedment or encasement of a structural steel member.

Prestressed piles have conventionally been designed with moderate values of prestress to meet the needs of handling and driving. The increasing application of prestressed piles to resist combined direct load and moment requires that the designer be alert to the potential structural value and economy of using much higher values of prestress.

11.2.1 Design for Resistance to Driving Stresses

The stress-wave theory explains many of the phenomena observed during the driving of prestressed piles. When a prestressed pile is struck by a hammer blow, the impact lasts from 0.004 to 0.08 second. The compression wave travels at the speed of sound in concrete, about 13,000 fps (4000 m/s) for a typical concrete used in prestressed work. (For structural lightweight aggregate concrete, this velocity is reduced 15 to 30%.) The wavelength is normally 50 to 150 feet (15 to 45 m), depending on the duration of the impact. The maximum compressive stress in the pile head will be from 1000 to 3400 psi (70 kg/cm^2 to 240 kg/cm^2), but can reach over 4000 psi (300 kg/cm^2). (For structural lightweight aggregate concrete, the maximum compressive stress is reduced about 20%.) When this compressive stress wave reaches the pile tip, it reflects. If the tip is on hard bottom (e.g., rock), it will reflect as a compressive wave. If the tip is on soft bottom (e.g., mud), it will reflect as a tensile wave. (See Fig. 27).

The toe stress may thus reach a compressive stress approaching twice the head stress, when driving through water onto rock. Conversely, when driving the tip into soft material, a tensile stress up to 50% of the head compressive stress may be reflected back up the pile. Actual tensile stresses have been measured by strain gauges up to 1400 psi (100 kg/cm^2) and even higher.

There are a great many variables (21 or more) involved in computing these stresses. A number of computer programs have been set up. The varia-

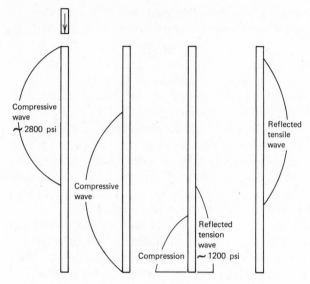

Compressive wave ~ 2800 psi

Compressive wave

Reflected tensile wave

Reflected tension wave ~ 1200 psi

Compression

(a) Reflected tensile stress—wave: soft driving conditions at tip

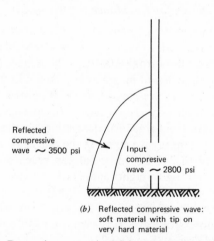

Reflected compressive wave ~ 3500 psi

Input compresive wave ~ 2800 psi

(b) Reflected compressive wave: soft material with tip on very hard material

Fig. 27. Dynamic stresses in driving prestressed concrete piles.

bles include weight and length of pile, modulus of elasticity, damping properties, weight and velocity of ram, stiffness of cushion, and soil resistance. Actual observations tend to confirm these mathematical solutions, although the values of maximum stress observed tend to be somewhat lower than computed, due to the influence of the surrounding soil, and the dissipation of energy into the soil and into heat and sound. The location of maximum tensile stress appears to remain close to the top one-third point (in most, but not all, cases) while the mathematical solution may indicate an equally high stress at the bottom third point. This, again, is probably due to the frictional resistance of the soil along the sides of the pile.

As discussed later under "driving," the values of maximum stresses can be effectively reduced by reducing the velocity of ram impact, and by increasing the amount of cushioning.

According to the stress-wave theory, with a pile driven through water, or very soft soil, onto rock, a high compressive stress can reflect from the toe, travel up the pile to the head and, if the ram has left the head by this time, the wave will be reflected as a tensile wave. This is a rather uncommon condition in actual practice, due to soil restraint and damping in the pile itself; however, a number of cases have arisen, requiring modification in the driving procedure.

These "rebound" tensile stresses are resisted by the tensile strength of the concrete and the effective prestress, up to the point of cracking. After cracking, these stresses are resisted by the steel tendons, up to their yield point. When the stresses equal or exceed the yield point of the steel, low-cycle fatigue ensues and the tendons may fracture in brittle failure. The adjoining concrete becomes heated and fails in bond and compression under repeated impact.

The steel tendon area, at yield, should have an equal or greater force than the prestress plus concrete tensile strength; otherwise, the first crack will shortly lead to failure. Microcracks that form serve to increase the damping. Theory plus experience indicates the desirability of providing a steel tendon area of 0.5% or more of the concrete area.

It is interesting to note that this same problem of rebound tensile stresses had been noted with reinforced concrete piles, although numerous cracks due to shrinkage and the large steel area usually prevented failure. For these piles reinforced with mild steel, an area of 3% was often prescribed. The ratio of yield points of mild steel to prestress steel is about 1 : 6, which agrees with the inverse ratio of steel areas recommended (i.e., 0.5% : 3% = 1 : 6).

There is no general theory for design of the spiral binding for prestressed piles. In practice, steel spiral areas of about 0.2% have been widely employed without serious difficulty. (For this purpose steel area ratios are determined on a vertical section through the pile axis; Fig. 28).

However, longitudinal cracking has occurred with increasing frequency, especially with hollow-core and cylinder piles. The Concrete Society of Great Britain recommends a greater area of spiral for hollow-core than solid piles; values of 0.3 to 0.4% are recomended.

During driving, the bursting forces at head and tip are large. There are

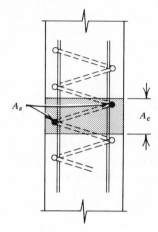

Fig. 28. Vertical section through pile axis for computation of ratio of spiral steel area to concrete area.

transverse tensile forces due to prestress and radial bursting forces due to the hammer blow.

Specifications for the lateral binding of the end zones of girders generally set values for A_s/A_c of 0.4%, solely to resist the transverse tensile stresses in the end zone. The bursting stresses due to driving are apparently of the same or greater magnitude (extrapolating from our known values of direct stress). This indicates a requirement of 0.8 to 1.0% spiral steel area in the top 24 inches. The Concrete Society of Great Britain recommends a minimum of 0.6% spiral over a length equal to 3 times the least dimension of the pile. This is based on observed (empirical) data.

The author has determined in practice that the provision of 1% steel area in the top 12 inches of cylinder piles will effectively prevent splitting. Some engineers provide bands around the top for this same purpose. It is interesting to note that these bands sometimes break during driving, indicating that the stresses are real. Hoops of mild steel are recommended in preference to bands. They can be placed very close to the outside wall of the pile, since cover for durability is usually provided by the pile cap.

Chamfering of the circumferential edges and top corners of the pile head will reduce bursting and spalling tendencies.

After installation, piles may be subjected to shearing forces due to lateral loads from soils and seismic forces. During driving, shear may occur due to restraint or obstructions. Therefore, some spiral should be provided throughout the length of the pile, not just at head and tip. Decreasing the spiral spacing is much more effective than increasing the wire size.

In summary, therefore, the following amounts of spiral are recommended for prestressed piles:

Top 12"   1.0%
Next 24"   0.6%
Body of pile 0.3% (solid piles)
               0.4% (hollow-core piles)
24" above tip 0.6%
Tip 12"   1.0%

11.2.2   Design Notes

The following design notes relate to details for prestressed piling. Inattention to these details has led to difficulty in installation and subsequent behavior: like all mass-produced items, the basic details are of great importance.

Lifting Holes.   Lifting holes through the pile are undesirable because they cause local stress concentrations. Piles should be lifted by slings around the pile, or by lifting eyes embedded in the pile. A loop of short pieces of strand, with each end embedded at least 24 inches (60 cm) into the concrete, makes a good picking eye. Special picking inserts are also manufactured for this purpose.

If the picking eye or loop is permanently embedded in soil, no removal or corrosion protection is usually required. If the picking eye is exposed, it may be desirable to cut it off and coat it with an epoxy patch.

Pile-to-Cap Connections.   The following types have been successfully employed:

(a) Mild-steel dowels are grouted into holes in the head. These holes may be preformed, with flexible metal duct. Removable tubes, such as hose or inflated rubber tubes, tend to leave too smooth a surface for bond.

Alternatively, the holes may be drilled in after the pile is driven, provided a small drill is used, so as not to break out, crack, or otherwise damage the pile.

The bars should be embedded a sufficient distance (i.e., 30 or 40 diameters for deformed reinforcing steel) to develop full bond. The bars may also be used to develop additional bending moment in the head of the pile.

Bars should be grouted in. Dry pack is not nearly as satisfactory as grout. The grout should be designed for low shrinkage. Epoxy grout may be used, but is more expensive.

(b) The prestressing tendons may be extended into the pile cap a distance

Shape	Advantages	Disadvantages
 △	Highest ratio of skin-friction perimeter to cross-sectional area. Low manufacturing cost.	Low bending strength.
□	Good ratio of skin-friction perimeter to cross-sectional area. Low manufacturing cost. Good bending strength on major axes.	
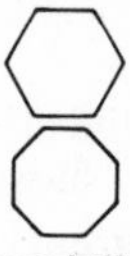 ⬡ ⬢	Approximately equal bending strength on all axes. Good penetrating ability. Good column stability (l/r ratio)	Surface defects on top sloping surfaces as cast are hard to avoid.
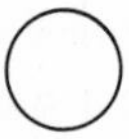 ◯	Equal bending strength on all axes. No sharp corners—aids appearance and durability. Minimum wave and current loads. Good column stability (l/r ratio).	Manufacturing cost generally higher. Surface defects on upper surfaces as cast are hard to avoid.
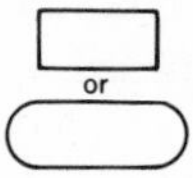 ▭ or ⬭	May be used where greater bending strength is required around one axis, especially if minimum surface to lateral wave and current forces is desired.	Difficulty in maintaining orientation during driving.
 ✚	High bending moment about both axes in relation to cross-sectional area.	High cost of manufacture. Difficulty of orientation.
 x-⊐⊏-x	High bending moment about axis x - x in relation to cross-sectional area.	High cost of manufacture. Difficulty of orientation.

Note: Hollow cores may be employed with most shapes. Varying cross-sections may be employed along the length of the pile—such as enlarged tips for bearing, enlarged upper sections for moment, or a change to a circular upper section in order to eliminate corners, etc. These changed sections may either be cast monolithically or spliced on at any stage. Change in cross-section should employ a transition section at least twice the length of the radial change.

Fig. 29 Cross-sectional shapes for prestressed concrete piles.

of 18 to 30 inches (4.5 to 7.5 cm). They should be splayed. While such embedment will develop the full strength of the strand, there may be an excessive rotation of the pile-cap connection if the stresses allowed are greater than those used for mild steel of the same cross-sectional area. This is only of importance where the cap is small, as in high railroad trestles; for normal footings and caps, rotation is generally of no importance.

When piles must resist repeated tension, then strand embedment should be increased to 30 to 40 inches (75 to 100 cm) to provide safety against bond failure.

(c) The pile may be extended up into the cap a distance of 18 to 30 inches (45 to 75 cm). The surface of the pile should be cleaned prior to concreting. It may also be roughened by bush-hammering or sand-blasting.

(d) For hollow-core or cylinder piles, a cage of reinforcing steel or a structural steel beam may be concreted into the top portion of the core. To insure bond, the inside surface of the core should be cleaned and roughened, as by sand-blasting, and epoxy bonding compound applied. The concrete in the plug should be designed for low shrinkage.

Precast plugs, containing the steel cage, are particularly useful with large-diameter cylinder piles. The small annular space between the inside of the pile shell and the outside of the precast plug can be filled with grout. This minimizes the problem of shrinkage.

Note. When filling a hollow-core pile with concrete, be sure that there is sufficient hoop strength in the pile (spiral) to withstand the hydrostatic head of the fresh concrete. Rates of pour should be adjusted to prevent cracking or bursting the pile.

(e) Post-tensioning from the pile head to the cap is theoretically an excellent means of developing full moment, but has not been widely used because of the high labor cost. If bars are used, these may be embedded and grouted into holes which are formed or drilled into the pile. Because the tendons to be stressed are very short, the anchoring system must effect the seating without loss, or else permit subsequent shimming to offset this seating loss.

Instead of multiple bars, a single large tendon may be embedded in the core of hollow core piles and later post-tensioned through a sleeve to the cap.

An as yet untried method that offers promise is to group the strands in the four corners of a square pile, bundling them at the head, and letting them protrude as four bundles during driving. Each bundle is then led through a sleeve through the cap and later post-tensioned. (Fig. 30)

Distribution of Tendons. The matter of distribution of the tendons within a pile has been given careful analysis in Japan and elsewhere. Current United States practice uses uniform distribution around a circular pattern, even for square piles. Swedish practice has been to group the tendons near

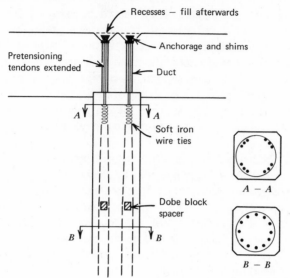

Recesses — fill afterwards

Pretensioning tendons extended

Anchorage and shims

Duct

Soft iron wire ties

Dobe block spacer

$A - A$

$B - B$

Note: Spiral wrapping not shown

Fig. 30. Head detail: pretensioning tendons extended and stressed to pile cap.

the corner of the square. In the USSR, short foundation piling have been employed with the tendons concentrated at the center. Elastically, and for concentric stresses, it seems to make little difference as long as the center of stress coincides with the center of gravity. After cracking in bending, however, the strand pattern does have an effect on the ultimate strength. The Japanese tests have concluded that the ultimate bending moment is not affected appreciably as long as there are at least four tendons around the periphery.

A circular pattern, with circular spiral reinforcement, is used on round, hexagon, and octagon piles. With square piles, both circular and square patterns are used, with spiral shaped to suit. However, there appears to be no significant difference in behavior between the circular and the square pattern, and no evidence that the corners of square piles are more likely to break off when a circular pattern is used. Therefore, because of the savings, increased efficiency of circular spiral, and ease of manufacture, the circular pattern for square piles predominates in current practice.

Pile Shoes.   Practice and opinion vary widely from country to country, based largely on early experience during the developmental testing of concrete piles. In the United States, pile shoes are generally not used at all, whereas in Scandinavia and Great Britain, pile shoes have generally been standard practice. The reason may lie in the geology of the regions. High toe stresses occur mainly when driving through extremely soft materials (e.g., peat) onto hard rock (e.g., glaciated granite).

Extensive recent experience confirms that shoes are unnecessary when driving into sands, silts, clays, and even soft shales, etc. Pile shoes are of help when the piles must penetrate buried timbers, coral (limestone) strata, and rock.

These shoes may take the form of plates, points, or stubs. Plates should be sufficiently thick to withstand local deformation (i.e., 1 inch (2.5 cm) thick or so) and anchored into the pile tip by anchor rods of sufficient embedment to develop bond even under repeated loading (i.e., a minimum of 40 diameters). Reinforcing steel, if used for these anchors, should be mild steel, and plug welded with full penetration welds to the plate. Intermediate or hard-grade bars often fail at the weld, although use of low-hydrogen electrodes helps. Fillet welds tend to fail under the repeated impact.

Fabricated or cast steel points may be used. These are useful in obtaining penetration, but tend to make the pile run off line; thus, pile alignment will

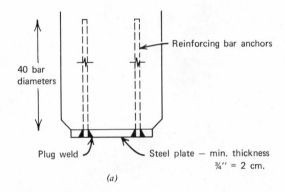

(a)

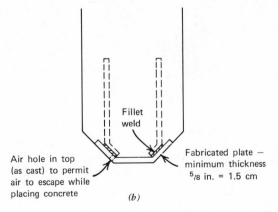

(b)

Fig. 31. (a) Steel plate pile shoe; (b) "blunt" steel pile shoe.

not be as accurate as with flat plates. A rather blunt point, therefore, is often chosen as a reasonable compromise.

Such points also serve to confine the edge of the concrete at its tip, preventing corner splitting or spalling. They need not be of as thick steel as a flat plate, since their shape serves to stiffen them; thus, 5/8 inch or thicker plate is usually satisfactory. In manufacture, some means must be made to prevent air from being trapped in the shoe while placing the concrete; an air escape hole is usually provided for this purpose. (See Fig. 31).

Stubs consist of H-pile sections, solid dowels of diameters of 3 inches (7.5 cm) or more, or fabricated crosses. They may be welded to a plate which is in turn anchored to the pile. More usually, they are anchored by embedment of the stub section into the pile, a distance of 4 to 5 feet (1.25 to 1.5 m). In

Fig. 32. Prestressed concrete bearing piles for major San Francisco building utilize steel H-pile stubs for penetrating bedrock.

early applications, shear lugs were welded to the embedded portions; most recently, dependence has been placed on bond alone, with satisfactory results. Again, means must be provided to permit placement and consolidation of the concrete around the stub and to permit escape of air; a hole through the web is usually provided.

Steel stubs must be of sufficient thickness, stiffness, and high yield to prevent their own distortion. (See Fig. 32) One of the advantages of prestressed concrete piles in penetrating hard material is their rigidity and resistance to local deformation. Nothing is accomplished by installing a long, slender, steel stub if that stub bends, tears, or otherwise distorts.

Stubs can be fitted into the strand pattern by forcing: even if this deflects the strands or the spiral, it has no ill effect.

Heavy spiral or stirrup reinforcement must be provided around stubs to confine the concrete. The proper design of the entire tip assembly deserves careful consideration, so as to permit placement and consolidation of the concrete.

In Norway, hollow-core piles are driven to the surface of sloping rock, then a hole is drilled through and into the rock, through which a steel pipe dowel is placed and grouted. This is called "the Oslo point."

In Singapore, prestressed piles have been anchored into rock by first driving them so as to seat on the rock, then drilling through a hollow core, inserting a prestressing tendon, grouting it throughout its lower (rock) portion, then stressing it and finally grouting it into the hollow core. These piles have then been used as tension (uplift) piles.

The shape of the tip, like the shape of the shoe, may be varied to suit driving conditions. A square tip, with chamfered corners, is most common today in practice. In former times, points and wedges were formed. These tend to cause the pile to deflect. High bending stresses are induced in the point, tending to break it off or to break the pile. However, rounded or heavily chamfered tips are useful when driving through old rock seawalls, etc. to prevent the edge or corner from catching on a rock.

Hollow-Core and Cylinder Piles. With hollow-core and cylinder piles, when driving through soft material, or under water, soil and water may be forced up the core, causing a hydrostatic head inside that may burst the pile. This is particularly serious if water completely fills the core so that the hammer blow causes a hydraulic ram effect.

Vents should be provided to allow the water to escape. A series of such vents will be adequate for the hydrostatic head effect, but not for the case when the pile head is driven to or below the water surface. In this latter case, a driving head should be designed to provide full venting, that is, an area approximately equal to that of the core; otherwise, the hydraulic ram effect may destroy the pile.

Underwater vents through the walls are also useful in preventing internal hydrostatic pressures during sudden drops in the outside water level (as in a river) and in preventing internal freezing.

Cover. Adequate cover must be provided over the tendons and spiral to insure durability. Current European practice calls for 1 inch (2.5 cm) cover in fresh water and soil, and 1-1/2 inches (4 cm) in salt water. Current US practice is to use 1-1/2 inches (4 cm) in fresh water and soil, and 2-1/2 to 3 inches (6 to 7.5 cm) in salt water. Actually, there appears to be little reason for placing the strands close to the edge, since elastic bending resistance is unchanged and ultimate bending resistance is only slightly lowered; for example, with an 18-inch octagonal pile, increasing the cover from 2-1/2 to 3 inches decreased the ultimate moment by approximately 5%.

Structural Lightweight Concrete. Structural lightweight aggregate concrete is increasingly employed for bearing piles, sheet piles, and fender piles. The weight for transporting and handling is reduced. The deadweight, submerged in water or soil, is greatly reduced as compared with conventional concrete. This may give important benefits when large, long piles must achieve support in weak soils. By using lightweight concrete, substantially all of the frictional resistance of the soil is available for support of the superimposed structure.

During driving, prestressed lightweight concrete piles are more flexible and can deflect more without cracking when driven through obstacles. They can absorb more energy from impact and dynamic loads. From a practical viewpoint they behave approximately the same in driving as a conventional pile: the lighter mass, lower modulus of elasticity, and greater damping tending to offset each other.

There is a greater tendency for the heads of prestressed lightweight concrete piles to spall. This can be countered by providing increased spiral wrapping at the head and perhaps several layers of cross-bars (i.e., a grid) or of mesh.

Changes in Section. Abrupt changes in section or properties (transformed section) should be avoided. When solid heads or tips are used with hollow-core piles, a gradual transition should be made, in the form of a cone (with rounded tip) or hemisphere. Additional mild steel, especially circumferential, may help to contain the stress concentrations across the transition.

Similarly, at splices or head zones where heavy longitudinal mild-steel bars are installed, the locations of the ends of the bars should be staggered longitudinally and extra spiral provided in this zone.

11.3   Manufacture

Prestressed concrete piles are manufactured by a wide variety of processes. The essential considerations of the manufacturing process are the following:

(*a*) Uniform, dense, thoroughly consolidated concrete should be produced, of high strength.

(*b*) Tendons should be in relatively accurate location and be thoroughly surrounded by dense mortar.

(*c*) No cracks should be permitted to occur, whether due to shrinkage or thermal expansion, etc. The manufacturing process should ensure that a minimum of surface tensile stresses are "locked into" the concrete.

(*d*) The pile must not be cracked in bending during picking, handling, or transporting.

(*e*) The surface of piles which will be exposed to salt spray should be relatively dense and free from deep pockets or deep "bug holes" due to bleed.

(*f*) The head should be truly normal to the axis and free from protuberances or ridges.

(*g*) The cross section should be sufficiently true to maintain the prescribed cover.

(*h*) Economy should be obtained in manufacture and handling.

11.3.1 Long-Line Pretensioning Method

This has been employed so far for piles ranging from 20 feet (6 m) up to 260 feet (80 m) in length, and from 10 inches (25 cm) up to 72 inches (180 cm) diameter. Solid piles, hollow-core piles, and cylinder piles are extensively manufactured by this process. All shapes and cross-sections can be made by this method.

Forms are most commonly fixed forms with a slight taper. The precautions on lifting piles from the forms were discussed earlier under "Design." Removable forms are also used, and may be of the hinged, slide-back, or lift-off type. Removable forms are frequently used to form the upper half of octagonal or round piles.

If the top sections are removed shortly after concreting, the top surface may be finished to remove bleed holes and surface imperfections.

The typical production sequence is as follows:

1. Set out precut bundles of spiral in the forms, one bundle to each pile location.
2. Pull strands down bed, through the coils. Take up slack, using a dynamometer to equalize length of each strand. Affix strand ends to stressing block, stress, and anchor. Alternatively, each strand may be tensioned individually.
3. Spread out coils to proper spacing, and tie them to strands.
4. Place segmental end gates and clamp them to strands.
5. Place concrete and vibrate.
6. Cover and keep damp for about three hours, then start steam-curing cycle.

7. At some point during the above cooling period, release prestress into piles, and re-cover.
8. Lift piles from forms to storage.
9. Supplemental water cure in storage (optional).
10. Supplemental drying period (optional).

Supplemental water cure should be employed for piles which will be exposed to aggressive environment in service, for large, thin-walled, hollow-core or cylinder piles, and whenever extremely drying atmospheric conditions will result in surface crazing.

The prestress should be released into the piles (Step 7) before exposure to the atmosphere; otherwise, the piles may develop horizontal shrinkage cracks. While these will probably close later when the prestress is released, the tensile strength of the concrete will be impaired.

A slow-cooling cycle is important for large, solid, or massive piles. If they are exposed to cold atmospheric conditions while the inside core is still hot, thermal stresses may cause cracking or crazing. Thermal and shrinkage strains are additive.

Steam covers should be sufficiently tight to provide protection from cold winds, so as to insure relatively uniform temperature conditions inside. If, because of wind, for example, there is a substantial difference in curing temperature inside, then when the prestress is released into the member, the strength, effective "age," and modulus of elasticity will vary, plastic flow will occur, and the resulting pile may have a decidedly permanent "sweep."

For piles which will be exposed to salt air or the salt-water splash zone, or to a freeze-thaw zone, if bleed holes ("bug holes") have occurred due to the trapping effect of vertical and overhanging forms, and if they are more than a nominal depth (say, 3/8 inch), then they should be filled with regular mortar (dry pack) or epoxy mortar in that portion of the pile which will be exposed. Because there appears to be no way to completely eliminate these air and water bleed holes from the upper surface of octagonal and round piles when poured with the dry mixes used for prestressed piling, the use of removable top forms followed by finish trowelling appears to be the best solution.

Forms for long-line-pretensioning must be constructed so as to prevent transverse fins or obstructions, as, for example at end gates. Any such fins will lock to the freshly set concrete while the form is expanding under the rising steam temperature, and will cause transverse tensile cracks, which may later be aggravated under driving.

Sliding or slip forms are also used, particularly for the upper half of octagonal and round piles. The method of extruding the concrete and of moving of the form must not "pull" the concrete so as to impart built-in tensile haircracks or residual stresses in the surface.

A typical sliding form consists of a section about 20 feet long, surmounted by a hopper, and fitted with several form vibrators. It is slid or pulled along rails at a rate of from 1/2 to 3 fpm. The concrete consistency, vibration, and rate of travel must be carefully controlled to prevent sloughing of the surface. Sometimes the hopper is equipped with a screw-type feeder or internal vibrator. Occasionally, a small second hopper trails the first, feeding mortar to the surface.

Hollow-cores are formed by several methods:

(a) Permanent tubes of waterproofed paper or light-gauge corrugated metal.

(b) One method, used in England, wraps the waterproof paper around a core formed of stressed tendons. The tendons, in turn, may be held in position, at frequent intervals, by spacer rings.

(c) Removable inflated cores consisting of rubber, with or without metal stiffeners. These may require auxiliary spacing or stiffening, especially in the larger diameters, to prevent becoming egg-shaped or "wobbling" longitudinally.

(d) Removable rigid forms, which collapse mechanically or hydraulically. All mandrels and cores must be held accurately in place to prevent sagging, dislocation, and especially to prevent flotation as the concrete is vibrated. Hold-down devices must be spaced at frequent intervals to hold the core within the specified tolerance, usually 3/8 to 1/2 inch (1 to 1.25 cm). The use of the stressed tendons to hold the core in place is usually unsatisfactory because, as the core floats upward, it also displaces the tendons.

(e) Sliding internal mandrels must be designed so that they do not drag the concrete, causing internal spalling, nor permit haircracks or built-in tensile stresses in the interior walls which may later cause trouble in driving. The mandrels must be guided or supported ahead of the concrete being cast, by beams, rollers, or the strands. The system must be designed so that the strands in turn are held in position to ensure that neither they nor the spiral coils are dislocated during the travel of the mandrel. The tail support for the mandrel may be in the freshly placed concrete. These sliding mandrels are usually fitted with form vibrators. Sometimes heating units are mounted in the mandrel in order to give a quick set to the concrete. To reduce drag, cores are usually machined; they may even be of polished stainless steel. While Teflon has not yet been used, so far as is known, it would appear to offer an even lower drag coefficient.

The internal surfaces of a slip-formed core must be kept moist, especially in the period before regular steam curing.

Extrusion machines have been developed which form both the inside and outside surfaces of hollow-core and cylinder piles at one pass.

186 PRESTRESSED CONCRETE PILING

In Malaya and Fiji, large-diameter cylinder piles have been manufactured by using precast rings as spacers and as supports for the formwork. The rings were spaced at about 10-foot (3 m) centers and the tendons were run through preformed holes. The forms were then placed and the concrete cast. Particular care must be taken to insure bond between the precast spacers and the cast concrete, so as to insure full concrete tensile strength during driving. Use of an epoxy bonding compound would appear particularly desirable, along with water curing of the pile up to the time of stressing, so as to prevent shrinkage cracks at the joints.

11.3.2 Segmental Construction

Segments are precast, either by centrifugal spinning or cast in vertically forms. This method of manufacture has evolved from precast concrete pipe technology. After the precast segments have hardened and cured, they are placed end-to-end, jointed, and post-tensioned.

Spun segments offer the advantage of thorough consolidation of the concrete. The centrifugal force may be augmented by vibration or by an internal roller. Excess water is drained from the center. The outer surface is dense and smooth.

The spinning method requires that techniques be instituted to hold the duct formers in true position. They naturally tend to belly out from the centrifugal force. If restrained, a void may form outboard of each tube. To overcome this, the duct formers are usually tensioned against the end forms and, after spinning, the duct formers are removed, so that subsequent grouting of tendons will fill any voids.

Vertical casting of segments permits the economical forming and casting of any size pile, up to the largest diameters. Shear keys may be easily formed in the top and bottom edges. Form vibration may be used effectively as a supplement to internal vibration.

After casting, stripping, and curing, the vertically cast segment must be tipped to horizontal. For very large diameter pile segments, this is usually done in a tilting frame, which allows rotation without imposing excessive crushing loads on the bottom edge. Another method is to set the segment in sand, then tip it while lifting. The sand serves to distribute the crushing load on the edge. A third method is to rotate it in air, using a double sling arrangement.

After segments are turned and accurately aligned on a horizontal bed, the joints must be made. If the ends as cast are very true, the joint may be made with epoxy mortar. With one spinning process, true ends are achieved by using an absorbent form liner (to draw off excess water and prevent bleeding). With vertical casting, the end forms are usually machined cast steel.

Wider joints may consist of cast-in-place concrete, with a minimum width of 3 inches (7.5 cm). With joints thinner than 3 inches it is difficult to consolidate the concrete, with the result that stress concentrations and spalling may occur during driving. "Fine concrete," using a 3/8-inch (1 cm) coarse aggregate and high cement factor [say, 8 sacks (430 kg/m^3)], is usually employed. The ends of the segments may have previously been painted with an epoxy bonding compound. In any event, the surface film of mortar should have been removed by light-sandblasting or bush-hammering. After pouring, steam curing may be beneficially employed to accelerate the gain in strength of the joint.

After jointing, tendons are inserted, stressed, and grouted. In some systems, the end anchorage is removed after the grout has hardened, the tendon stress being transferred by bond alone. This is satisfactory, particularly with strands, up to a reasonable concentration of force per tendon. With higher forces, bond slip may be a problem, but more serious is the bursting force that may crack the thin wall outside the duct. Adequate spiral reinforcement, particularly at head, tip, and throughout the zone of possible cut-off, is essential.

Sections of piles may be manufactured in standardized lengths, for example, 40 feet (12 m) or so, and then spliced during driving. This method is extensively used in Japan, Sweden, and England. It is particularly well adapted to locations in which transport of long piles would be a serious problem, in which crowded conditions make the lifting (pitching) of long piles difficult, and in which the desired or available equipment for driving is limited in size and height.

Such standardized sections can be manufactured in a highly mechanized plant and stored in inventory. In driving, the lengths can be varied by simply adding or leaving off a section.

For such a system to be successful, splice details must be strong (in compression, tension, and moment), durable, rapid of execution, and economical.

The Japanese have adopted a cast-steel end piece that facilitates anchoring of the tendons during manufacture and permits rapid jointing by welding. An electrical connection is available for those installations where cathodic protection of the splice is needed.

The sections themselves are manufactured by a variety of methods: centrifugal spinning with tendons post-tensioned against the end pieces; centrifugal spinning with tendons inserted later and post-tensioned; and horizontal casting with pretensioning.

In addition to welded splices, a number of mechanical connectors have been developed; these engage and lock mechanically, as by a screwed joint, or wedge effect. These are, of course, very rapid in execution. They generally are designed to transmit full compression and tension, but cannot be considered full-moment splices.

Friction splices are also available, in which a sleeve is driven over a male casting, locking itself by wedging. These are generally fully effective in compression, but tension values are erratic, particularly under the first few hammer blows after a splice. Therefore it may be necessary in soft driving to restrain the lower section so that the upper section may be driven onto the lower. This restraint may be by means of a timber clamp resting on the ground.

Fig. 33. Prestressed pile segments are spliced with epoxy-dowel splice for full-moment connection. Use of segments permits driving under high-voltage power lines.

Experimental and developmental work has been directed toward a pre-stressed splice. To date, practical difficulties of tensioning have prevented widespread use.

Epoxy-doweled splices have been very successful structurally and econom-ically, but require a time for set that may range from 30 minutes to 12 hours, depending on the setting characteristics of the epoxy and the temper-ature. Usually four dowels of deformed reinforcing bar are extended from the top section into corrugated metal tubes in the head of the bottom sec-tion. Bar extension lengths of 20 to 30 diameters are recommended. The size of the four bars should be selected to give a steel area of 3% of the concrete area across the joint. (See Fig. 33) Epoxy may be poured into the holes before inserting the dowels or injected afterward. Other proprietary com-pounds are also available. In Norway, unsaturated polyester resin is used. The set is accelerated by internal electrical resistance heating. A small cop-per wire is wound around the dowel tubes. After the resin has been poured into the holes, the top section is set and an electric current from a generator set (about 24 V, 100 A) is used for five minutes to start the reaction. Once started, the reaction proceeds rapidly, even in cold weather, and sufficient strength is often generated to permit driving to re-start within 15 minutes or so. Since copper in concrete is potentially a source of electrochemical corro-sion of the prestressing tendons, use of a steel wire conductor of proper size and electrical properties would appear to be a safer practice. (See Fig. 34).

Steam jackets have similarly been clamped onto the pile at the splice and used to accelerate the reaction.

Because of the structural and durability benefits of the epoxy-dowel splice (or polyester resin), developmental work is now underway along these lines:

(a) Substitution of a single pipe stub in lieu of 4 dowels. If two nipples are run to the pile surface, steam can be circulated to accelerate strength.

(b) Use of epoxy as an addition to one of the several mechanical splices now on the market: then commencing driving immediately, letting the epoxy set after completion of the installation.

It is important that with any splice, the sections be essentially center-bearing, and that the concrete immediately above and below the splice be contained with heavy spirals. If the splice is edge bearing, then the driving stresses will not be transmitted uniformly to the concrete, and a fatigue condition, under alternate compressive and tensile waves, may develop. With sufficiently rigid steel, as in the Japanese standard joint, this is appar-ently not critical; however, it is always desirable to keep the driving stresses as uniform and concentric as possible, and the author prefers a center bear-ing splice.

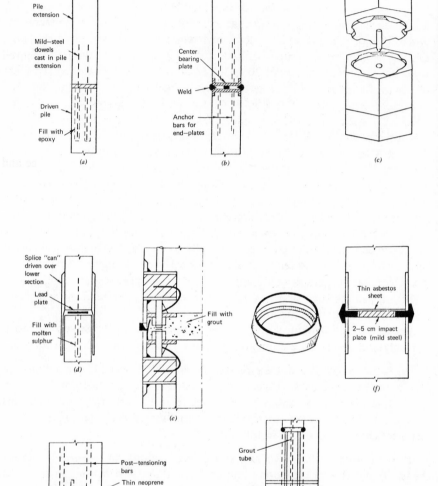

Fig. 34. Typical splice details for prestressed concrete piles: (a) epoxy-doweled splice —United States, Norway; (b) welded splice; (c) mechanical splice (Swedish patent); (d) steel splice sleeve or "can"; (e) welded splice—Japanese patent; (f) "Brunsplice" joint—United States patent; (g) post-tensioned splice—Great Britain; (h) steel pipe splice for hollow piles—Norway.

11.4 Installation

Prestressed concrete piles are installed by a wide variety of techniques, including driving, jetting, drilling, vibration, and weighting.* However, the great majority of prestressed piles are installed by driving, either alone or in a combination with other techniques. Hundreds of thousands of prestressed piles have been successfully installed by driving under the widest possible variations and combinations of subsurface conditions, ranging from penetration into silts, sands, clays, and soft rock, to founding on hard rock, and to penetration through overlying hard material into lower bearing strata. From this experience, some general rules have been developed to ensure successful installation. The rules have independently emerged in ramarkably similar form from many countries, including Japan, Britain, Australia, United States, Sweden, and Canada. These rules, as taken from the author's "General Report to the F.I.P. Symposium on Mass-Produced Prestressed Precast Elements" (Madrid, June 3, 1968), are the following:

1. The pile and hammer should be in alignment.
2. The pile head should be square and free from protuberances or projecting wire or strand. Binding should be avoided in the pile helmet.
3. Avoid excessive hammer (ram) velocity. Reduce ram velocity in soft soils.
4. Provide adequate cushioning at pile head: 6 to 14 inches (15 to 35 cm) of softwood has consistently proved most effective in reducing internal stresses in the pile while facilitating maximum penetration of the pile. Provide a new cushion for each pile. The Japanese use corrugated paper board packing; other materials used are rubber (including crude rubber sheets), felt, sacking, asbestos fiber, coiled hemp rope, and plywood.
5. Protect pile splice rings, or other fittings, with a ring cap.
6. Avoid excessive restraint to pile, either in torsion or bending. The pile position and orientation, once the pile is well embedded in the soil, cannot be corrected from a position at or near the head without causing distress.
7. For cutting off, preferably make a circumferential cut first with a diamond saw or small rotary drill. This will ensure a neat, un-spalled head. Other means include clamping with a steel or wood band, then cutting with a small pneumatic hammer.
8. When predrilling or jetting, the tip must be well seated before full driving energy is applied.
9. The greatest driving efficiency is obtained by using a heavy ram, a soft head cushion, and low-velocity impact.
10. With very long slender piles, guides or supports should be employed to prevent "whip", vibration, and buckling during driving.

*For driving with "impact rammers," (directional vibration), see Postscript, section iv.

11. With batter or raking piles, supports must be provided for the over-hanging lengths—both those extending above and below.

12. The lifting and handling points for handling and picking (pitching) must be properly marked and used. With long piles and multiple (two to six) pick-up points, the angle which the slings make with the pile affects the stress in the sling; pick-up points may therefore be different from those used for handling at the plant and in transport.

13. In construction over water, pile heads are sometimes pulled into correct alignment. With long unsupported lengths, it is easy to over-stress or even break the pile. Thus, specifications should limit the movement or force of pulling to ensure against overstress. However, specifications should be written in such a way as to be practicable of accomplishment and enforcement. The allowable amounts vary consid-erably, depending on the depth of water, soil conditions, etc. For example, with a standard 16-inch prestressed pile at the outer edge of a wharf in 40 feet of water and a soft mud bottom, a force of 500 pounds at the head will move the head about 3 inches and reduce the compres-sive prestress at a point 10 feet below the mud-line to about 200 psi. This may be acceptable in many cases.

    In the same wharf, at the inboard edge of the wharf, in 6 feet of water and a rocky fill bottom, a force of 1700 pounds is necessary to move the pile 1/3 of an inch and reduces the prestress by the same amount, i.e., down to 200 psi compression. Thus, for such a pile, pulling the head is impracticable.

    One practicable means that can be enforced is to specify that the tool used for pulling be set so it cannot exert a force greater than, say, 500 pounds.

14. After piles have been driven in rivers or harbors, they may require temporary lateral support against tidal current, wave forces, etc., and, in the case of batter piles, against dead-weight deflection.

15. Diesel hammers are widely used to install prestressed piles. Because of generally light ram and high velocity of impact, particular care must be taken to provide a proper head cushion. Fortunately, the inherent performance of a diesel hammer prevents development of full ram velocity in soft driving, thus automatically reducing dynamic stresses.

16. Jetting is often employed to aid driving. This may be pre-jetting (pilot jetting) to break up hard layers or it may be jetting during driving. In the latter case it is hazardous to jet at or below the tip during driving as this may create a cavity, and produce a "free-end" condition, lead-ing to excessive tensile stresses and cracking in the pile.

17. Predrilling, dry, with water, or with bentonite slurry, may be effectively used to aid penetration through hard and dense upper

layers. This is becoming increasingly necessary due to the fact that soil engineers are requiring deeper penetrations for settlement control.

The pulling and removal of prestressed concrete piles is very likely to produce cracks because of inadvertent bending during pulling. It is very difficult to take a truly axial pull, even with a pair of slings which have equalized loops on both sides of the pile. Extensive jetting should be performed on all sides of the pile prior to pulling. The pull should be gradual and maintained, since soil tends to fail slowly in friction under a sustained pull.

Sometimes prestressed concrete piles are damaged during or after installation by accidental impact, etc. If the damage consists of spalled concrete or cracks, it may often be satisfactorily repaired by one of the following methods:

(a) Underwater-setting cement, which does not contain calcium chloride, may be placed below the water surface by a diver. This is mixed in small batches, placed in a sealed can and lowered to the diver, where it is forced into place by the diver.

(b) Larger voids or spalls may be patched with tremie grout. A metal form is placed around the pile, and grout is poured into the void, with the pipe always immersed in the fresh concrete. The form should be overflowed to get rid of any laitance.

(c) Patching with underwater-setting epoxy.

With all of these methods, the spalled area should be thoroughly cleaned of damaged and broken concrete, marine growth, silt, etc., and then, if possible, a key should be chipped either by a hand chisel or very small air chisel. (Use of a large chisel may cause further cracking.) Patches should be protected from wave erosion, etc., until they have hardened.

Cracks may best be sealed with underwater epoxy, although in some cases underwater-setting mortar can be forced into the crack. With a serious structural crack, a small tube may be drilled into the crack and epoxy pumped in until it exudes from the crack.

The above repairs have been generally successful when carefully performed for the correction of structural damage. They are not necessarily applicable nor successful for repair of durability failures, i.e., disintegration due to corrosion or chemical change. This chemical and electrochemical degradation usually requires special, far more extensive techniques (see Chapter 4, Durability and Corrosion Protection).

For the installation of very large and long piles, special combination techniques are often employed. For example, on the Oosterschelde Bridge in The Netherlands, the piles, which were 13 feet (4 m) in diameter, were sunk through silts and sands by a combination of internal dredging and weighting. The internal dredging was performed by a cutter-head on a special articulated arm. The derrick barge was equipped with an outrigger that was clamped onto the head of the pile; through the rigging the barge was then lifted so as to impose several hundred tons of force on the pile. After the proper tip elevation had been reached, a bottom plug of underwater concrete was placed. While initially the piles were unfilled, provision was made for later filling with sand, if needed, to dampen vibration. Freezing of the inside water was prevented by means of vent openings near the sand line, well below the ice zone.

On the San Diego-Coronado Bridge in California, 54-inch (135 cm) diameter cylinder piles were installed in three successive stages through dense clay and sand strata to the desired consolidated-sand bearing stratum. In stage 1 they were prejetted with high-pressure, high-volume jets, then driven to within 6 feet (2 m) of design tip. The jetting was controlled so that it did not affect the sands within this 6-foot zone. In stage 2, only side jetting to relieve friction was permitted, keeping the jets well above the tip, but the penetration was achieved by driving with a large hammer. This driving was continued until the pile reached a point 2 to 3 feet (0.6 m to 1 m) above design tip. In stage 3 the hammer alone was employed to drive the pile to final bearing. The specifications required at least 200 blows at refusal, to insure consolidation of the sand around the pile. After all piles in a pier footing had been driven, the hammer was again used for a specified number of blows on each pile to insure that the overlying sands were re-vibrated and consolidated. Piles were then cleaned out, down to the hard plug of sand that had been driven up into the tip. Then a tremie concrete plug was placed.

In driving prestressed piles on some of the Arkansas River projects, involving friction piles in sands, jetting plus driving was found to be most effective. It was found that driving should be continued after jetting ceased to ensure re-consolidation of the sands. In these sands the jetting actually proved to be beneficial to the piles' bearing capacity because it washed out the fine silts, giving greater frictional and lateral support.

In the section on Pile Design, considerable discussion was included regarding rebound tensile stresses. When these become excessive, they manifest themselves in a rather "mysterious" fashion, usually occurring in the soft driving, and especially in the first few blows when the pile breaks through the surface and runs. Cracks show up as puffs of dust, usually about one-third of the length of the pile from the top. If driving continues, other cracks will form about 2 feet apart. They usually go all the way around, and are characterized by surface spalling that looks much worse than the actual crack. Repeated driving produces fatigue in the concrete next to the crack; the concrete gets hot, and the aggregate-paste bond is destroyed. Eventually the tendons fail in brittle fracture.

This type of failure is made harder to diagnose by the fact that the first crack usually occurs at a weak point, such as a lifting eye, insert, form mark, or honeycomb. Such a local stress concentrator may determine the location of the first crack, but is usually not the cause of the crack.

The theory and preventive design for rebound tensile stresses was discussed earlier. However, should the rebound stresses prove to be excessive and cracking be noted, then one is faced with a situation in which the prestressed piles are already manufactured and a given piece of driving equipment is on the site. Several steps can then be taken to reduce rebound tensile stresses and prevent further damage:

1. Increase the thickness of head cushion and reduce the stiffness. The best material is unused, green, softwood, made up of rough 1-inch or 2-inch thicknesses to a total thickness of 12 or 14 inches. These provide the maximum cushioning during the early, soft, driving, and become much stiffer (twice as stiff or more) during driving. For this reason, they should not be re-used, even if still in apparent good condition.
2. Reduce velocity of impact. If the driving energy must remain the same, use a heavier ram with less stroke. Usually, however, the existing hammer may be used by reducing the velocity of impact during the early, soft, phases of driving. With a differential-acting hammer, it is often only necessary to throttle down, so as to slow the hammer. With a single-acting hammer, a small reduction can be accomplished with throttle; significant reduction requires the use of an adjustable slide bar to reduce the stroke in the zone of soft driving.
3. Change jetting practice as necessary to prevent any tendency to wash out a hole or soft spot under the tip while the sides are still wedged in hard strata or gripped by side friction. Limit jetting during driving to the relief of side friction.
4. Some of the worst cases of rebound tensile failures occur when driving through an overlying fill, into a very soft mud stratum. If problems persist despite steps 1 and 2, break up the overlying hard layer by predrilling or prejetting, so that the pile will push through the soft strata under its own self-weight plus that of the hammer.
5. Ensure that the pile head is not binding in the driving head; that it is free to turn. Any torsional stresses are additive to the rebound tensile stresses and permit the start of tensile cracking on the faces of the pile.

11.5   Sheet Piles

Prestressed concrete sheet piles offer the advantages of durability, rigidity against local deformation, and excellent appearance. (See Fig. 35) They do not necessarily represent a savings in first cost over steel sheet piles, because sheet piles must usually take both positive and negative moments. To

Fig. 35. Precast prestressed sheet pile wall in San Francisco Bay.

provide for this, prestressed concrete sheet piles must be essentially uniformly prestressed to a fairly high degree. However, if a computation of design moments indicates greater positive than negative moments, then a degree of eccentricity of prestress may be used. Alternatively, mild steel can be added at points of high negative moment. Another means of varying the center of prestress throughout the length is to deflect the strands to match the points of maximum moment or to embed an unbonded tendon for stressing after the pile is in place.

The amount of eccentricity that may be tolerated without affecting the ability to withstand driving stresses will, of course, vary widely, depending on the many variables in soil, pile, and driving equipment. However, eccentricities which give a variation of a few hundred psi in the concrete have been successfully employed.

Sheet piles are frequently installed by jetting. With jetting, it should be possible to use greater eccentricity of prestress. Another method of installation is vibration, with or without the aid of jetting. With vibration, both compressive and tensile stresses are traveling up and down the pile. We do not know the minimum prestress required to withstand the tensile stresses due to vibration but, presumably, they are similar in degree to those for driving, that is, 700 psi (50 kg/cm^2).

In the design of sheet piles some tension may be permitted to exist, provided the ultimate strength requirements are satisfied. Tension up to one-half the modulus of rupture is commonly permitted. A limit condition in design is cracking; obviously it is essential that the sheet piles not be cracked in the salt-water splash zone.

The details of lateral mild-steel reinforcement of sheet piles are very important. The mild steel serves the function of confining the concrete in order to insure transverse bending strength, of providing shear strength, and of making the tongue and the wings of the interlock grooves function as an integral part of the pile. These wings in particular tend to break off during driving, due to the wedging effect of the tongue of the adjoining pile, or due to the wedging effect of the soil. For this reason, it will generally be found best to drive the sheet piles with the tongue leading, so that there is no soil plug formed in the groove. When a double groove joint is used, a steel pipe is usually inserted in the leading edge of the groove as a temporary tongue, then later withdrawn. (During such withdrawal, grout may be injected.) Because of these wedging stresses, it is therefore desirable to detail the mild steel with bars extending into the wings.

In driving prestressed sheet piles, it will often be found best to use auxiliary jetting, either several internal jets, or an external gang jet. A roller may be placed over the leading edge and pulled back against the preceding piles; this will tend to keep the pile up tight at the ground line. By sniping off the leading toe of each sheet pile, the toe also will be wedged back against the preceding sheet pile.

To drive each pile to the same grade, a special driving head may be needed, with a slot so it can pass by the head of the preceding pile. Alternatively, a short, narrow extension may be cast onto the head of each pile to serve as a driving block for the hammer. After driving, this extension may be cut off, and the tendons tied into a capping beam; or the extension may be left on and the slot between adjoining extensions used for tie-backs. (See Fig. 36)

Prestressed concrete sheet piles are utilized for waterfront bulkheads, cut-off walls, groins, wave-baffles, and to retain soil during excavation for foundations.

In some recent building foundations, the prestressed concrete sheet piles were installed by a combination of predrilling and driving. After excavation, the joints were welded and filled with non-shrink grout, so that the sheet-pile wall served as the permanent foundation wall of the building.

When prestressed sheet piles are used for waterfront bulkheads and cut-off walls, the joints must be sealed. The most common method of sealing is to fill the joints with grout. In sand, a pipe may be advanced by jetting which washes the sand out of the joint, then grout may be pumped through the pipe as it is withdrawn. In grouting through water, a polyethylene, burlap, or canvas bag may be pushed down the joint by means of a pipe, then the grout pumped into the bag as the pipe is withdrawn.

The grout used should be a rich, cohesive, workable mix. For most cases, Portland cement, sand, and water form the mix, preferably with an admixture to promote cohesiveness and reduce the water/cement ratio. For joints

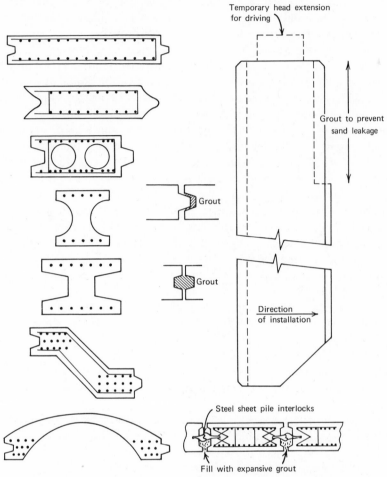

Fig. 36. Typical cross sections and details of prestressed concrete sheet pile.

which will be exposed to drying shrinkage, a low-shrinkage mix or shrinkage-compensated cement should be used. It would appear that epoxy grouts, poured or pumped into the joint, may offer excellent possibilities, particularly in the wave zone.

A number of interlocks have been developed for prestressed sheet piles, to give both structural strength and a degree of sand-and-water tightness. The ordinary tongue-and-groove interlock transmits shear but not tension. Steel sheet piles can be cut in half and embedded in the prestressed sheet piles. These will, therefore, be as tight and have the same tensile strength as the steel sheet-pile interlock. A polyethylene interlock has been developed by

Belden Concrete Products, Louisiana, which is embedded in the concrete and which acts both as an interlock capable of some tension and as a water-stop. (See Fig. 37)*

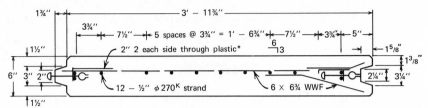

No. 2 bar reinforcement for 6—in. pile

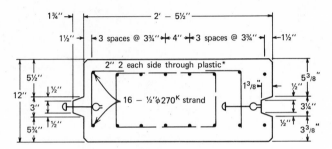

No. 2 bar reinforcement for 12—in. pile

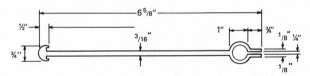

Cross section of plastic interlock

Fig. 37. Plastic interlock for prestressed sheet piles.

11.6 Fender Piles

Prestressed piles have many advantages to offer as fender piles, dolphin piles, and other piles primarily resisting lateral forces. Prestressed fender piles are durable and economical. They can be designed to maximize deflection and energy-absorption, by keeping the moment of inertia low in relation to strength (by choice of section), by use of concrete with a low modulus of elasticity (such as structural lightweight aggregate concrete), and by proper choice of prestress levels.

One use of fender piles is for wharves and piers, where ships must

*For use of prestressed concrete sheet piles in Japan, see Postscript, section v.

constantly berth without damage to ship or fender system. Prestressed fender piles have been installed on major cargo wharves in Kuwait, Singapore, Los Angeles, and San Francisco and have given very successful performance in service. (See Fig. 38) A serious fire at one installation destroyed the adjoining treated-timber fender but left the prestressed fender piles undamaged, save for slight surface spalling.

Structural lightweight-aggregate concrete appears ideal for fender and dolphin piles because of its greater deflection and energy absorption.

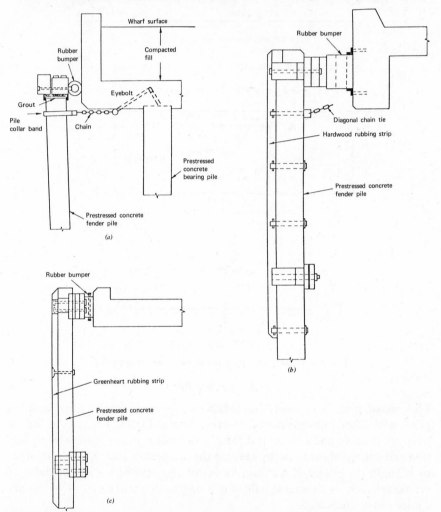

Fig. 38. Fender systems for general cargo wharves utilizing prestressed concrete fender piles: (a) Port of Los Angeles, California; (b) Port of Jurong, Singapore; (c) Port of Kuwait.

The greatest problem with prestressed dolphin piles is their lack of ultimate strength when design loadings are greatly exceeded. They tend to snap off under excessive impact. ·Therefore, it is suggested that greater reserve ultimate strength be provided in the form of additional unstressed strands or mild steel.

Prestressed fender piles are usually designed for a maximum tension of up to one-half the modulus of rupture. It is important to prevent cracking that will reduce durability; on the other hand, it would be wrong to make them too stiff. The design should emphasize energy absorption. As regards dura-

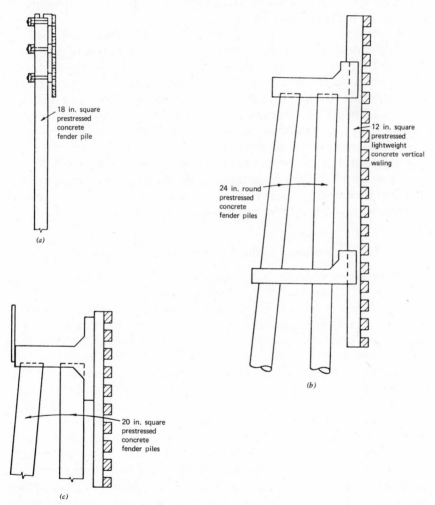

Fig. 39. Bridge pier fenders utilizing prestressed concrete fender piles. (a) Eureka Slough bridge (b) Sacramento River bridge (c) Benecia-Martinez bridge.

bility, they can usually be economically replaced at intervals during the life of the structure itself, so they do not require as great a factor of safety as do foundation piles.

On most of the wharf fender piles which have been installed to date, it has been felt necessary to provide a timber rubbing strip so that the ship would not abrade and spall the concrete fender pile. Alternatively, timber camels or hanging timber pile butts have been used for this purpose. Since the energy absorption of the timber in this case is relatively small, its main purpose has been to prevent local spalling. This might be more economically and effectively accomplished by a steel pipe or box section, wrapping around the concrete in the critical zone.

Another type of fender is the protective fender for bridge piers. (See Fig. 39) This is designed to protect the pier, and the prevention of damage to ships and the cost of repairs are not of primary concern. Here, ultimate strength plus durability are paramount. To increase the ultimate strength under excessive lateral load such as collision, unstressed strand or mild-steel bars should be added. Alternatively, the number of strands can be substantially increased, with the stress per strand reduced proportionally.

Fire resistance is of great importance in the protective fenders of bridges, especially steel bridges, where a fire in the fender might weaken and drop the steel span. This is a major advantage of prestressed concrete for this application. For these reasons prestressed concrete fender piles have been selected for the fenders of a number of major bridges in California.

11.7 Economical Evaluation of Prestressed Concrete Piles

The construction engineer is particularly concerned with the economical evaluation of alternative piling materials and systems because industry practice frequently offers or permits piling alternatives to the contractor or, otherwise, makes him responsible for their selection. The true economy of piles, as with other structural elements, is, of course, their ability to perform their assigned structural function, with adequate ultimate strength for overload conditions, for the period of the design life, and at minimum cost. Thus it is necessary to consider the structural behavior of the pile, the durability, and the costs of transporting, handling, driving, and connection, as well as the costs of manufacture.

Prestressed concrete piles have proven to be the most economic solution for a wide range of piling installations, especially when the inherent structural properties can be fully utilized, when durability or fire resistance are important, and when here is sufficient total volume in a geographical area to justify a proper manufacturing set-up and mobilization of proper driving equipment.

For marine structures the criteria usually includes a life of 25 to 50 years, good column strength, high load-bearing capacity, strength in bending, ability to be handled and driven in long lengths, and low first cost. For all of these parameters, prestressed concrete piles excel and, thus, they have been widely adopted for marine installations throughout the world.

It is interesting to note that the new super-tanker terminals for 250,000 ton tankers and larger, in 100 to 120 feet of water, with extremely high lateral loads due to berthing, wind, and sometimes ice, place an over-riding consideration on strength in bending. Here is one parameter for which steel piles, especially high-yield alloy steel, are apparently superior to conventional prestressed concrete piles. For most such super-tanker terminals at the current time, therefore, high-yield steel piles are being selected, despite less favorable properties to meet the remaining criteria. Developmental efforts are being made, therefore, to adapt prestressed piling to this new requirement; e.g., square hollow-core or cylinder piles of high strength prestressed concrete and increased prestress levels, may be combined with unstressed strand, mild steel, or a steel core to give higher ultimate strength.

For foundations the usual criteria are low first cost, high capacity as a short column, and ability to be driven to the required penetration. In general, what is sought is the lowest cost per ton of carrying capacity. Since there are many types of piling available in the market for use as building foundations, a careful analysis should be made, comparing the load-carrying capacity of the pile, the soil capacity for the sizes and penetrations under consideration, the costs of furnishing, transporting, and driving, and the size, depth, and cost of footings. Time required for installation should also be evaluated.

These last two items, footings and time, give great weight to the choice of high-capacity piles. In concrete piles, high structure capacity can most economically be obtained by use of the highest-strength concrete available. In most localities, building codes establish a formula for computation of the maximum permissible bearing load, which is usually directly proportional to the concrete strength. However, high capacity is of no use if the pile cannot develop a corresponding capacity in the soil. Soil support is furnished either by side friction and adhesion, or by end bearing, or by a combination of both. Side friction-adhesion is primarily a function of the surface enveloped in the bearing stratum. (In cohesionless soils of uniform density, such as beach sand, pile shape (e.g., taper) may also be a factor as may surface roughness in some sands.) Prestressed piles are usually of constant cross-sectional area; therefore, they offer a large surface-friction area. Thus, for the great preponderance of friction piles, standard prestressed concrete piles of uniform cross section have the highest bearing capacity in the soil. In many soils, such constant cross section prestressed piles will develop the required bearing value at substantially less penetration than timber, steel, or tapered concrete piles.

For special cases in which friction may be very critical, for example, when it is desired to found the piles in a rather thin sand stratum between two clay layers, the friction support may be increased by special techniques. One of these consists of corrugations on the surface of the pile in the bearing zone. Special types of corrugations have been developed to maximize the frictional effect (U.S. patent applied for). Another technique consists of increasing the cross-sectional area of the pile through the bearing zone. This is much practiced in the Netherlands. This enlarged tip may be cast monolithically with the rest of the pile, or may be spliced to the column portion of the pile during manufacture. A third technique, not yet fully developed, is to inject grout through an internal grout pipe, after installation. In this case, the pipe is usually also used as an internal jet to aid in installation and to keep the pipe clear for grouting. This method has been successfully utilized to increase tension frictional resistance and, presumably, could increase bearing frictional values as well. The difficulty is in controlling the process, so that reliable values can be obtained. Use of a cement grout with high penetrating ability is desirable; there are a number of admixtures made which reduce surface tension, and one process which uses colloidal mixing to aid flowability. At the present time this technique is primarily adapted to increasing the ultimate capacity in tension and bearing of piles which have already developed their design capacities by more conventional means.

End-bearing piles depend primarily on the size of the tip, for which constant-section prestressed piles are well suited, and on the ability to be seated into the material and mobilize the resistance of the soil. In soft rock, hard clay, etc., the soil capacity, as measured by deformation under load, is improved by the "prestressing" from the pile being forced into it. Prestressed piles have demonstrated this ability of penetration and mobilization to a high degree, because of the high strength concrete held in a homogeneous (uncracked) condition by prestressing.

The author has found it useful to prepare tables of cost-per-ton of load carried, evaluating piles of other maerials and capacities with prestressed piles of several different sizes and capacities. Such a table will quickly point up those factors which predominate. Above all, it will destroy the all-too-commonplace illusion that piles should be selected on the lowest first cost per foot.

Prestressed piles turn out to be very competitive in many such analyses. They must, of course, be compared with other forms of concrete piles. As compared with cast-in-place concrete piles, they offer high soil capacity, greater durability, better combined moment-load capacity, and usually, reduced footing size. As compared with conventionally reinforced (mild steel) precast concrete piles, they offer major savings in column strength, durability, ability to be driven to deeper penetration, and savings in steel cost. Prestressed piles require about one-sixth as much steel weight as conventionally reinforced piles; since the unit cost of prestressing steel is double, the net cost is only $1/6 \times 2 = 1/3$. Due to this steel economy, and manufacturing economies, made possible by mass production in modern plants (usually long-line pretensioning with multiple forms), the first cost of prestressed concrete piles is very favorable in comparison with other types of concrete piles.

For prestressed concrete sheet piles, the requirement for double-bending, that is, positive and negative moment-resisting capacity, offsets, to a large degree, the other economic advantages of prestressed concrete, as compared with steel. A recent study indicated that prestressed sheet piles have a first cost about 10% high than unprotected steel sheet piles of the same structural capacity, but about 5% less than steel sheet piles that are painted or coated for protection against corrosion.

Prestressed concrete fender piles are roughly equivalent in first cost with treated timber. They may have a higher cost of installation. Their selection depends on an evaluation of their durability and fire resistance. In evaluating durability, consideration should be given to the high cost of replacement of fender piles, timber or concrete, and the out-of-use cost of the wharf; on such a comparison, prestressed concrete will be found very favorable.

11.8 Problems and Failures

Proper design, manufacturing techniques, and installation practice can be achieved only by a careful analysis of problems and failures. These represent an extremely small proportion (considerably less than 1%) of the total number of prestressed piles which have been successfully installed. These known problems are presented as specific cases, with assigned cause, and corrective action taken.

11.8.1 Horizontal Cracking under Driving

This is the most common problem. Puffs of "dust" are seen about one-third of the length from the top and horizontal cracks appear at 1-1/2-foot intervals, with considerable spalling at the surface. If driving continues, the concrete will disintegrate in fatigue failure, and grow sensibly hot just above the crack. Eventually, the tendons will break in brittle fracture.

Cause: (*a*) Tensile rebound stresses from free-end condition at toe (soft driving).

Cause: (*b*) Tensile rebound stresses from free-end condition at head when driving through water or soft material to rock.

  Cure: (1) Increase amount of soft-wood cushion.

         (2) Reduce height of stroke (fall) of ram.

         (3) Do not jet or drill below tip of pile during driving.

    (4)   Increase weight of ram.
    (5)   Predrill through overlying fill.

11.8.2   Inclined Cracking with Extensive Surface Spalling

Cause: (a)   Torsion plus rebound tensile stress
Cure: Steps (1), (2), and (3) of Subsection 11.8.1, plus:
    (4)   Make sure driving head cannot restrain pile from twisting.
    (5)   Do not restrain pile in leads or template.

11.8.3   Vertical Splitting of Cylinder Piles or Hollow-Core Piles during Installation

Cause: (a)   Soft mud and water forced up inside to a level above outside levels, creating an internal hydrostatic head.
Cure: (1)   Provide adequate vents in pile.
Cause: (b)   Hydraulic ram effect of hammer hitting water column.
Cure: (1)   Do not drive pile head below water level *or* provide a special driving head with very large openings (greater than 60% of void area).
Cause: (c)   Shrinkage cracking in manufacture.
Cure: (1)   Adequate water cure, especially for the first few days, starting immediately after conclusion of steam curing.
    (2)   Re-design mix to minimize shrinkage.
    (3)   Increase area of spiral.
Cause: (d)   Piles split while being filled with concrete, because of hydrostatic head of liquid concrete.
Cure: (1)   Reduce rate of pour to ensure it will obtain initial set without exceeding tensile strength of pile concrete and also within acceptable strains of spiral, whichever may be lower.
    (2)   Increase area of spiral steel.
Cause: (e)   Vertical splitting at top during driving due to bursting stresses.
Cure: (1)   Increase spiral at head.
    (2)   Increase cushioning at head.
Cause: (f)   Vertical splitting at or above toe due to plug of soil wedging in void.
Cure: (1)   Prejet or predrill to prevent plugs.
    (2)   Jet or drill inside to break up plugs during driving.
    (3)   Increase spiral.
    (4)   Use solid tip.
Cause: (g)   Jetting inside produces internal hydrostatic head.
Cure: (1)   Provide vents in pile walls.

11.8.4   Vertical Cracking of Cylinder Piles after Installation

Cause: (*a*) Freezing of water or mud inside.
Cure: (1) Provide sub-surface vents to allow circulation of water.
(2) Place wood or "styrofoam" log inside.
(3) Fill pile with frostproof material.
Cause: (*b*) High river level suddenly fell, leaving water inside higher than out, combined with drying shrinkage on outside and colder temperature outside.
Cure: (1) Ensure adequate vents.
or
(2) Fill piles with concrete core.
and
(3) Increase spiral.
Cause: (*c*) Piles cracked from logs, debris, etc., carried by river
Cure: (1) Fill with sand or concrete

11.8.5   Breakage of Prestressed Vertical and Batter Piles in Wharf by Ice Masses during a Spring Thaw

Cause: (*a*) Ice masses slid down batter piles, wedged against adjoining vertical piles, broke both.
Cure: (1) In such an environment, design pile layout so ice masses cannot wedge between piles.
(2) Increase bending strength of piles by greater prestress, or by larger-sized piles (greater section modulus).
(3) Increase spiral to provide greater shear resistance.
(4) Add additional longitudinal mild steel reinforcement for greater ultimate strnegth.

11.8.6   Excessive Spalling at Head under Driving

Cause: (*a*) Hammer impact on unrestrained concrete.
Cure: (1) Chamfer head.
(2) More cushioning.
(3) More spiral at head.
(4) Make sure head is square and plane.
(5) Make sure strand and other reinforcements do not project above the head of the pile.
(6) Be sure driving head (helmet) does not fit too tight (1 to 2 inches clearance should be provided all around).
(7) Make sure driving head cannot become cocked: it should ride in leads in axial alignment with hammer and pile.

11.8.7 Disintegration in Service of Corners of Prestressed
Piles below Water Line

Cause: (a) Reactive and unsound aggregates.

Cure: (1) If caught very early, seal entire pile surface with underwater-setting epoxy, after chipping out damaged zones to undamaged concrete, and patching with epoxy mortar or underwater/setting cement mortar.

(2) Jacketing, with a structural jacket.

(3) Replacement.

11.8.8 Vertical Hair-Line Cracks along Center of Faces of Solid Piles

Cause: (a) Differential shrinkage of outside of pile plus differential cooling, especially in cold, dry, windy weather.

Cure: (1) Provide graduated cooling period during last phase of steam curing.

(2) Cure in water immediately after removal from steam cure.

11.8.9 Horizontal Hair-Line Cracks on Top Surface at Lifting Points

Cause: (a) Negative moment exceeds concrete strength plus prestress.

Cure: (1) Provide mild steel at lifting points.

(2) Use more lifting points.

(3) Provide greater prestress.

11.8.10 Horizontal Cracks in Bending over Support Points
while Installing Batter Piles

Cause: (a) Negative moment exceeds concrete tensile strength plus prestress, and is aggravated by rebound tensile stress during driving.

Cure: (1) Increase effective prestress.

(2) Increase pile section modulus by increasing size. In some cases, the pile section may be made rectangular.

(3) Provide support for pile over a longer length (either above or below water or both), as by use of leads.

(4) Make sure hammer is held at proper angle by slings or leads so that its weight does not induce additional bending on pile.

12

Prestressed Concrete
Marine Structures

12.1 Introduction

Marine structures rank among the foremost applications of prestressed concrete. It was early recognized as the optimum material for harbor and coastal structures because it combined durability, strength, and economy. This chapter will discuss various aspects of this application. There is an obvious inter-relationship with two other chapters, Prestressed Concrete Piling (Chapter 11) and Prestressed Concrete Floating and Submerged Structures (Chapter 14). Therefore, this chapter must be read in conjunction with the others for the full presentation of marine applications of prestressed concrete.

Prestressed concrete began to be generally applied to wharf and pier construction about 1955 and, within the short span of ten years, had gained acceptance for use in harbors throughout the world. It has been adopted as the principle material and technique for the harbor development of such diverse ports as San Francisco, Los Angeles, Long Beach, Seattle, Galveston, Charleston, and Norfolk in the United States; Kuwait, Singapore, and ports in Java, Malaya, South America, England, Norway, etc. The climatic environment has ranged from tropic to Arctic, and the piles have been subject to attack from warm sea water, salt fog, and freeze-thaw sea water conditions.

Among the many types of marine structures for which it has been adopted are:

(*a*) Cargo and petroleum wharves.
(*b*) Bulkheads.

Fig. 40. Navigational light standard for U. S. Coast Guard, San Francisco Bay.

(c) Overwater airport runways and taxiways.
(d) Offshore platforms.
(e) Trestle roadways and pipeways.
(f) Bridge piers.
(g) Navigation structures. (See Fig. 40)
(h) Dolphins.
(i) Protective fenders.
(j) Coastal jetties.
(k) Small boat harbors.
(l) Groins.

Special features pertaining to these are discussed later in this chapter.

The structure of the above applications may consist of the following prestressed concrete elements:

A. Piles   Bearing piles
Batter (raker) piles
Cylinder piles
Fender piles
Sheet piles
Soldier piles

B. Deck units   Full-depth deck slabs
Half-depth deck slabs (for composite action)
Pile cap girders
Marginal beams
Deck beams and girders
Firewalls

C. Wharf girders   Craneway girders
Pipeway beams
Pipe bridge girder
Trestle approach span girders
Catwalks

D. Substructures Jackets
Crossties and bracing
Skirt slabs
Distribution slabs and girders
Columns

E. Bulkheads   Sheet piles
Curved slabs
Flat slabs
Ties
Struts

12.2 Prefabrication

Marine construction implies over-water construction, with all its inherent problems. Provision of supports is difficult; transport of and access for men and materials is expensive and time-consuming. The difficulty of access and support, frequently combined with salt-water splash below and hot drying (or freezing) conditions above, makes it extremely difficult to obtain high-quality concrete, possessing both strength and durability. Many marine structures are located in areas remote from the optimum location for construction. Premium wages are exacted in some locations. Labor may be in short supply at other sites.

All of the above factors emphasize the need for prefabrication and precasting. Precasting, however, is usually most efficient and effective when it employs pretensioned reinforcement. Moreover, the connection of precast units so they will act monolithically can often be best done by post-tensioning. (See Fig. 41)

Fig. 41. Prefabrication, utilizing precasting, pretensioning, composite construction, and post-tensioning, enabled rapid and economical extension of overwater runways and taxiways, LaGuardia Airport, New York.

Therefore, marine structures are best constructed by prefabrication. Prestressing makes the prefabrication of concrete more efficient and practicable, and facilitates the tying together of the structure to act as a whole.

12.3 Durability

This feature has been a dominant one in the selection of prestressed concrete for marine exposures. Chapter 4 treats extensively of the subject of durability, so only special consideration relative to marine structures is examined here.

Prestressed concrete fully immersed in sea water is completely saturated. Oxygen contents are relatively low; therefore, most forms of corrosion of reinforcement are minimized. On the other hand, chemical attack on the concrete is maximized. If the aggregates lack in soundness, their deterioration will be accelerated. Reactivity between the alkali in cement and reactive aggregates will be increased. With cements whose chemical constituents have been improperly selected (e.g., high C_3A), the cement constituents may be subject to replacement by softer compounds, leading to severe loss of strength. Therefore, for the underwater zone, particular care must be given to:

1. Soundness of the aggregates under sodium-sulfate tests.
2. No alkali-aggregate reactivity.

3. Use of a Portland Cement, such as ASTM Type II.
4. Low water/cement ratio.
5. Curing. Optimum results will be obtained with a three-stage cure:
    (*a*) Atmospheric steam cure.
    (*b*) Supplemental water cure.
    (*c*) A drying out period.

Not all of these curing stages are always necessary. As a matter of practice for most regions, steps (*a*) and (*c*) or steps (*a*) and (*b*) are usually adopted.

If the complete cycle is to be abbreviated, the author prefers to cut the supplemental water cure [step (b)] to three days and allow one week of drying [step (c)]. With dense concrete, internal curing will still take place after the termination of the water cure, and the drying of the surface will improve the durability in sea water.

Near the sand line, concrete may be subjected to abrasion from moving sand and gravel. Abrasion can also be caused by ice, particularly where the ice contains embedded sand and rocks. Abrasion can be partially countered by providing a dense concrete, of as high quality as possible; that is, high cement factor, low water/cement ratio.

Aggregates should be selected for hardness and resistance to abrasive loss in the rattler test. It may be desirable to increase cover in a zone of known abrasion.

Cavitation may occur in a surf zone. Cavitation is due to the collapse of bubbles of water vapor. It is very difficult to give complete protection against cavitation; a hard trowelled finish will usually prevent serious cavitation loss. In extreme cases, or when repairs are required, an epoxy coating is recommended.

Attack of marine borers on concrete, by mollusks and sea urchins, is usually of no consequence when a normally strong, dense concrete is employed.

Above low water, in the so-called splash zone, lies the region of greatest exposure. In addition to the sea water itself, there is ample oxygen for corrosion of the reinforcement. Chlorides may be deposited by evaporation in permeable concrete and lead to salt-cell electrolytic corrosion. This zone is also subjected to freezing and thawing and, possibly, to ice abrasion.

Emphasis must be placed in this zone on:

Low water/cement ratio.
High cement factor.
Thorough consolidation of concrete.
Good curing.
Adequate cover over reinforcement.
In freeze-thaw exposures, use of air-entrainment (6% to 8%).

In the zone above the splash zone (e.g., the decks of wharves) the concrete is still subjected to moist salt air.(from spray or fog). The most predominant form of attack is salt-cell electrolytic corrosion of the reinforcement. The concrete, therefore, must be designed for impermeability. Freeze-thaw attack must also be considered where applicable.

Emphasis for this zone should be on the following:

High cement factor.
Low water/cement ratio.
Adequate cover.
Air-entrainment for freeze-thaw exposure.

Coatings and jackets may be applied in special cases, to seal the surface and to protect it. It should be emphasized, however, that prestressed concrete has been extensively and successfully employed in marine exposures of tropical, desert, temperate, and Arctic climates with no coatings or jackets whatsoever.

The simplest and most common form of coating is a bitumastic paint, applied cold. As such, it serves primarily as a sealant. Hot asphalt coatings give a thicker and more continuous protection and have proven effective in completely sealing concrete subject to extreme salt-cell electrolytic attack.

Epoxy coatings have been used in recent years, some applied to precast concrete while still at the manufacturing yard; others applied in place, using underwater-setting epoxies.

Wood lagging has been applied for many years in the tidal range and splash zone in Western Norway. The more recent trend, however, is to eliminate it and rely on the use of high-quality prestressed concrete with 8% air-entrainment.

Wrought-iron jackets have been applied in areas of extreme ice abrasion. Where practicable, natural stone facing has been used similarly. The high cost of facing of bridge piers with granite could be substantially reduced by preparing precast prestressed panels with the facing embedded, then setting these panels up as side forms for the concrete, rather than the stone-by-stone method commonly employed. Use of ceramic facing might then be a more economical way to provide the abrasion-resistant facing, with the prestressing acting to hold the individual facing pieces locked in. Considerable success has been had on dam facings by using precast panels of extremely high-quality dense concrete, again set up as forms and tied into the main body of the concrete pier. Some of these panels have been prestressed, and this would seem to offer the added advantages of freedom from shrinkage and handling cracks, the assurance of concrete continuously under compression, the ease of handling, and improvement of the tensile quality of the adjoining cast-in-place concrete.

Recent research in the USSR has indicated that such panels, prestressed to a high degree, e.g., initial prestress equal to 50% of concrete strength, possess substantially increased durability.

Jackets of precast concrete, sometimes prestressed, have been extensively used in the past, set around or over the previously constructed concrete. These have then been secured by grouting, either tremie grout or by sealing, pumping, and grouting in the dry. Although not employed so far, a combination of a highly prestressed jacket with epoxy mortar injection would seem to offer technical advantages, which would have to be weighed against the higher costs.

Unfortunately, jackets have sometimes proven less durable than the concrete they were supposed to protect. This is because jackets by nature are relatively thin. To place the concrete in such sections has required more workability, hence a higher water/cement ratio, and more permeable concrete. Cover over the reinforcement is also reduced.

The present trend, therefore, is to eliminate jackets, putting primary reliance in the quality of the basic concrete, and providing more cover in zones of high exposure.

12.4 Repairs

Where concrete has suffered corrosion, disintegration, or damage, considerable study must be given to the proper method of repair.

Localized damage, as from impact, is most easily repaired. The damaged area should be thoroughly cleaned and keyed. The surface should then be coated with epoxy. The cavity can then be filled with concrete or grout, either poured in the dry or underwater. Use of a shrinkage-reducing admixture is definitely indicated above water, and good water curing is essential. Underwater concrete, placed by tremie or grout injection methods, does not shrink.

For relatively small underwater repairs, an underwater setting cement may be used. For prestressed concrete this should not contain calcium chloride. There are several such products available on the market which have been used successfully in repairs. Underwater epoxies have been similarly employed.

Where concrete disintegration and/or extensive corrosion of reinforcement is occurring, the first step is to determine the cause and extent. In a few notorious cases (e.g., the original Hayward-San Mateo Bridge and an iron-ore loading pier in South America) the process had proceeded so far that permanent restoration was impracticable. In the first case, all of the concrete was permeable; it was in the splash zone, and salt cells had formed throughout the structure. Localized repairs consisted of chipping, replacement of reinforcing steel, encasement in shotcrete, and bitumastic coating.

However, the anodes and cathodes just shifted and the electrolytic process continued.

In the case of the pier the disintegration was below water and was occasioned by a combination of unsound aggregates, possible alkali reactivity of the aggregates, and high alkali cement. Once again, attempts to apply coatings in localized areas were unable to stop the progressive disintegration beneath.

In an apparently similar example to the bridge, a wharf deck was found to be suffering salt-cell corrosion of the reinforcement. Extensive repairs were made by cutting out bad concrete, keying surfaces, replacing corroded steel, rebuilding the members by shotcrete, and *coating of the entire pier substructure* with hot asphalt. Since the pier was covered, the concrete itself was basically dry, and the complete coating sealed off the electrolyte (water) from re-entering. The results of the repairs were completely successful. Another factor was the much better quality of the original concrete; thus, salt-cell formation had not penetrated so deeply.

12.5   Increased Span Lengths

This has an important implication for marine structures, where live loadings are increasing all the time, and where economy depends so heavily on reducing the number of pile-to-cap-to-slab connections. Prestressing extends the efficient structural length of slabs and girders. This, in turn, permits the use of larger, more heavily loaded prestressed piles which, in turn, have greater bending strength, thus reducing the need for batter piles.

12.6   Economy

Prestressed concrete has resulted in substantial economies in marine structures to the point that some large modern structures are being built today at comparable costs per square foot to those of 20 to 25 years ago, despite the steady rise in construction costs. These economies are due to two causes:

1. Greater structural efficiency, as demonstrated by the following:
    (a) More heavily loaded piles.
    (b) Better column strength.
    (c) Greater lateral bending strength.
    (d) Ease in handling and driving.
    (e) Longer spans.
    (f) Better connections for monolithic action.
    (g) Integration of structural function so entire structure acts to reduce lateral, vertical, and torsional loads.

2. Economies in production, arising from the following:
   (a) Prefabrication, especially precasting.
   (b) Plant manufacture—savings in mass production.
   (c) Standardization and modular design.
   (d) Economy of prestressing steel in relation to conventional mild reinforcing steel.
   (e) Ability to use higher-strength concrete.

12.7   Effective Utilization of Prestressed Concrete in Marine Structures

12.7.1   Wharves

A typical marine wharf is comprised of piles, both vertical and batter (raking), cap girders, deck slabs, and marginal beams. Certain wharves may have additional elements, such as pipeway beams, firewalls, catwalks, craneway girders, etc. Fender systems are employed to absorb the impact of berthing vessels, to minimize damage to ship and wharf, and to prevent vessels of low freeboard and debris from getting in under the wharf deck.

For efficient structural performance, the entire wharf should function as a unit to resist vertical and lateral loads.

Vertical loads arise from cargo, cranes, trucks, railroad cars, storage, snow, and people. Lateral loads generally arise from ship-mooring forces, waves and currents, earth slope pressures, including the lateral earth pressure due to surcharge behind the wharf, wind on moored vessels and buildings, and seismic forces on wharf, and on the live load and earth slope. These lateral forces may be applied in such a way as to cause torsion in the structure as a whole. A special loading condition may be produced by tsunamis and should be considered in those harbor locations where tsunami effects have occurred in the past.

A wharf deck, if properly tied together, can act as a huge girder in the horizontal plane, supported against buckling by numerous piles. In considering the individual elements comprising a wharf structure it must be emphasized that they usually are designed to perform as part of a complete structural system.

a.   Bearing Piles

The special considerations pertaining to wharves are that they usually involve long unsupported lengths, hence they require consideration of column behavior. They are exposed in the splash zone where greatest attention must be given to durability. Bending stresses occur with batter piles, and also when a vertical pile is first seated on a sloping surface or driven through old riprap, timbers, sunken barges, etc.

In driving on a slope, it will usually be found best to set the pile a foot or so inboard of its final location, as it will tend to "walk" downhill during driving.

In driving through old riprap, etc., it has been shown best not to restrain the pile horizontally during driving. The pile is held at the top only, and the tip tends to find its way through the obstructions. (Any attempt to provide a third point of restraint will cause high bending stresses and possibly break the pile.) Such a procedure has been used satisfactorily in installing thousands of piles through old rock seawalls with a minimum amount of pile damage. However, this procedure will result in greater deviations in pile location, requiring that the pile cap be made wider, so as to permit a tolerance of the order of 12 inches or so.

Whenever piles are driven in a marine structure, some deviation in head location is inevitable. Common practice in the past has been to pull pile heads into true location. This, unless properly guided by engineering analysis, has resulted in many broken and cracked piles. The cracks are usually below the mud line, at the point of fixity, thus unseen, but they have reduced the pile bending capacity seriously. On the other hand, the blanket prohibition of all pulling of heads into position is unrealistic. The factors involved are the depth of water, soil properties, particularly near the surface, section modulus of the pile, and design bending loads.

Enforcement of restrictions on pile head realignment is very difficult. The author has successfully used the applied force as a reasonable solution which can be enforced. Pulling of pile heads is permitted only by use of a hand jack (such as a Coffin hoist) which is so set that the force applied cannot exceed a specified number of pounds, e.g., 500 pounds. This procedure has been more practicable to administer than a detailed survey and row-by-row calculation, particularly since the outboard piles may be constantly moving with tidal current, waves, etc.

In deep water, or when currents are swift or wave forces severe, it will be necessary to stay piles as soon as they are driven to prevent excessive deflection and breakage.

For marine structures, pile heads should always be notched, as with a saw cut, before cutting them off with pneumatic tools. Use of a timber clamp may also help to prevent cracking of the pile heads. This zone is under transverse tensile stress from the prestressing so, unless care is taken, cracking can extend below the point of cut-off. This, in turn, may lead to accelerated corrosion. For the same reason, it is desirable to embed the pile in the deck slab a few inches, to the extent that it is structurally feasible.

b. *Batter Piles (Rakers)*

These are extensively utilized in wharf structures to resist lateral forces. During installation, these piles undergo reversals of stress, from being cantilevered over a support during setting, to becoming a continuous beam during driving, to being simply supported in final position. These conditions must be analyzed. In deep water, with flat batters, it may be necessary to adopt one or more of the following steps during setting:

(*a*) Set the pile on the bottom in a vertical position, then lean back to a support above water level.

(*b*) Provide underwater support, such as below-water leads.

(*c*) Hold in slings on specified batter, and set until tip rests on bottom and top rests on support.

(*d*) Provide an above-water frame or leads to hold top of pile during installation.

Many batter piles, set properly, have been severely cracked or broken during driving by the weight of the hammer being placed on the overhanging pile head. The hammer must be constantly supported, either in inclined leads, or hanging in a bridle at the correct angle.

Many batter piles, properly installed in deep water, have been damaged or broken during the time of releasing from the leads and transfer of support to the previously-built structure. This is always a complex procedure that requires careful planning. Fortunately, the force required to support the pile head on release is generally very small, and can usually be provided by small timber blocking and small (1/2 inch or so) wire line.

Batter piles are generally designed to function in both tension and compression with an adjoining vertical or opposing batter pile. The transfer of tension loads from a pile cap into a pile is usually accomplished by mild-steel bars, grouted into the pile in drilled holes to a depth sufficient to develop the full value of the bars in the fully prestressed portion of the pile, that is, below the transfer zone, which is usually 20 inches (50 cm) long. Thus, embedments of 3 or 4 feet (1 meter) or more are common. Drilling of such deep holes after installation is difficult and may damage the pile. Therefore, it is better practice either to form these holes in manufacture or to embed the mild-steel dowels during manufacture. In both cases there must be sufficient excess length to take care of driving tolerances.

Embedment of the extending prestressing tendons, for example, strands from the pile into the cap, is adequate to develop strength. However, because of the small steel area, this may permit excessive rotation of the pile head. With wide wharves this is not a factor; in such a case the strand embedment is a cheaper and generally more satisfactory solution. Strand should be embedded a minimum of 24 inches into the pile cap.

Tension connections can also be made by use of a stressed tendon such as an alloy bar, anchored into the pile and stressed to the pile cap. This is a very effective structural means, but not extensively employed at present due to higher cost. It has many potential advantages, particularly with precast cap girders.

Tension piles usually develop their tension in the soil by skin friction on the bottom portion of the pile. When driven into soft rock, such as shale, steel stubs on the tips of the piles have effectively developed high pull-out values.

Another scheme, particularly useful when driving onto hard rock, is to anchor it by prestressing. A pile with a hollow core or tube is driven to bearing and framed into the superstructure; in some cases, even the deck may be poured, leaving a hole formed over the pile. Then a drill, working off the deck or framing, drills through the pile and into the rock to form a socket. A prestressing tendon is then anchored, grouted in the rock, and stressed from the top; then the pile core is grouted. In some cases, the tendon may be only partially stressed, or unstressed, or a dowel may be used. This depends on the relative values of design compression and tension on the pile so that the total compressive force in the pile (initial prestress plus anchoring stress plus live load compression) does not exceed the allowable compressive stress in the pile.

c. *Cylinder Piles*

Cylinder piles and other large cross-section piles are being increasingly employed in wharf projects, with the lateral loads being taken in bending. (See Fig. 42) Usually the top of the pile is fixed to the pile cap girder by use of a heavy cage of mild steel reinforcement. With a hollow-core or cylinder pile, this cage is set in the pile head and secured to it by a concrete plug. To insure that the plug will provide full shear transfer to the pile walls, several alternate techniques have been employed:

(*a*) Roughening of the inside walls of the pile shell top by sand-blasting or bush-hammering.

(*b*) Provision of corrugations in the walls of the pile; these are formed in the top few feet of the pile during manufacture.

(*c*) Painting the inside head with epoxy bonding compound prior to pouring the plug.

(*d*) Use of a low-shrink concrete mix for the concrete of the plug.

(*e*) Use of a precast concrete plug encasing the reinforcing cage; this plug is 1 to 2 inches (2.5 to 5 cm) smaller all around than the core. After placing, the bottom of the annulus is sealed with manila rope or paper pushed down, and the annulus is filled with grout. The precast plug may have a roughened or corrugated surface.

For cylinder piles 36 inches in diameter and larger, that is, having hollow-cores larger than 24-inch diameter, the combination (*a*) and (*e*), or (*a*), (*b*), and (*e*), is recommended wherever this can be incorporated into the struc-

Fig. 42. Prestressed cylinder piles for Port of Baton Rouge, Louisiana.

tural scheme selected. For hollow-cores smaller than 24 inches; then the combination (*a*), (*b*), and (*d*), or (*b*), (*c*), and (*d*), is recommended.

Where cylinder piles must take lateral loads in bending, special consideration must be given to shear and torsional resistance. Greatly increased shear and torsional resistance can be provided at minimum cost by increasing the number of spirals and, to a lesser extent, by increasing the diameter of spirals or hoops. In the typical cylinder pile with 5-inch thick walls, a maximum spiral spacing of 3 or 4 inches is recommended throughout the entire length that is subjected to bending and torsion.

d.   *Pile Caps (Pile Cap Girders)*

These are normally subjected to heavy shear forces as well as positive and negative bending. With comparatively short spacing between piles, therefore, prestressing is usually more or less uniform over the section (concentric). With long spans between piles, the center of prestressing force may follow the moment curve, with tendons raised over the piles and near the bottom of the cap between piles. Post-tensioning after pile cap connections are made tends to lock the cap to the piles, and permits the embedded portion of the pile head to act monolithically with the cap in resisting compression.

Because of the heavy shear, pile caps usually have very heavy stirrups of mild reinforced steel.

Much modern design is aimed at providing continuity of the deck over the caps and at utilizing the upper portion of the cap as a part of the slab. This may readily be done with a monolithic cast-in-place deck. However, the provision of temporary supports, etc., is costly, so there is a trend to pouring the bottom half of the cap first, then setting precast deck slabs on this cap, then pouring the top half of the cap.

Thus the advantage of precast deck slabs can be combined with the provision of continuity for the deck, and with common use of the concrete of the upper portion of the cap. The lower cap obviously requires minimum support during pouring, and the quantities of cast-in-place concrete are held to a minimum.

There is high horizontal shear across the mid-depth of the cap. This means that the heavy stirrups must protrude from the bottom half-cap to tie the upper portion to it. The concrete surface should be as rough as possible, preferably with the cement past sandblasted (or air-and-water-jetted off, immediately after the pour), so as to expose the aggregate. An epoxy bonding compound may also be used as an added insurance of bond. Shear keys may be stamped in the surface of the green concrete.

This bottom half-cap may be precast. The connection to the piles is then made in one of three ways:

(a) Reinforcement protrudes below the cap and is inserted into the hollow core of the piles. A small hole permits pouring of the grout plug.

(b) A large hole is left in the precast half-cap. The reinforcing cage is set through this hole into the hollow core of the pile and the plug is concreted.

(c) A large hole is left in the precast half-cap and the pile head penetrates sufficiently far to develop the connection. The pile head is roughened so as to bond with the grout pour in the annulus.

When setting a precast half-cap over batter piles, an elliptical hole will provide more tolerance.

With precast half-caps, the bottom reinforcing must be in the side walls. Since the holes in the cap must permit tolerance for pile head location, and the caps in turn must support deck slabs, this often requires widening of the cap beyond that required for structural reasons. While adequate side cover must be provided over the bottom reinforcement (prestressed or conventional), only bare minimum cover need be provided next to the hole as this will be filled with concrete. In such a case, positive steps must be taken to prevent a shrinkage crack between the grout fill and the half-cap. Widening of the cap is an economical way to provide tolerance for setting of the precast deck slabs despite small inaccuracies in pile bent location.

With the half-cap scheme, post-tensioning tendons, in rigid ducts, may be placed immediately on top of the half-depth cap, and post-tensioned after the top half is poured. This is an economical way of providing concentric post-tensioning and of locking piles, cap, and precast slabs together. (See Fig. 43)

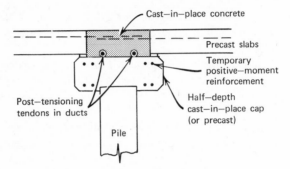

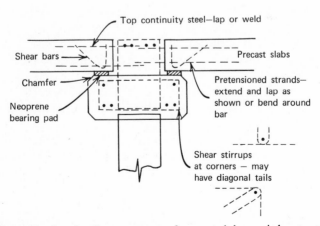

Fig. 43. Reinforcing and other details at supports of precast slabs or girders to resist shear and tensile stresses due to creep, shrinkage, and temperature.

When prestressed precast deck slabs rest on the cap, they tend to shorten due to shrinkage and creep and, thus, shear off the sides of the cap. The precast slabs, therefore, should be tied across the cap by means of extending strands or reinforcement. Stirrups should be provided to resist the shear in the caps (and the end of the slab). A strip of asphalt-impregnated fiberboard or neoprene will provide uniform bearing and permit rotation.

Full-depth caps are frequently employed where spans are long. This eliminates the necessity of providing for horizontal shear across the joint and facilitates pile cap to pile connection. However, it requires that the deck be structurally independent of the cap, thus preventing common use of the cap concrete. Continuity in the deck slabs is more difficult to achieve.

e. *Wharf Slabs*

Precast prestressed concrete is at its best here: the economy of precasting and the structural efficiency of prestressing combine and augment each other in this application.

In most cases, simple span precast prestressed slabs are used. These may later be made continuous by the use of mild steel over the pile cap. Lateral monolithic action may be provided by transverse post-tensioning or mild-steel reinforcement.

Both continuity and transverse monolithic action are facilitated by the scheme of half-depth slabs. These are pretensioned slabs, usually having tie steel (stirrups) protruding for tying the two halves of the slab together. The surface is roughened by a jet of air and water while the concrete is still green, or by sandblasting. Mild-steel reinforcement is then placed for transverse load distribution and for negative moment over the cap; then the top concrete deck slab is poured in place.

This same continuity over the cap may be accomplished with full-depth deck slabs, by welding projecting reinforcement or by stepping or tapering back the top half of the slab for a few feet from the end. Transverse ties may be made by welding inserts in pockets in the joints between slabs, then grouting. (See Fig. 44)

With very wide but thin slabs, it is essential to provide a reasonably uniform seating by means of asphalt-impregnated fiberboard strips or neoprene.

An occasionally used scheme is to set the precast slabs, then lay post-tensioning tendons in the joints betwen slabs, raising these in low saddles over the pile caps, concreting the joints, then stressing the entire deck longitudinally in sections of several hundred feet of length.

For providing longitudinal continuity steel over caps and also transverse distribution steel, the use of tensioned elements (Chapter 3, Subsection 3.1.6) would seem particularly appropriate.

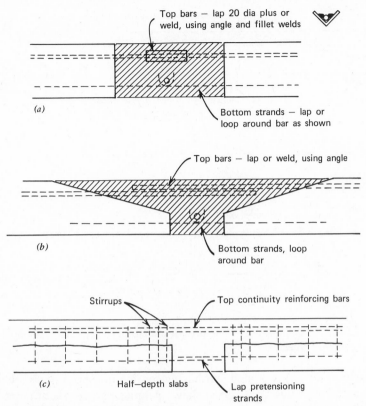

Fig. 44. Methods of obtaining continuity over supports when employing precast pretensioned slabs.

Another means of providing continuity was utilized on a large wharf in Kuwait. Full-depth slabs were desirable in this location because of the problems associated with the curing of large exposed surfaces in this environment. The slabs were made in double-span lengths, set alternately in checkerboard fashion so that every other slab was continuous over the support. Transverse post-tensioning tied the entire deck together. (See Fig. 45)

These particular deck slabs were haunched, permitting the use of straight tendons, providing more shear next to the supports, and a greater effective depth for the prestressing to compensate for the fact that only half the tendons cross each cap.

Another scheme utilizing precast slabs for continuity is to provide diagonal post-tensioning in criss-cross fashion. This is most readily achieved with half-depth slabs, with rigid ducts laid diagonally on top and stressed after concreting. (See Fig. 46)

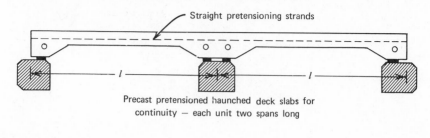

Straight pretensioning strands

Precast pretensioned haunched deck slabs for
continuity — each unit two spans long

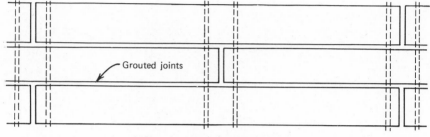

Grouted joints

Plan view showing units erected in
staggered pattern and tranversely
post—tensioned for continuity

Fig. 45. Scheme for continuity using precast pretensioned deck slabs as employed at Kuwait Harbor Project.

A concept not yet exploited, but having considerable promise, is the use of long lengths of precast prestressed slabs, spanning perhaps six or more bays. The joints would be staggered so that only one slab was jointed in any one bay. Slabs would be post-tensioned transversely. These slabs would be concentrically prestressed if spans are short, or with tendons deflected up-and-down if spans are long. Such slabs would be relatively inexpensive to manufacture, and could readily be set with a large modern floating derrick. The effect of creep would also have to be considered and flexibility provided in the pile bents. Such a scheme might be particularly economical with half-depth slabs.

A scheme originated in The Netherlands obtains continuity by tapering the end of each slab, and allowing it to overlap the cap several feet. Transverse post-tensioning then distributes the load transversely between overlapping points.

Where the wharf deck will be subjected to uplift due to tsunami effect, the slabs and their connections may require special design considerations. These are discussed more fully under "Coastal Structures," Section 12.7.3.

f. Combination Cap-and-Slab Decks
Much of the cost, time, and difficulty in precast systems for wharf decks is

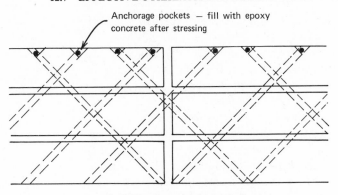

Anchorage pockets — fill with epoxy
concrete after stressing

(a) Use of double—diagonal post—tensioning to
obtain continuity and monolithic action
with precast prestressed slabs

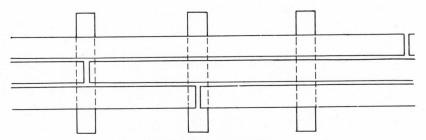

(b) Precast pretensioned deck slabs, full—depth,
uniformly prestressed, each unit extending
over four or more spans. Post—tension
transversely for monolithic performance

Fig. 46. Schemes for assembly of precast deck slabs for monolithic behavior.

in the joints and connections. Prestressing is an extremely beneficial technique for making these joints structurally efficient. However, the cost and time remain. Therefore, considerable attention is being given to monolithic cap-slab deck sections, cast-in-place or precast in large combined sections.

With cast-in-place slabs, one such development is the use of a constant-depth thickness that is a continuous flat slab. This greatly simplifies the forming of the soffit. The cap is reinforced (or prestressed) in the line of the piles; it, in effect, a wide, thin cap. The deck is reinforced (or prestressed) longitudinally right through the cap.

This same flat, soffit can be maintained in ballast-deck (filled) wharf structures, even if structural requirements require that the cap be a few inches thicker, by raising the top of the cap above the general level of the deck slab. (See Fig. 47)

For precast schemes one solution is to build the cap and slab as one deck unit, allowing this deck unit to extend a few feet beyond the pile bent. This permits the development of a degree of continuity and, in effect, cuts out one-half of the number of joints.

Another scheme is to widen the pile cap so that in effect it spans half-way to the next cap, thus becoming a double cantilever. Longitudinal post-tensioning and/or mild-steel overlap at the center of the span can be used to tie the structure together longitudinally. The soffit can be tapered in such a way as to improve the structural efficiency, and the main prestressing steel kept in the top of the slab, completely protected from the splash zone.

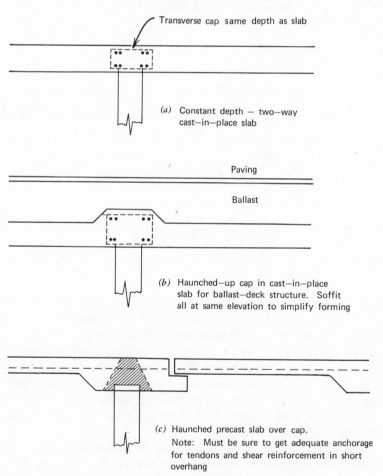

Transverse cap same depth as slab

(a) Constant depth — two—way cast—in—place slab

Paving

Ballast

(b) Haunched—up cap in cast—in—place slab for ballast—deck structure. Soffit all at same elevation to simplify forming

(c) Haunched precast slab over cap.
Note: Must be sure to get adequate anchorage for tendons and shear reinforcement in short overhang

Fig. 47. Some schemes for deck slabs for concrete marine structures.

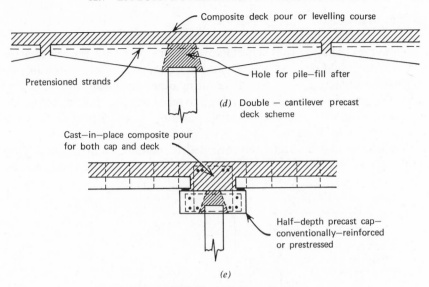

Fig. 47. *(Continued)*

When this scheme is superimposed on a substructure of cylinder piles, with their wider spacing and elimination of intersecting batter piles, then the deck unit becomes a tapered block, requiring minimum reinforcement for vertical loads. It can have its connecting plug cast monolithically, and all tolerance adjustment can be made at mid-span. Transverse and longitudinal post-tensioning can then be provided in the joints between the slabs, or over the top, with a cast-in-place topping slab.

g. *Marginal Beams and Fender Beams*

Modern wharf decks require edge beams to take the increased reaction from vertical concentrated loads, and to support fendering inserts, bumpers, etc. In some cases, with a relatively thick deck slab, these beams can be incorporated right in the same slab thickness by providing additional mild-steel reinforcement. In ballast-deck (filled) wharf structures, the side wall can serve these other functions.

With many wharf decks, however, separate edge beams are required. These can be cast-in-place and tied into the pile cap girder with mild steel. Alternatively, they may be precast in relatively long lengths, set on steps or brackets on the ends of the pile caps, and post-tensioned to and through the cap. At the center portions of each span, the edge beam should also be post-tensioned transversely to the deck slab in order to serve its function as an edge beam support for vertical loads.

Obviously, the transverse post-tensioning in the case of both cap and deck slabs can be used for the dual purpose of tying on the edge beam and transverse prestressing of the deck.

h. *Firewalls*

Firewalls are utilized to keep fire from spreading longitudinally under long wharves. They were much employed in the past with timber wharves where a fire could develop a horizontal draft chimney effect. Firewalls are still used in concrete wharves where oil or other petroleum products may catch fire and the fire spread unchecked to adjoining ship berths, endangering the ships as well as any combustible portions of the concrete docks, such as creosote fender systems.

Firewalls should be preferably oriented to minimize wave forces. They should extend below lowest low tide. Lightweight concrete is more effective

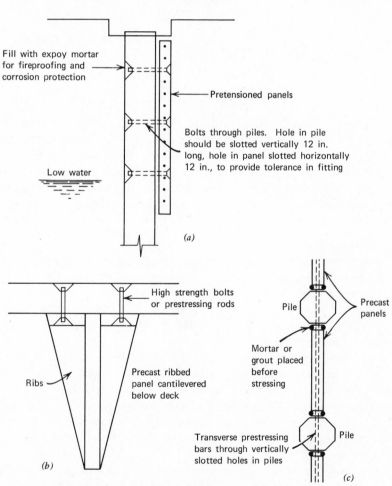

Fig. 48. Three schemes for fire walls in marine structures.

in preventing heat transmission. Because of their location near and in the tidal zone, precasting is usually the most economical method of construction.

Firewalls tend to work back and forth in the waves; therefore they must be secured properly. In many wharves they are tied to the piles by bolting or with dowelled and concreted joints. This puts an extra horizontal bending on the piles and may not be desirable structurally. An alternative method is to suspend the firewall from the deck, and to post-tension it vertically to the deck slab. Haunched ribs can be used to provide the necessary moment capacity. (See Fig. 48)

i. *Craneway Girders*

Many modern wharves must support large gantry cranes for container handling, ore loading, etc. Heavy, deep girders are required. Past design practice was to employ a large number of closely spaced piles under these girders, in order to minimize the shear and moment in the girder. With present wheel spacing and multiple wheels on gantry cranes, it may often be more structurally efficient today to use the same span as the remainder of the wharf, and make the craneway girder act monolithically with the rest of the deck slab, distributing the heavy wheel loads transversely to adjoining piles in the bents.

In some cases these girders have been cast in long lenths, heavily prestressed, and set across multiple caps. The prestressing force may be deflected up and down for positive and negative moment, or the soffit may be haunched. (See Fig. 49)

Particular attention must be paid by the designer to the reversal of moment that may occur with the crane wheels in an adjoining span, thus producing negative moment in the center of the span.

Similarly, the very heavy concentrated loads may produce axial shortening in the pile under the wheels, causing positive moment over the support. Mild steel should therefore be provided in the bottom of the girder over the support. Such long, prestressed girders are subject to shortening under creep. Particular attention should be paid to providing adequate stirrups in the caps under the girders and to chamfering the edge of the cap. A good detail is also to provide an asphalt-impregnated fiberboard or neoprene bearing strip at the edge of the cap. At the joints where two girders join, there will be high forces trying to tear the cap apart. Longitudinal expansion should be permitted, as by a double bent, or else this jointing cap must be tied together with heavy mild steel, or post-tensioned, for example, with short bars.

When craneway girders are precast, they can be tied to the adjoining deck slab by transverse post-tensioning to distribute the load and to make the entire deck work as a whole.

Fig. 49. Prestressed concrete craneway girder for container terminal in San Francisco Bay.

j. Pipeway Beams

Many wharves carry extensive pipeways on trestle-type bents, connected and braced by pipeway beams. The forces on such beams include not only the usual loads, but also the longitudinal forces due to pipe expansion and contraction. These beams lend themselves to precasting because of the long spans usually employed; the problem has been how to connect them in a three-dimensional pattern with full moment connections.

Post-tensioning, usually with bars, has proven the most satisfactory method. Provision must be made to accommodate tolerances in pile location, either by moving the pile head to correct location (suitable for slender piles and deep water), or in the joint details. The jointing may be made over the pile, or a capping beam cantilevered a short distance out from the pile and the joint made there.

If the longitudinal pipeway beams are prestressed, consideration must be given to their shortening under creep and prestress. The usual practice is to take all longitudinal movement to an anchor bent, then provide expansion joints every 600 feet or so with a double bent.

k. *Catwalks and Pipe Bridges*

Many marine structures employ a central loading platform, separate or combined breasting dolphins, and detached mooring dolphins. These are connected by catwalks. In some cases pipeway bridges are similarly used to connect adjacent loading platforms.

The loadings on the catwalks are very light; therefore, steel truss spans have been largely used in the past. Railings must be provided, both for safety of personnel and to drag a ship's line along.

Prestressed lightweight concrete, usually in the form of a channel section with up-raised legs, can be made to work very efficiently. The up-raised legs serve as both railing and girder; the bottom slab serves as walkway and bottom flange. The legs will have to be thick enough to serve not only as web but top flange of the girder as well. Such a catwalk will have low maintenance. It is usual with catwalks to fix one end and allow the other to expand and contract. With a prestressed catwalk, care must be taken to provide for shortening due to creep, and the expansion shoe and adjoining concrete must be designed to take both the shear and bearing.

For pipe bridges, the standard highway I-girders will often be found the best section. Rectangular sections are also used. Cross-struts can be post-tensioned to the I-girders, or the connection made by steel brackets and bolting.

A mechanical safety-securing device should be provided to prevent one end from dropping completely off its support when the support moves under ship impact or wave or seismic force. This device must, of course, permit movement but restrict its magnitude. Among the devices employed are dowels in enlarged holes, filled with asphalt (New Zealand). Another method is to use chains; a third consists of keeper rods.

l. *Fenders*

Prestressed concrete is eminently suited for the fender piling of typical cargo wharves. Extensive installations have been in service for many years at Kuwait Harbor and the Jurong Wharves in Singapore (See Fig. 50). Smaller installations and test structures are in use in Los Angeles and San Francisco harbors. Prestressed lightweight concrete has the advantage of greater deflection and energy absorption. These fender piles are discussed in more detail in Chapter 11, Prestressed Concrete Piling.

Connecting wales for fender systems are conventionally made of laminated treated timber. This timber is not only a potential fire danger but is increasingly expensive. Its maintenance cost is high, due to the attack of marine borers. Even treated timber is vulnerable to marine borers in this zone after abrasion, local damage, or limnoria have penetrated the treatment.

Fig. 50. Prestressed concrete fender piling for wharves at Jurong Harbor, Singapore.

Prestressed lightweight concrete planks can be effectively used for fender wales. They should be thin (1-1/2 to 4 inches thick), with a high degree of concentric prestressing, and wire mesh reinforcement. They should be allowed to cure 30 to 60 days to get rid of as much creep as possible. The thinner planks can be laminated; the thicker planks can be used singly. For the thinner planks, galvanized mesh and galvanized strand would provide an extra degree of durability; however, this is usually not necessary if the lightweight concrete mix is placed with great care.

Although existing installations of prestressed concrete fender members have employed timber rubbing strips to protect the pile surface from abrasion, studies are being made on alternative, cheaper ways. A heavy coat of epoxy or an embedded steel plate may prove a better solution than the bolted-on timber strip.

12.7.2 Bulkhead Wharves

Bulkhead wharves are much used along inland waterways, serving to form a vertical wall between the land working area on one side and a navigable depth on the other. A great deal of engineering study has been directed toward analysis of the lateral earth loads, including those due to surcharge, and the failure circle which, in turn, determines the required penetration and the bending moment in the bulkhead. Having determined the loads and resistances, the structure is then designed as a wall with suitable anchors or ties. The rigidity of the wall will have an effect on the pressures developed.

The wall is exposed to mooring impact and forces, and the environmental attack of the water and splash zone.

Prestressed concrete has been extensively employed for such bulkhead wharves because of its durability, strength in bending, and economy.

Walls are usually built of sheet piles or a combination of soldier piles and panels. Some walls have utilized large cylinder piles with connecting panels.

These walls are usually tied back by ties and deadmen anchors. Prestressed concrete is particularly well suited for use as ties.

a.   *Sheet Piles*

The design, manufacture, installation, and jointing of prestressed concrete sheet piles is discussed in Chapter 11, Prestressed Concrete Piling.

When used as a bulkhead wall, the point at which the tie acts determines the moment pattern in the sheet pile. Thus, the pile may require eccentric or

Fig. 51. Precast prestressed soldier beams are utilized to construct graving dock.

deflected prestressing. Alternatively, mild steel may be placed in the relatively short region of high negative moment.

To tie the sheet piles together along the wall, and to span between ties, some sort of wale (longitudinal beam) is required. If the tie is at the top, then a capping beam may be poured in place, either conventionally reinforced or, preferably, post-tensioned.

If the tie is at some distance down from the top, one solution is to excavate and dewater down the necessary distance in back of the piles, and pour a cast-in-place horizontal beam. This again may be conventionally reinforced or post-tensioned.

The post-tensioning will help to tie the wall together, so as to act as a continuous membrane, and also to keep the joints tight.

b. *Soldier Beams and Slabs*

These soldier beams are usually of H cross section or else rectangular with suitable grooves. They may be steel or concrete. If prestressed concrete, consideration must be given to the points of positive and negative moment as discussed above for sheet piles, except that the magnitude of moment will be higher with the soldier beams. (See Fig. 51)

The connecting slabs serve as a wall to transmit loads laterally to the soldiers. These slabs may be prestressed planks or flat slabs, reinforced transversely, with vertical prestressing for durability and handling.

The slabs may relay on shell action to transmit the loads from the earth; cylindrical shells are sometimes used which may have vertical prestressing for handling, durability, and edge restraint. They exert a side thrust on the soldier beams. This is resisted in the lower zones by the passive pressure of the earth, but at the top a longitudinal tie must be provided, such as a post-tensioned capping beam. (See Fig. 52)

c. *Ties*

Ties are used in tension to hold the wall back to the deadman anchor. These ties are exposed to corrosive action because of their usual location at about tide line in filled soil. They also must have bending strength to resist settlement of the fill, and heavy concentrated loads, etc. Steel tie rods, heavily coated and wrapped, have been much used in the past, but highly stressed prestressed concrete tie beams appear to be superior in performance and economy. Connections are always a problem with any material. With prestressed concrete, the connection at the wall may be by direct shear on a T-head, or by dowelling or post-tensioning to a capping beam. One interesting solution provides for the connection to be made at the joint between adjoining sheets, this joint then being concreted. (See Fig. 53)

Ties, being subject to both tension and bending, should be designed for no more than 0 tension at the worst point. The available tensile strength in the

Fill behind bulkhead wall

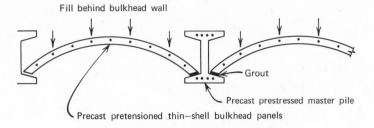

Grout

Precast prestressed master pile

Precast pretensioned thin–shell bulkhead panels

This system can be used only in sand or soft soil
(jetting plus light driving to install), since accuracy
and verticality are essential. Underwater fit should
be verified by diver

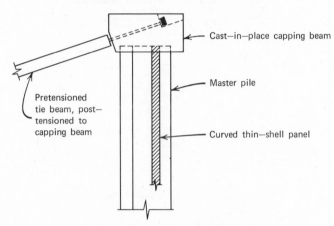

Cast–in–place capping beam

Master pile

Pretensioned
tie beam, post–
tensioned to
capping beam

Curved thin–shell panel

Fig. 52. Precast prestressed bulkhead system.

concrete and the ultimate strength of the strands will usually then provide
the necessary margin of ultimate strength.

12.7.3 Coastal Structures

These are defined herein as structures built on beaches, reaching out into
the ocean. Thus they penetrate the surf zone, passing from an onshore environment into a deep water, ocean environment.

Structures included are fishing and recreational piers, pipeline piers for
intakes, discharges, and products; piers for the support of offshore oil drilling and production, offshore loading terminals and facilities, groins and jetties for control of sand movement (erosion and deposition), and breakwaters. Completely submerged structures, such as intakes, outfalls, and
pipelines, are discussed in Chapter 14, Prestressed Concrete Floating and
Submerged Structures.

The increasing emphasis on the ocean for transport with special carriers,

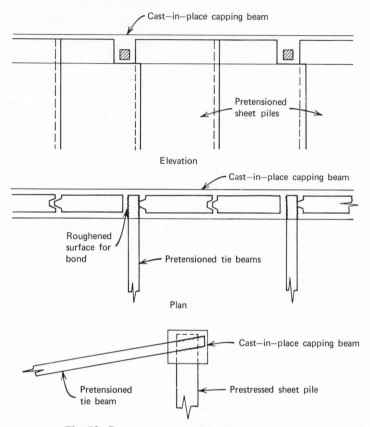

Fig. 53. Precast prestressed bulkhead system.

and the growing industrialization of coastal areas has given a great impetus to the development of coastal structures and it is anticipated that there will be a greatly expanded activity in the future. Great strides have been made in coastal engineering. Specific reference is made to the Conferences on Coastal Engineering and their published *Proceedings*.

Coastal structures must serve in a wide variety of environments, from tropical to Arctic, and from gently sloping sand beaches to exposed rocky shores.

Environmental aspects to be considered include:

(*a*) Storm waves, surf, (cavitation.)

(*b*) Sand movement.

(*c*) Current.

(*d*) Seaweed (kelp) and growth (barnacles).

(*e*) Mooring forces, especially the forces imparted by waves on vessels moored to the structure.

(*f*) Wind, including wind on moored vessels.

(*g*) Seismic, where applicable.

(*h*) Tsunami and storm surge, where applicable.

(*i*) Ice build-up on structure.

(*j*) Floating ice pressure against structure and moored vessels.

(*k*) Erosion and deposition of sand.

(*l*) Effect of extreme temperature and salt water.

Prestressed concrete is extremely well suited to the construction of coastal structures and has, therefore, been given perferential selection in many parts of the world. It must be emphasized, however, that the environmental aspects (*a*) through (*l*) above, as applicable, present an extremely severe set of parameters, especially as many interact to intensify and augment each other. Judgment is often required to assess the probability of forces acting simultaneously; to design for the worst possible combination may be beyond economic feasibility in many instances. Similarly, an evaluation must be made of the consequences of failure, including contamination, shut-down of service, destruction of adjoining facilities, etc. Consideration must be given to aesthetics, fire protection, and the effect of the coastal structure on the adjoining beach areas (sand migration, currents, wildlife, etc.).

Prestressed concrete used in coastal structures offers the advantage of durability, resistance to abrasion from sand and ice, structural strength in bending, maintenance of an uncracked section under alternating loads (as from waves) and high column strength. In arctic installations, it offers excel-lent impact resistance at low temperatures, an especially important point under repeated impact loading from ice.

To resist storm waves, it is desirable to present a mimimum area and a minimum volume to the waves. Use of highly stressed piles of circular cross section will reduce wave forces to the minimum.

Surf tends to produce cavitation in concrete surfaces. High density, well-compacted concrete, and provision of slightly increased cover, will generally give satisfactory results.

Sand movement on the beach is extremely abrasive. Use of high-density concrete, in a circular or octagonal cross section (or a square pile with large chamfers), reduces the abrasive damage to a minimum.

Sand migration will also tend to give greater unsupported column lengths at times, followed by unequal build-up later. The high column strength and bending resistance of prestressed concrete can be beneficially employed to counter these changes in the sand line.

Seaweed and barnacles tend to increase the front presented to wave, surf, and current action.

The repeated action of waves, particularly those acting on a vessel moored to a structure, tends to produce fatigue-type stress reversals. Fortunately, prestressed concrete behaves excellently in resisting fatigue due to the low stress change in the steel tendons in relation to the change in loading.

However, stress reversal does sometimes have to be considered in sheet-pile groins, where the waves on the two sides of the grin are typically out of phase. This constant "working" has destroyed steel sheet-pile groins; prestressed concrete is, therefore, a more suitable material but must be properly designed.

Tsunamis and storm surges and, sometimes, extreme storm waves, produce violent upward forces on the undersides of decks. These have torn apart many timber docks in the past. Some modern coastal structures are designed to lose their decks under such conditions; they are only slightly fastened down, with connections designed to fail, so that the substructure will remain undamaged.

A minimum of bracing and similar obstructions should be presented below the deck. This favors the use of prestressed concrete piles. Concrete deck slabs have inherent weight and can be designed to resist the upward force, provided they have sufficient mild-steel reinforcement to take the reversal in loading. Openings can be designed so as to minimize and relieve the wave entrapment in a particular area.

There has been a general tendency to neglect the consideration of tsunamis in design of marine structures, due to the unpredictability of occurence and force. However, some harbors and coastal areas are notorious for focusing these waves and surges. A design for tsunamis can be based on a rational appraisal of probability, consequence, and practicability, including cost.

Ice build-up occurs in areas of tidal change, and often includes beach ice that is dislodged and wedged in the piling. It adds both lateral resistance to current and weight. The downward weight on piles, particularly on batter piles, may be sufficient to break them. These ice masses also may slide down between batter and vertical piles and break them by wedging action. Thus the configuration of the structure must take ice build-up into consideration.

The force from moving ice is of major consideration to the designer and often exceeds all other lateral loads combined. There is a great difference in force, depending on whether the ice fails in tension or in compression. This, of course, is a function of the configuration of the structure and of the ice conditions. Prestressed concrete offers strength, rigidity against local buckling, and good fatigue strength.

Construction

The installation of piles in sandy beach areas is frequently aided or accomplished by jetting. Internal jets have been found very effective in beach sands because they eject the water right at the pile tip. For most coastal structures, it will be prudent to use a standard steel jet pipe in the pile, with screwed or welded fittings; this prevents the possibility of cracking of the pile due to a leak in the jet pipe. Care should be taken not to jet a large hole under the pile tip and then drive on the pile as this free-end condition will cause excessive tensile stresses and possible breakage.

In rocky beaches prestressed piles are often concreted into drilled holes. After drilling, the hole may partially fill with sand. The use of an internal jet pipe may permit jetting in of the pile, and the pipe can then be used for grout injection.

Prestressed sheet piles are usually installed in a sandy beach; here the use of external "gang" jets, or several internal jets, are efficacious.

Many coastal structures are constructed from "over the top," working progressively seaward out over the completed structure. Therefore the time

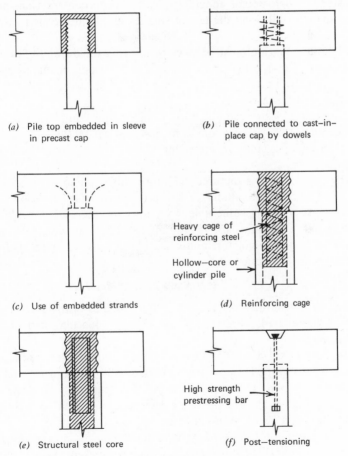

(a) Pile top embedded in sleeve in precast cap

(b) Pile connected to cast-in-place cap by dowels

(c) Use of embedded strands

(d) Reinforcing cage

Heavy cage of reinforcing steel

Hollow—core or cylinder pile

(e) Structural steel core

(f) Post—tensioning

High strength prestressing bar

Fig. 54. Methods of connecting piles to pile caps.

required for completion of a span must be kept as short as possible. With precast elements, these may be temporarily connected by welding or bolting, and the construction equipment moved out over them. Later the permanent connections are made. Alternatively, prestressed connections may be employed, using short bars. Dry-fit joints between superstructure elements are particularly expeditious. Provision must be made, however, to accommodate tolerances in pile position and elevation. Extremely clever schemes have been developed for this purpose: sleeve joints, jackets, etc. Use of extra-wide cap girders permits greater tolerances in pile location. Particular care must be taken not to overstress or break a pile by attempting to pull it into position.

Quick-setting grout and epoxy compounds help to speed completion of these connections.

Prestressed lightweight concrete may facilitate construction, permitting longer spans or allowing the use of smaller and lighter construction equipment.

Particular attention must be given to insuring the durability of joints and connections. These connections are frequently made under adverse conditions. The painting of all joints with epoxy is an excellent way of sealing them against moisture entry.

Furthermore, it is recommended that, on the completion of a coastal structure, a detailed inspection be made of all connections to make sure of their completeness and tightness, for it is at these connections that most problems occur.

Obviously, coastal structures are more subject than other structures to salt spray and the possibility of salt-cell electrolytic corrosion. Pile caps have been particularly vulnerable. Extreme care, therefore, should be taken to follow the procedures for maximum durability in the splash zone (See Section 12.3 in this chapter, under Durability; also Chapter 4, Durability and Corrosion Protection).

12.7.4 Pile-Supported Bridge Piers

Caissons and similar structures which are used as bridge piers are covered under Chapter 14, Prestressed Concrete Floating and Submerged Structures.

This section will discuss the use of prestressed concrete piles, especially cylinder piles and the larger-diameter caisson piles, with specific reference to their incorporation in overwater bridge substructures. Reference is also made to Chapter 11, Prestressed Concrete Piling, and to Cylinder Piles under Subsection 12.7.1 in this chapter.

Pile supports for bridge piers must resist heavily concentrated vertical loading, plus lateral loads from traffic, wind, seismic forces, currents, waves, debris, and ice. This combination of vertical load plus bending requires careful consideration and calculation in design. It affects both the shape of the pile cross section and the degree of prestress. In many cases, it is desirable that piles for bridge piers have higher prestress values, say, up to 0.15 to 0.20 of f'_c, in order to most efficiently resist these combined loadings.

Previous sections have stressed the need for adequate spiral reinforcement to resist tension and bursting stresses.

In debris-laden harbors, or when impact from small boats and ice, is possible, cylinder piles may be strengthened against local damage by filling with sand or concrete. Chapter 11, Prestressed Concrete Piling, sets forth the precautions to be followed in such filling so as to prevent longitudinal cracking. In cold climates, the sand fill should be graded to be non-frost-susceptible. Cylinder piles must also be designed so as to prevent damage from internal freezing in cold weather. (See Chapter 11, Prestressed Concrete Piles, for techniques.)

Large-diameter cylinder piles and cylindrical caissons (up to 14 feet diameter) have been used to support major bridge piers.

They have been installed by combinations of jetting, dredging of the interior, weighting, vibration, and hammering. Bentonite injections may be used to reduce the skin friction in the upper zones. The dividing line between cylinder piles and caissons is one of degree only, as the two blend together in both structural and constructional consideration.

When high-capacity piles are employed for bridge piers, it becomes necessary to install them within a minimum tolerance to prevent eccentricity in the pier itself. Thus, there is a trend to the use of templates and guides in order to enable the accurate initial setting of the pile.

In deep water, cylinder piles for bridges may require underwater bracing in the form of tie-strut connections or jackets. (See Subsection 12.7.7, Offshore Platforms.)

12.7.5 Navigation Structures, Dolphins, and Bridge Fenders

Prestressed concrete has been rather extensively used for navigation structures, because of strength and durability, and because of simplicity of structural concept, thus permitting maximum prefabrication, simplified connections, and overall economy (See Fig. 40, page 210).

Prestressed lightweight concrete has been chosen for many of these structures; among other reasons, because of its greater flexibility, that is, deflection under load. White cement and, on occasion, white quartz aggregate, has been used to give greater visibility. It would seem that panels could be made with exposed glass fragments or beads to improve the reflective power.

Such structures are frequently moored to by small boats, and may be hit by boats and barges. Thus the design must provide adequate ultimate strength to prevent the pile from snapping off. This can best be done by using additional unstressed strands and ample spiral.

Dolphins are subjected to horizontal loads only. The piles should be highly prestressed, and can use from 0.20 to 0.30 f'_c as effective prestress, while maintaining a balanced design. Such high prestress values require more spiral binding.

Dolphin piles should be selected, as fender piles are, to give maximum energy absorption. This means a compromise selection of high-strength concrete with low modulus of elasticity, criteria which structural lightweight aggregate concrete best fulfills. The cross section similarly needs maximum strength with minimum moment of inertia—a square section is often selected.

Prestressed concrete is increasingly used for the protective fenders of bridges. This use differs in purpose from wharf fenders: for bridge fenders, the primary aim is to provide maximum protection to the pier. Prestressed concrete is fire-resistant, a very important fact to be considered when selecting the fender for a steel bridge. It is durable, resistant to abrasion and impact. As for all structures subjected to predominantly lateral loads, use of lightweight aggregate concrete and higher degrees of prestress are desirable. Hollow-core piles and cylinder piles may be filled to protect against local damage from debris and ice (see previous subsection).

For bridge fenders, the criteria of minimizing damage to the vessel, and ease of repair of the fender after damage, are secondary. Therefore, the top portions of the pile may be encased in heavy ring girders of concrete, either prestressed or conventionally reinforced. These may be precast sections connected to the piles by mild steel with poured concrete joints, or by post-tensioning.

12.7.6   Overwater Airports

These may be constructed either as floating structures of prestressed concrete (See Chapter 14, Prestressed Concrete Floating and Submerged Structures), or as fixed structures supported on piles. The overwater extension of runways and taxiways at LaGuardia Airport, New York, is a notable example of the latter type. The prestressed concrete deck structure consisted of precast cap girder segments, pretensioned half-depth deck slabs and cast-in-place deck. The cap girders and deck slab were then post-tensioned to achieve monolithic behavior capable of supporting the extremely heavy design loading.

The deck slabs were shallow, inverted double-tee sections, with holes through the webs, through which the post-tensioning tendons were later run. The cap girder segments were jointed over the pile heads, with a poured-in-place joint. Ducts crossing these joints were spliced and taped to prevent grout in-leakage. Stage stressing was employed; the first stage to support the deck slabs and the concrete topping, and the second stage to support the live load. This meant that a considerable time interval occurred between placement of tendons and stressing and grouting. VPI powder was dusted on the tendons during their insertion and the ends sealed. The precast cap segments were located just about at the high-tide line, so prior to insertion of the tendons, they were flushed with fresh water. In the light of subsequent knowledge, this particular environment is so adverse for possible salt contamination and salt-cell formation that more positive steps would probably have been justified; however, no corrosion had been reported four years after construction.

This combination of precasting, prestressing (both pretensioning and post-tensioning), and composite construction proved extremely satisfactory and showed its great value for decking large areas for heavy impact loadings. (See Fig. 55)

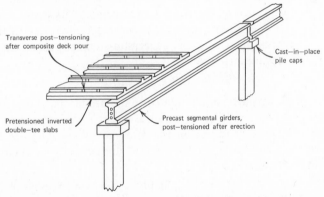

Fig. 55. Precast prestressed deck concept utilized for overwater extension of LaGuardia Airport.

The interaction of piles and deck under heavy concentrated loads was of great importance in design, in minimizing individual pile loads and in absorbing the impact energy under the design parameter of a collapsed landing gear. Lightweight concrete piles may offer advantages in this regard for future construction because they combine low modulus with high strength, permitting greater shortening of the pile under the concentrated load and, thus greater load distribution.

Experience gained on this huge structure would indicate the following lessons for the future:

(a) The precast concept is sound, economical, and rapid. It can be economically extended even to non-typical areas.

(*b*) The combination of pretensioning for precast members with two-directional post-tensioning of the composite structure is a brilliant design technique and highly practicable.

(*c*) The large tendons in the cap should be of a system that requires minimum size ducts; that is, the tendons should be capable of being inserted prior to placement of the end anchorage.

(*d*) Special detailing and specifications should be directed to insure the utmost in durability.

(*e*) Reinforcing steel detailing should aim at standardization and simplicity: on this project, the labor costs of fabricating and tying the mild steel were very high.

(*f*) Joints are the costliest item; therefore, the deck slabs should be as wide as possible in order to reduce the number of joints.

(*g*) Adequate depth of concrete (cover) above the stems of double tees and reinforcing steel must be provided to prevent shrinkage cracks in the deck surface.

Other schemes of deck will undoubtedly be developed and used for overwater airports, such as the combination cap-slab units discussed under Subsection 12.7.1 **f.** of this chapter. This construction concept at La Guardia, however, represents a bold and successful step forward in construction concepts for overwater airports.

12.7.7 Offshore Platforms

Offshore platforms are mainly related to the provisions of services for the oil and gas industry, although the sulfur industry utilizes them also. As means are found further to exploit the resources of the continental shelves and adjacent coastal water, platforms will undoubtedly be required for a wider variety of industries.

Most offshore platforms consist of piling, jackets or other below-water framing, deck sections, and appurtenances. Prestressed concrete has been so far utilized extensively for piling in locations where design wave forces are not extreme (e.g., Lake Maracaibo), for deck sections of moderate size, and to a very limited extent for jacket framing. Prestressed concrete offers several fundamental advantages wherever construction engineering has developed means to utilize it on offshore platforms: durability, economy, fire-resistance, and maximum use of local materials and labor.

a. *Decks*

Offshore platforms are used to support production drilling equipment and processing equipment, including compressors, storage and pumping equipment. Many of the dead loads imposed, therefore, involve vibratory and dynamic loading. Mass is, therefore, a desirable feature in the deck structure.

Platforms are subject to extremely heavy lateral forces primarily from storm waves, but also from wind and current. The pin piles normally used to support the structure must function in both tension and compression. In many locations, the soils are such that the determining factor for pile penetration is the uplift requirement, thus requiring drilling and other very expensive and time-consuming methods for pile installation. In such instances, heavier deck structures may be an advantage.

On the other hand, weight of a deck structure is a disadvantage under seismic loadings. It is also a disadvantage in handling and setting, requiring more pieces than lighter-weight structures do.

Prestressed concrete does offer substantial advantages for deck structures where the loads are relatively fixed in location. The platform structure can be manufactured and assembled ashore, transported and set in segments, and connected by post-tensioning. This application would seem to lend itself to the use of dry joints with epoxy coating (See Chapter 3, Subsection 3.2.4). Connections can thus be effectively and expeditiously completed and will have excellent performance under dynamic loading. Such deck segments can be highly refined in both design and construction; use of high strength concrete in certain members combined with lightweight concrete in others is a promising approach. Deck girders may be framed or Virendeel trusses, deck slabs may be waffle-ribbed or hollow-core. Deep deck girders may be post-tensioned longitudinally and, in addition, have their webs post-tensioned vertically with bars, to resist high shear.

To meet the restrictions of lifting capacities in construction yards and offshore, one system is to make up the deck structure in reasonable-sized segments of, say, 30 tons maximum weight. These can them be assembled and connected on a barge into units of approximately 250 to 350 tons. Thus trusses may be formed of precast chord and web segments. Final setting in location can therefore utilize the 500-ton capacity of many existing large offshore construction derricks.

The most-used material for decks of offshore platforms is, of course, structural steel. However, the growing tendency in many areas to specify steels that are impact-rated for low temperatures, the increasing sophistication of welding requirements, and the cost of painting for this adverse exposure, all make the cost of steel deck sections extremely high. Thus, there is a steadily increasing interest in prestressed concrete.

One of the problems in fitting any prefabricated deck structure, steel or prestressed concrete, is the accommodation of pile tolerances. Fortunately, most offshore structures employ templates to keep piles within acceptable tolerances, and sleeves in the deck structure or other means are provided for final adjustments. The problem is no more severe with prestressed concrete than with structural steel.

b. *Piling for Offshore Platforms*

Piles are normally subjected to very high axial loads, both compression and tension, as well as bending moments. For these reasons, steel pipe piles have been used very extensively. In shallow water and where the wave criteria have not been so extreme, prestressed concrete cylinder piles have been employed, especially in Lake Maracaibo. These piles offer the benefits of economy, durability, and adequate strength for combined loading. Prestressed cylinder piles have been used for conductor piles.

Such piles are usually friction piles. Their installation may be accomplished by weighting and driving, aided by progressive removal of the soil plug and by jetting. In many soils, jetting weakens the capacity of the soil and is thus severely restricted. On the other hand, the author has encountered a number of cases where the jetting improved the soil capacity, washing out some of the finer silts, allowing the coarser sand grains to be consolidated around the pile by the vibration due to driving.

If the piles are essentially in cantilever or rigid frame action, fixed at the deck and fixed at some distance below the mud line, then the maximum moment will occur at a point slightly below the mud line and at the deck. Similarly, when a bracing frame or jacket is employed, the maximum bending moment occurs in the zone near and just below the bottom. This is the zone of pile-soil lateral interaction.

Additional moment capacity can readily be provided at the deck by either an internal steel core or an external sleeve. At the mud line, however, increased bending resistance requires careful planning. Increased moment resistance can be provided for this zone by one of the following means:

(*a*) Mild-steel bars added to the reinforcement of the cylinder pile wall, extending through and beyond the zone of maximum moment.

(*b*) An inner steel core, lowered down the cylinder pile to proper elevation and grouted.

(*c*) An external steel sleeve, jetted down around the pile and connected by grout.

(*d*) Tapered pile sections. These have been extensively employed with large rectangular conventionally reinforced piles in Lake Maracaibo, but so far not with cylinder piles, mainly because the early cylinder piles were assembled from uniform cylindrical segments. Actually, the manufacture of cylinder piles by the pretensioning method lends itself to the use of tapered piles, giving maximum sections at point of maximum moment. Tapered piles could also readily be formed by the post-tensioning together of precast segments, provided the segments are manufactured so as to fit the taper.

(*e*) Installing a prestressed cylinder pile to moderate depth, then driving an inner pile of prestressed concrete or steel pipe through and beyond it, overlapping the two in the zone of maximum moment, and connecting the two shells with grout. (See Fig. 56)

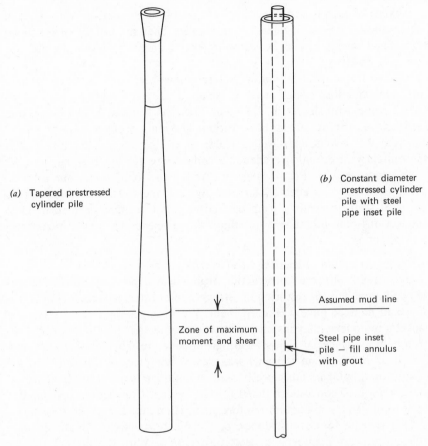

(a) Tapered prestressed cylinder pile

(b) Constant diameter prestressed cylinder pile with steel pipe inset pile

Zone of maximum moment and shear

Assumed mud line

Steel pipe inset pile — fill annulus with grout

Fig. 56. Special prestressed piles for offshore structures.

Abrupt changes in pile stiffness as a result of steps (a) to (e) above must be given careful consideration in location and detail so as not to create a new weak point.

c. Jackets and Bracing

The typical offshore platform employs jackets made of tubular steel sections. Such jackets can be very large and heavy even in steel (up to 3000 tons and more). The larger ones are usually transported by special barge, launched over rocker arms, following which they barely float. They are then up-ended by the derrick and controlled flotation, and set on the bottom. So-called tower jackets utilize much larger legs and are usually transported to the site floating on the lower legs. At the site, they are up-ended by flooding and derrick control.

Jackets of prestressed concrete have so far been used only to a small extent and have been of relatively small size. They are usually of the tower type. They have been built in basins or graving docks, floated to the site, and sunk into position.

Bracing frames of steel and of concrete have been used for both offshore platforms and deep bridge piers. These are usually confined to a single horizontal plane, with sleeves for the piles, They can be lifted off the barges and set in place, or else floated into position. The main problem with the use of concrete for jackets and bracing fraes, other than weight for handling, is the difficulty in developing adequate connections for the bracing, whether truss or girder type. Post-tensioning through the joints offers one solution which is particularly effective for truss web members. Another solution is to use steel stubs protruding from the ends of the truss members, so that the connections may be made by welding, then encasing the joints in concrete for protection.

d. *Reconstruction of Exisitng Steel Platforms*
One major offshore construction firm (Brown & Root, Inc.) has developed a method for restoring the strength of old steel jackets, where corrosion and damage have weakened them. Working generally from inside the jacket's main legs, they install post-tensioning ducts in the bracing tubulars, then pump in a special high-strength grout-concrete. The main legs are reinforced as necessary at the anchorage ends, tendons are inserted, and the composite tubular member prestressed. Advantage is taken of the compressive strength of concrete confined in steel tubing. Connections are detailed so as to obtain the maximum reinforcement from multi-axial prestressing.

This scheme for reconstruction may indicate means of obtaining more efficient tubular members in new construction. With the present tubular design, the force from waves is proportional to both the projected area and the displaced volume of water (i.e., virtual mass). Relatively thin-walled steel tubes filled with concrete and post-tensioned may offer a more economical technique in some cases for obtaining the maximum axial strength and most efficient connection details, particularly in cold weather environments.

e. *Caissons*
Caissons of conventionally reinforced or prestressed concrete have been proposed for many years for offshore platforms. Interest is being revived because of additional requirements for offshore storage, and the newer criteria for resisting ice pressures in the Arctic Ocean and other northern waters. Such caissons would be manufactured in basins in warm water ports,

then towed to the site and sunk. They would be secured in place by filling with gravel ballast, by piles, or by post-tensioning to underlying soils or rock.

Prestressing could be advantageously used in caissons for ring reinforcement and for vertical reinforcement, depending on the specific requirements involved.

Caissons of this type have been proposed for Offshore Terminals in the Arctic Ocean and for the piers for the proposed Intercontinental Peace Bridge joining Alaska and Siberia across the Bering Straits. Smaller but similar caissons, employing a varying extent of prestressing, have been installed for lighthouses offshore Scandinavia and Eastern Canada.

13

Prestressed Concrete Bridges

13.1 General Provisions

Many of the most notable successes of prestressed concrete have been in the construction of bridge superstructures. Its use offers advantages of low first cost, low maintenance, high durability, and attractive aesthetics.

Prestressed concrete has been successfully employed for bridges ranging from short spans to long spans, in all environmental conditions, from tropical to sub-Arctic and desert to rain forest. It has been widely utilized in crowded city viaducts and for isolated structures in underdeveloped lands.

Prestresed bridges may be precast or cast-in-place. Precast elements and segments may be joined with other elements or with cast-in-place concrete for composite, monolithic action.

Prestressed concrete is also, but less widely, utilized for piers, columns, abutments, struts, suspenders, etc.

For short-span bridges (up to 140 feet) precast prestressed bridge superstructure units have been standardized. These include I-girders, T-girders, flat slabs (both solid and hollow-core), box units, and channels. They are usually single spans, designed to act in composite action with a cast-in-place deck slab, and may be made continuous over the supports by reinforcement in the deck slab. (See Fig. 57)

A Joint Committee of the American Association of State Highway Officials Committee on Bridges and Structures (AASHO) and the Prestressed Concrete Institute (PCI) has produced a set of standard I-girders, of six types, for spans from 30 to 140 feet. Similarly, a joint committee of the American Railway Engineering Association (AREA) and the Prestressed

252

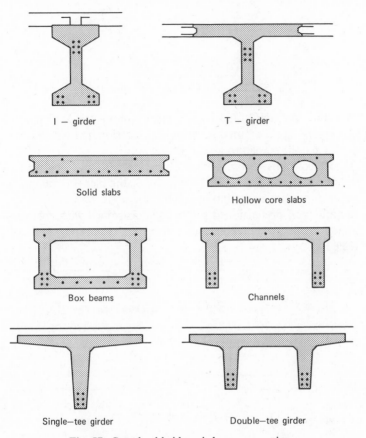

Fig. 57. Standard bridge girder cross sections.

Concrete Institute has produced a set of standards for bridge deck units for railroad trestles and bridges.

For medium-span bridges (up to 250 feet or so), these same units may be employed by using the cantilever-suspended span scheme, the hammerhead scheme, or inclined struts, etc. Segmental construction techniques are also frequently utilized to extend the range of the standard girder sections.

A recent study was made by the Prestressed Concrete Institute with Lin, Kulka, Yang, and Associates as consultants, to select and present methods of obtaining spans from 100 to 170 feet in length for highway overpasses utilizing standard precast girder segments. It assumes the maximum transportable length is 100 feet. The study is entitled "Prestressed Concrete for Long-Span Bridges," published by the Prestressed Concrete Institute, Chicago, 1968.

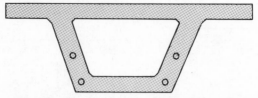

Fig. 58. Box girder section for rapid transit.

The standard AASHO-PCI I-girders are among the segments studied. Where necessary to accommodate large post-tensioning ducts, the standard girder forms may be spread to give a thicker web.

The report contains recommended splice details, together with calculations for splices at mid-span, third points, over-the pier (maximum negative moment), and at inflection points.

Combinations of pretensioned precast girders joined with post-tensioning were found to offer practicable and economical solutions to these longer spans. (See Fig. 59).

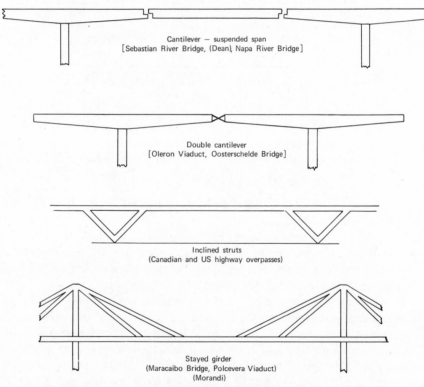

Fig. 59. Means for extending span ranges in prestressed concrete (a).

Cast-in-place bridges may be constructed on scaffolding and prestressed. These may be box girders or tee girders. Box girders have especially favorable action for torsional resistance for bridges built on a curve. Attractive aesthetical results may be achieved by sloped webs.

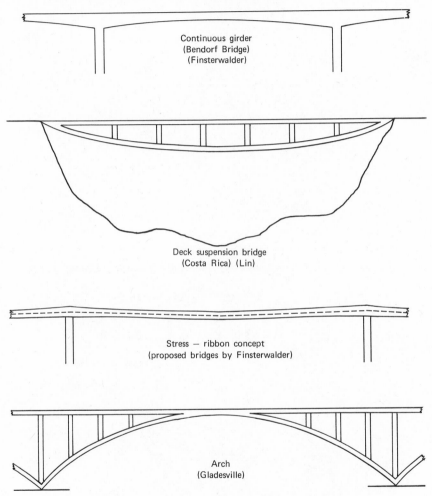

Continuous girder
(Bendorf Bridge)
(Finsterwalder)

Deck suspension bridge
(Costa Rica) (Lin)

Stress — ribbon concept
(proposed bridges by Finsterwalder)

Arch
(Gladesville)

Fig. 60. Methods for extending span range in prestressed concrete (b).

Longer-span bridges, up to 700 feet, have been built of cast-in-place concrete or precast segments, cantilevered out section by section, each being prestressed back to the balancing extension on the other side of the pier. Some of these bridges have been quite spectacular achievements and have, justifiably, received fame. (Figs. 59, 60, and 61 and Postscript Section viii).

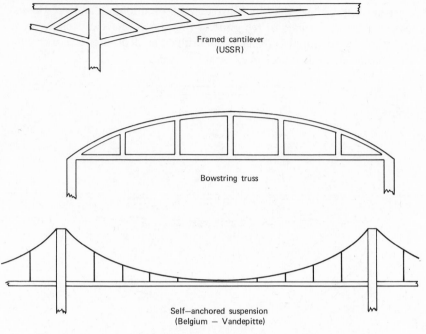

Framed cantilever
(USSR)

Bowstring truss

Self—anchored suspension
(Belgium — Vandepitte)

Fig. 61. Methods for extending span range in prestressed concrete (c).

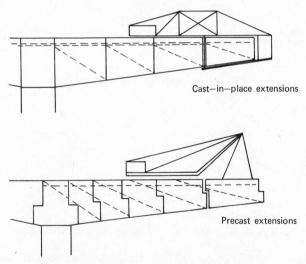

Cast—in—place extensions

Precast extensions

Fig. 62. Method of progressive cantilevering.

Extremely long spans have been built (750 feet) and designed up to 1000 feet, using prestressed suspender struts. The Lake Maracaibo Bridge and the Polcevera Viaduct in Genoa are of this type.

Finsterwalder has proposed a stress-ribbon concept in which a long-span bridge would consist of the concrete slab encasing the highly-stressed tendons.

A similar concept, evolved from the cable-suspended roof concept, has been utilized by T. Y. Lin for a bridge across a gorge in Costa Rica. Precast units are hung on the cables, and a precast viaduct with level roadway posted up from this.

For columns and shafts, precast elements may be constructed by erecting precast segments on top of one another or cast-in-place, using sliding or panel forms. Prestressing is employed to counter bending moments due to lateral forces.

Cap girders have been manufactured in precast segments, set and stressed together and to the pier, so as to act monolithically.

There is, and will undoubtedly continue to be, a debate over the relative merits of cast-in-place and precast concrete.

Cast-in-place has these advantages:

1. Fewer joints.
2. Easily constructed to horizontal and vertical curves and flares.
3. Large and complex cross sections and varying span lengths are readily accommodated.

Precast segments have these advantages:

1. Control during manufacture and inspection before erection permits higher strength, closer tolerances, thinner sections.
2. Shrinkage may be largely completed prior to erection.
3. Less creep, due to greater age at time of stressing.
4. Ability to easily compensate for dead-load deflection by wedging, jacking, and shimming precast segments to theoretical profile, before final jointing.
5. Minimization of falsework.

However, precast construction requires jointing and this must be performed with technical care and control.

The ultimate choice will depend on the final economy. It is to be emphasized, however, that adopting a single design, suitable for construction by either precast or cast-in-place methods, is not conducive to economy. Only by making full use of the properties and advantages of each method, and reflecting these in a specific design, will true economy be achieved.

13.1.1. Jointing and Connections

Properly detailed and executed joints and connections are essential to the success of concrete bridge constructions

Joints and connections must be so detailed as to meet the design requirements for transfer of bearing, shear, moment, etc., and, in addition, must be practicable of construction and inspection under the actual conditions at the site.

With stepped-end girders, as in cantilever-suspended span construction, high stresses develop due to the combination of direct bearing and anchorage stresses. Closely spaced stirrups are recommended and the tendon anchorage should be as low as possible.

Joints will generally be visible and thus will affect the appearance of the bridge. Well-designed and constructed joints may often be utilized and even emphasized, so as to enhance the appearance of the structure.

See Chapter 3, Subsection 3.4.1, Segmental Construction and Joints.

Consideration must be given to fatigue loading in all joint design, especially welded reinforcing steel splices and welded structural steel splices. Tendon splices should have satisfactory behavior under cyclic (fatigue) loading as shown by manufacturer's tests and guarantees.

Joints in precast segments which are subject to a reversal of stress at any section under design loading should be detailed so as to provide adequate restraint against movement, or else the joint shall be so detailed (with bearings, pads, or expansion joints) as to prevent hammering or other fatigue conditions.

Precast segments may frequently be combined with cast-in-place concrete to act in composite action, as a monolithic structure. Provision must be made for shear transfer. The precast units may serve as partial forms, and also as support for forms, for the cast-in-place concrete.

Consideration must be given to the various stages of loading to prevent over-stressing of the precast elements while the cast-in-place concrete is still wet. The constructor must consider the effect of differential shrinkage and of different moduli of elasticity.

See also Chapter 3, Subsection 3.5, Composite Construction.

Transverse post-tensioning may be beneficially employed in bridge decks to prevent sag, especially in cantilevered overhangs supporting a heavy curb.

13.2   Erection

Particular attention is directed to Section 3.4 (Lifting and Erecting Precast Elements).

The majority of erection methods and techniques which have been developed and employed for prestressed bridge construction fall into the following classes:

13.2.1   Crane Erection of Precast Girders

This includes erection by land-operating cranes, by derricks on water, and by cranes or derricks mounted on the structure itself.

When using inclined slings, the temporary buckling stresses due to increased compression in the top flanges of the girder must be considered, as well as the increased forces on the lifting loops or devices. Angles of force should be considered for each position during the lift. Lifting loops must be suitable for all angles of lift to prevent localized crushing or over-stress.

Land cranes must have firm under-support, adequate for the concentrated temporary loads under their tracks, wheels, or outriggers. The position of the crane, angles of lift, and working radii must be plotted on working drawings and accurately laid out and enforced in the field.

When two cranes are used to erect a single member, each should have capacity to take at least 66% of the total load, and precautions should be taken to prevent undue swinging and side-pull on the booms, and to insure that the girder does not hit one of the booms during the successive steps of rotation of the booms.

Derricks or cranes mounted on the structure must be properly secured and the temporary loadings imposed on the structure, including torsion, must be checked.

Care should be taken to prevent either the unit being lifted or another part of the structure from hitting the boom, as this may cause the boom to buckle.

Water-borne derrick barges should be checked for capacity during all stages of lift and placement, with due allowance for list due to load and wave action. (See Fig. 63.)

The list of a water-borne derrick or crane tends to surge it out of position laterally, putting an added strain on anchor lines. Also, the rotation of the revolving crane or derrick while listing puts added strain on the barge, the derrick base and roller path, and the swing engines, and also produces torsion in the boom. The list also increases the actual picking radius as the load drifts outward and, thus, may overload the crane. Before picking near-capacity loads, therefore, a thorough engineering check must be made for all phases of the pick.

13.2.2 Floating-In

Either entire spans or major portions thereof are built or assembled on scaffolding on a barge, then towed to the site, moored in exact position, and lowered onto the bearings.

Large single barges may be used, or multiple barges may be joined with trussing. The effect of differential movement, due to waves, must be considered in its effect on the precast span or element. Wind forces on the barges and spans must be taken into consideration.

Lowering may be by utilizing the tides, flooding of compartments in the barges, or jacks. In flooding, the effect of the free surface on stability must be considered. This usually requires that the barges or pontoons be compartmentalized.

Stability must also be carefully calculated during transport on the barges because of the great weights involved and the height of center of gravity.

With substructure elements, buoyancy may be provided within the element itself, and sinking accomplished by adding ballast, such as gravel, iron ore, concrete, or sand in compartments, or by flooding isolated compartments. Alternatively, positive buoyancy may be maintained and the element submerged by pulling or jacking against pile anchors.

Substructure units may also be transported *under* a barge or pontoons. In such a case, the unit may be constructed in a drydock or basin, which is then flooded, the barge floats in over it, picks it and carries it underneath itself until in position. This method takes advantage of the reduction in dead weight due to submergence, has inherent stability, and all lifts and lowerings are direct.

13.2.3   Erection on Falsework

A steel or aluminum truss is placed in position and the elements are lifted one by one, for example, by crane, onto the falsework. When an entire span unit is erected, the precast units are jacked and shimmed to the exact profile, then joined and stressed. This method is especially adapted to the case of parallel girders, because after one girder is erected and stressed, the stressing automatically decentering the falsework, the falsework span may be moved sidewise for the next parallel girder.

Alternatively, the falsework truss span may be above the final girder location, the precast units being raised from barges into position by hoists, and held until jointed and stressed.

13.2.4   Launching Gantry

This method involves the use of a special erection or launching gantry, which may include means for moving itself forward as portions of the bridge are completed.

Under one system precast segments are moved forward at deck level from one abutment, out over the completed superstructure. The segment is then picked up by the launching gantry and carried forward. To enable it to pass through the supporting legs, it is usually rotated at right angles to its final position during movement, then turned back and set in its final position.

The individual precast segments are usually jointed and stressed before the next segment is launched. See Chapter 3, Subsection 3.4.1, Segmental Construction and Joints.

When using a launching gantry to erect prestressed girders, provision must be made for lateral transfer of the girders after they have been moved into their span. Rollers, wheels on tracks, skidding with jacks, etc., are often employed to accomplish this lateral transfer. Positive stops must be provided to prevent the girder being moved beyond the end of the cap through accident.

Launching gantries are major steel bridges in themselves, subject to reversal of stress conditions as they are moved, and to impact as they handle the precast segments. Since the connections are usually field bolted, it is important that provision be made for frequent inspection of all joints, and repair or strengthening of any members accidentally damaged. If high-strength bolts are used, a clear identification marking must be placed on them to prevent careless replacement by a conventional bolt.

Safe walkways and, where applicable, moveable safety nets or platforms should be provided as an integral part of the launching gantry.

13.2.5 Direct Launching

This scheme has been employed to move precast girders lengthwise from a completed portion of the superstructure to their span location. A light steel or aluminum launching nose is overbalanced by a counter-weight or heavy rearward extension. Movement forward may be accomplished by jacking, rolling, tracked carriages, or cranes. The girder must be analyzed for temporary stress conditions as it is cantilevered forward and, if necessary, strengthened by external trussing or internal reinforcement. This method is particularly suitable for a single span in remote locations.

The same principle has been used in Venezuela, Germany, France, and the USSR, to launch an entire series of spans of prestressed concrete, the girders being approximately uniformly stressed to take care of moment reversal until they are in final location. Then the tendons are deflected up and down to their permanent profile, or additional curved tendons added. Special 'frictionless' bearings of teflon on chrome-nickel steel plates are used on top of the piers. The piers may require temporary guys or stays during the launching operations.

13.2.6 Cantilever-Suspended Span

The precast hammerhead section may be floated in or lifted in, supported by barges or cranes at each end, and set on the pier. Temporary stresses as a simple span must be countered by external or internal reinforcement. (See Fig. 63)

Since this is the section subject to maximum negative moment, its required final prestress force will generally be very high. During this stage, before the adjoining suspended spans are set, the tensile stresses in the bottom may exceed allowable limits. Stage stressing (See Chapter 2, Subsection 2.2.2) may be employed, or additional internal reinforcement provided

Fig. 63. Prestressed concrete girders being erected, Napa River Bridge, California.

in the bottom of the girders, or external structural steel beams bound to the segment.

Smaller hammerhead girders may be hoisted by one crane lifting at the center.

Stability may be provided by stressing temporarily or permanently to the pier shaft or by an inclined leg support from the pier base or by falsework towers at one or both ends.

Precast girders are particularly adaptable to suspended spans. They may be lifted in with one or two cranes working from below or from the cantilevered ends of the superstructure, or they may be moved forward on a falsework truss or by launching gantry.

Suspended spans may be assembled from precast segments on barges, and floated or lifted into place.

13.2.7 Progressive Cantilevering

This is an extremely useful method for construction of concrete bridges with precast or cast-in-place segments. As each segment is placed, it is jointed and stressed back to the completed portion of the superstructure.

The sequence of erection is chosen to keep the partially completed superstructure balanced about a pier, in double-cantilever.

To facilitate setting of a precast segment, a step or ledge may be provided on the previous segment so that the new segment can be readily set into exact position. Erection bolts should be provided so the segment can be pulled into exact position and held.

Dry joints and epoxy joints [See Chapter 3, Subsections 3.4.3(e) and (f)] are particularly adaptable to use with progressive cantilevering as they enable each segment to be joined and stressed as one continuous operation, usually in one day. Other types of joints may be employed, with accelerated curing so as to minimize delays between successive segments.

Temporary suspension of the cantilevered segments may be provided by external tendons, e.g., cables running up to a temporary tower above the pier. Stability during erection of the cantilevered arms may be provided by temporary vertical stressing down to the pier, or by inclined legs or falsework towers.

13.2.8 Sliding of Segments

Precast segments may be slid forward to their position in the span, sliding on skids or rails or rollers over falsework trusses, or falsework girders. Similarly, they may be slid along temporary or permanent wire rope cables to their correct position in the span.

Special erection techniques, usually associated with building erection, are set forth in Chapter 10, Prestressed Concrete Buildings. These include more detailed provisions for sliding and rolling in of girders, and provisions relating to the use of tower cranes, steel erection derricks, and helicopters.

13.3 Schemes for Bridge Construction Utilizing Precast Segments

Precast segments can be divided longitudinally, transversely, or horizontally. Transverse sections are generally, but not always, selected so that a fully stable cross section results. This may be the full transverse cross section of the bridge. Longitudinal sectioning results in girders, such as the solid or cored slab, channel, I, T, or box section, which may be joined in composite action by a cast-in-place composite deck. Alternatively, they may be jointed and stressed together transversely. Horizontal sectioning is generally used where precast flanges or deck sections are connected with cast-in-place webs, or conversely, where precast webs are joined to cast-in-place decks and bottom flanges. (See Fig. 64).

Bridge concepts utilizing precast segments have usually been one of the following types:

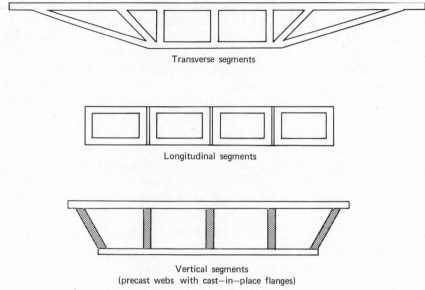

Transverse segments

Longitudinal segments

Vertical segments
(precast webs with cast—in—place flanges)

Fig. 64. Types of precast segmental bridges.

(*a*). Simple span.
(*b*). Continuous girder.
(*c*). Cantilever-suspended span.
(*d*). Double cantilever.
(*e*). Arch rib.
(*f*). Flexible suspension.
(*g*). External cantilever or rigid suspender type.

Trusses and framed construction are being increasingly utilized in Europe for long-span bridges. The precast segments are jointed at the ends and are welded, pinned, or post-tensioned through the joint. The latter appears to offer superior possibilities.

Precast deck sections may be assembled in such a way that the main longitudinal and transverse stressing is achieved by tendons located in the joints. The precast sections are first assembled on falsework or, in some cases, on cables. The tendons are then placed through the zone of the joint, the joint is poured, and the entire assembly stressed. (See Fig. 65)

13.4   Horizontal and Vertical Curves with Precast Concrete

Longitudinal precast girders, reinforced by mild steel or by prestressed tendons,can be fabricated on a horizontal curve. Bracing must be employed

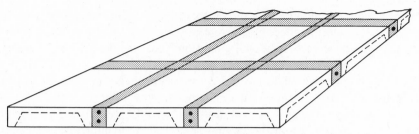

Fig. 65. Precast slabs with post-tensioned tendons in joints.

to ensure torsional stability during erection and until permanently secured in the structure. Both box and T sections have been employed—the wider flanges facilitate this type of curvature.

Vertical curvature can be provided during manufacture by adjustment of the soffit forms. (See Postscript ix)

With transverse segments vertical and horizontal curvature may be achieved by varying the joint width progressively. For dry joints, the change in curvature can be accomplished in manufacture by adjusting the already cast segment to its new angle before pouring the adjoining segment.

13.5 Integration of Design and Construction

The most effective use of precast concrete elements can be made when their use is contemplated in the original design and, thus, full advantage taken of the specific properties and characteristics of the precast method. Mere substitution for cast-in-place concrete will not normally develop the full advantages of precasting.

The erection of long-span concrete bridges involves a full understanding of all factors involved and analysis of stress conditions at all stages from manufacture through transport, erection, concreting, and final design load conditions. Thus, regardless of how achieved, design and construction must be integrated. The designer must follow through all construction stages, and the constructor must be aware of the effect of each of his operations on the final results as they affect the design. Whether the designer and constructor are in different organizations, as in conventional American practice, or in the same organization, as in European practice, there must be full collaboration if the best results are to be attained.

The most efficient use of precast elements can often be obtained by combining precast elements with cast-in-place concrete to work in composite action as a monolithic construction. By such an approach, the advantages of precasting can be employed in complex and variable-shaped structures.

Such a combination approach requires consideration of the different moduli of elasticity, and the effect of differential shrinkage and creep, and analysis must be made of each stage and condition; thus, the designer is inherently involved and concerned with the constructor's methods.

The ingenuity of both design and construction engineers has been responsible for the development of outstanding concrete bridge concepts, in which precast concrete elements were utilized to their full advantage. This ingenuity has demonstrated itself in the matter of design, manufacture, transport, erection, and jointing.

Since erection concepts will often determine the size and configuration of the precast segments and since erection is primarily a responsibility of the constructor, he should be permitted as much freedom as practicable in selection of the size and shape of precast segment within the general restrictions of the design. The importance of encouraging ingenuity on the part of designer and constructor should be given consideration during the preparation of plans, specifications, and criteria. This may be facilitated by the designer conferring with local fabricators and contractors during the design stage to discuss availability of erection equipment, special techniques, and cost comparisons. The thorough cooperation and coordination of both designer and constructor is essential if the maximum advantages are to be achieved from precast concrete.

13.6   Cast-in-Place Prestressed Concrete Bridges

Cast-in-place prestressed concrete bridges are extensively utilized for medium and long-span bridges, especially when the bridge alignment must incorporate horizontal curves. The cross sections most widely employed are T-girders, with the web and lower flange usually of the same width, box girders of rectangular cross section, and box girders of trapezoidal cross section. The latter is extremely efficient in its use of materials and lends itself to rather elegant and aesthetically pleasing solutions.

Most medium-span, cast-in-place concrete bridges are constructed on falsework scaffolding. Continuity is almost always employed for bridges greater than one span. Usually, such a bridge is made continuous for 4 spans or more. Because of maintenance problems and special costs associated with expansion joints, current practice tends to stretch out the length for continuity as far as possible, even around horizontal curves. In construction, therefore, a carefully planned sequence of tendon stressing must be followed, with the use of splices and overlapping tendons as detailed on the plans. Such a bridge lends itself to a sequential operation where, with careful planning, the prestressing crews may be utilized at a relatively constant level of work force over a substantial period of construction. The effect of expansion and contraction during construction must be considered for each stage.

The problems in cast-in-place prestressed bridges are usually associated with shrinkage, settlement of scaffolding, and differential properties of the concrete (modulus of elasticity and creep). Shrinkage can be controlled by careful selection of the mix, a low water/cement ratio (including use of a water-reducing admixture), and especial attention to curing. Sometimes, to prevent transverse shrinkage cracks, a small amount of prestress is imparted at an early age, such as 1 day. This is then increased at 7 days, and full prestress imparted at 14 to 28 days. During this period of partial stress, the tendons must be protected from corrosion, as by dusting on VPI powder during insertion, and sealing the ends.

Shrinkage can also cause horizontal cracks in the deep webs of girders. There is always a tendency of the lower concrete to consolidate and draw away from the upper concrete, and cracks will usually form at a discontinuity boundary, such as a tendon duct. Low water/cement ratio, thorough consolidation by vibration, and proper curing are the constructor's answers. Use of numerous small vertical reinforcing bars (stirrups) is the designer's answer. Some very deep webs have been prestressed by short inclined bars.

Differential moduli of elasticity are usually due to variation in the water/cement ratio and in the age at time of stressing. One solution used to overcome this and also to offset minor (elastic) deformations in the scaffolding is to provide 10% to 20% more tendon steel area than required: to then stress it to a point that raises the concrete spans to the required profile, using more or less prestress (within the limits provided in the design).

With long continuous spans and a complicated profile, friction losses during stressing may prove erratic. Use of a rigid steel duct, galvanized or leaded, will reduce the coefficients of friction and wobble. Such tendons should be stressed from both ends. A water-soluble lubricant has been used as an emergency measure to overcome excessive friction; it is then flushed out before grouting.

The trend is to use ever more concentrated prestressing forces, thus putting a greater load on the anchorage. This results in a highly congested area for proper concreting and consolidation, and so in a number of cases the anchorage block has been precast, then set in the forms and concreted into the cast-in-place concrete.

For medium and long span-bridges, an overhead gantry may be used to support forms and each cast-in-place segment until it is cured and stressed.

With continuous cantilevering, all movements (elastic deflection, creep, shrinkage, and thermal) must be constantly computed for each step; then properly countered by varying the prestressing force within limits provided in the design. The profile must be constantly monitored, in order to prevent cumulative movements.

Ducts for cast-in-place bridges usually cross construction joints. They must be grout-tight, either by a screwed or welded coupling or by wrapping with waterproof tape, or both.

When tendon ducts come close together in the vertical plane, such as at points of maximum moment, attention must be given to the possibility of one duct squashing into the adjacent duct when the tendon is stressed. With round ducts, especially rigid ducts, the concrete which encases them is usually sufficient to hold them in shape, especially if the tendons occupy a large portion of the duct cross section. With very large ducts and a very sharp bend, the duct material should be made thicker, e.g., use standard water pipe at these locations.

13.7   Short-Span Slab Bridges

For very short-span bridges (20 to 30 feet or so) pretensioned flat slabs may be the most economical solution. If these are cast in sequence against one another, they can be match-marked and placed side-by-side in the finished structure with perfect fit. Then they can be post-tensioned transversely, and no job-site concrete is required other than grout for the duct filling.

13.8   Foot Bridges

Foot bridges always pose a problem to the designer because they must be designed for the maximum possible load yet, normally, only realize a small fraction of this load. Channels, with legs upturned, while not as efficient structurally as other sections, may often prove the most economical solution, since the upturned webs may also act as handrails. The soffits may have a built-in camber in order to provide a proper appearance.

13.9   Conveyor Bridges

These have been constructed, in a number of cases, of precast prestressed units, such as channels. The section is chosen to serve other purposes as well, as a trough, or as a cover over the top. Because of their length, care must be taken to provide for longitudinal movements due to shrinkage, creep, and thermal change. Concrete serves this use well, as it is durable and corrosion-free, and has very low maintenance.

13.10   Pipe Bridges

These are usually built of steel, using the steel pipe as one of the carrying members of the truss or cable-stayed bridge. However, a number of recent pipe bridges have been built of prestressed concrete, using concrete pipe as a tubular girder. The pipe, either circular or rectangular in cross section, is prestressed in its walls, either uniformly, or with profiled tendons, to carry the dead and live loads. Circumferential stressing may also be employed, or mild steel may be used for the transverse reinforcement.

When a high degree of prestress is employed longitudinally, many of the advantages of prestressing of pipes are imparted. These include a failure mechanism (under excess pressure) that passes through successive stages of elastic behavior, cracking, opening, relief of pressure, and then partially or wholly re-closing.

13.11 Aqueducts

These are, in effect, pipe bridges, whether fully closed pipes or open troughs. They are extensively utilized in Italy (a modern day heritage from the aqueducts of the Roman engineers) and built of longitudinally prestressed concrete troughs. Their long lengths and constant cross sections lend themselves to mass-production. They are usually simple spans, set on T-head supports, with a bearing pad of neoprene to accommodate movement.

13.12   Railroad and Rapid Transit Bridges

These bridges must resist extremely heavy shear and side sway as well as direct load. Thus, rectangular cross sections are usually chosen: solid or hollow-core slabs, or box units. They are usually made by dividing the transverse cross section of the bridge into two or four boxes. These are then placed side-by-side and joined by epoxy grout, or Portland cement grout. Sometimes they are transversely post-tensioned; however, the present trend, especially with ballast deck structures, is to use just an epoxy-grout in the joint, with formed shear keys to transfer the load laterally from one unit to the other.

Because of the relatively high importance of shear, high-strength concrete is employed, and the webs are usually thick and heavily reinforced with mild steel stirrups.

For rapid transit and Metro viaducts, precast prestressed concrete girders offer long spans, esthetics, economy, reduction in noise and vibration, and savings in time of construction. The ability to achieve and maintain the track profile to extremely close tolerances is of great importance for high-speed operations, including monorails and air-supported vehicles.

14

Prestressed Concrete Floating and Submerged Structures

14.1 Introduction

Prestressed concrete is well-suited to the construction of floating and submerged structures, including deep-ocean and Arctic installations.

It possesses the important qualities of strength, water-tightness, durability, vibration damping, rigidity, adaptability to double-curvature, security and economy. Its apparently detrimental weight-to-strength ratio actually turns out, in many cases, not to be a serious drawback, even with floating structures, and may sometimes be an asset.

Prestressed concrete has a favorable mode of failure under accident or over-stress conditions, such as grounding, collision, and explosion. It develops localized cracks or disruption, but does not rip or tear as metals do. It is relatively easy to patch and repair after such accident.

Maintenance of concrete in sea water is low. High-quality prestressed concrete is durable and corrosion-resistant. Weeping and condensation have been virtually absent in concrete hulls in service.

The high impact resistance and fatigue strength of prestressed concrete at low temperatures make it the preferred material for service in the Arctic Ocean. (See Fig. 66)

Most floating and submerged structures must resist bending loads which tend to cause buckling of the skin. Concrete, in the thickness necessary to provide the requisite strength, has rigidity to resist local buckling and excessive distortion.

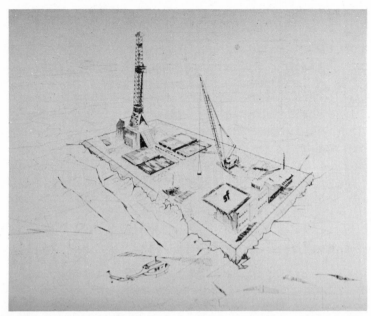

Fig. 66. Proposed prestressed concrete barge for support of oil drilling and production facilities in shallow water of Arctic Ocean.

Dead weight is generally an asset for bottom-supported structures, such as pipelines, subsea tanks, etc. In fact, steel pipelines are usually coated with concrete to add weight. In this case, the weight serves to increase stability under wave and current forces. For floating structures also, especially very large structures in deep water, additional dead weight only means an increase in draft and may provide increased stability.

Prestressed concrete offers both direct and indirect economies. The direct economies are in the more efficient use of the materials: high-strength concrete and high-strength steel, especially when subjected to multi-axial loading conditions. High-strength concrete is extremely efficient and economical as a compression-carrying member, whether the compression be from external loads or internal prestress. High-strength steel is likewise economical and efficient if it can be utilized in tension and protected from corrosion.

The indirect advantages include inherent durability, thus requiring little or no additional corrosion protection. Local materials and labor are utilized to a high degree, reducing the problems of foreign exchange, duties, and ocean freight.

Concrete has actually been already used for a very long time and to a very large extent in submerged and floating structures (See Fig. 67). Prestressing, relatively new to such applications, improves the structural performance, crack resistance, economy, and durability of these structures.

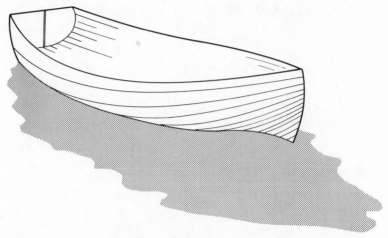

Fig. 67. "Seagull" built in 1887 to Monier's patents, now in Amsterdam zoo.

14.2   Problem Areas and Special Techniques

14.2.1   Durability

Earlier the statement was made that one of the great assets of high-quality prestressed concrete was its durability. Properly designed and constructed prestressed concrete is undoubtedly the most durable construction of all the practicable structural materials available for sea-water environments.

However, it must be emphasized that this durability is not automatically obtained. Careless or improper practice may lead to disastrous results in the form of corrosion or disintegration, or both. The frequency of such results is remarkably low, considering all the variables involved. However, when trouble has occurred, it has usually been general and progressive.

The environment for floating and submerged structures is of three natures. Completely submerged, with water on both faces, the concrete is completely saturated and little oxygen is available. Corrosion is inhibited, since no electrolytic cells are formed. The concrete itself, however, may be subjected to magnesium replacement of some of the calcium ions, leading to a general softening of cellular structure, and sometimes deleterious expansion. If the aggregates are unsound or reactive, swelling and disintegration may occur. Thus the concrete must be properly designed, mixed, and placed for maximum durability.

The second condition is complete submergence with air on the inside; e.g., a subaqueous tube. In this case, the salt water moves through the concrete and evaporates from the inner face. Salt crystals are left in the concrete and may set up electrolytic salt cells. Since oxygen is readily available, corrosion can take place. The greater the porosity and permeability of the concrete, the more serious the corrosion will be, particularly since carbon dioxide is available to reduce the pH of the cement paste.

For such structures, the concrete should be as dense and impermeable as possible, adequate cover provided over the steel, and consideration given to an impervious coating on the outside. Epoxy paint on the inside may help to seal against carbonation, and is often provided in vehicular tubes for light-reflecting purposes.

The third condition is in the splash zone, especially applicable to floating or partially submerged structures. Here the salt-cell problem is ever present, and in freezing environments, freeze-thaw action and ice abrasion must also be considered. Use of maximum air entrainment is essential in a freeze-thaw, salt-water environment.

When concrete structures are used for storage or passage of other fluids (oils, sewage, chemicals, etc.) specific precautions relating to those chemicals must be taken. The general rules of a dense, impermeable concrete, low water/cement ratio, high cement factor, and adequate cover are helpful in all cases, but special additives (e.g., air entrainment, pozzolans, sulfate resisting cement) may be necessary, or special coatings such as bitumens or epoxies.

14.2.2 Weight

The weight of prestressed concrete may be in some cases a detriment to its use. Despite the value of the concrete's increased volume providing structural rigidity and resistance to local deformations, etc., the strength/weight ratio is not as favorable as steel hulls of similar capability. Thus, for operations in shallow water, methods will have to be adopted to minimize the draft.

Since the wall thickness of concrete barges, ships, and hulls is usually fixed by other parameters than strength alone, the use of structural lightweight aggregate concrete will often show a substantial reduction in total weight and draft. The unit weight of high quality, impermeable, structural lightweight aggregate concrete is approximately 75% that of conventional concrete.

Another method of reducing weight is to reduce the wall thickness itself. Frequently wall thickness is determined by the need to fit in reinforcing steel and prestressing ducts. The substitution of central pretensioning for 2 layers of mild-steel reinforcement is one solution that has proven practicable. The substitution of pretensioning for post-tensioned tendons, with their sizeable ducts, is sometimes feasible.

Post-tensioned tendons and their ducts may be reduced in cross-section area by use of Dyform strand, higher-strength strand, or fewer strands per tendon, etc. Internal ribs may be provided paralleling the post-tensioned tendon path, so as to provide the necessary cover at this location, without increasing the overall thickness of the wall.

External post-tensioning tendons may be placed inside the box girder walls. They may be subsequently bonded to them by concrete encasement. For such an installation in floating structures, special and definite means should be taken to reduce the danger of corrosion. This may take the form of a metal sheath, with grout injection, as used by Dr. F. Leonhardt in his bridge constructions in West Germany. Alternatively, tendons may be plastic impregnated and coated, or the concrete encasement and adjoining concrete may itself be coated with epoxy paint.

Where the determining factor for wall thickness is local strength and rigidity, ribs may be used to provide increased stiffness and flexural strength to the sides, decks, hull, and internal bulkheads.

Cellular construction (hollow-core) walls have been proposed and are certainly worth careful consideration on very large vessels such as storage hulls and tanks. The cellular construction in effect gives a double wall with diaphragm connectors, which is a very efficient cross section in bending. The individual wall thickness (minimum thickness) must still be selected to give adequate cover over the reinforcement and space for its incorporation and proper concrete placement.

14.2.3 Joints

Prestressed concrete structures of the sizes contemplated usually must incorporate some joints, whether they be construction joints with cast-in-place concrete or joints between precast segments. This is no different from other large structures, except that joint details are more critical, and the consequences of poor joints more serious, when floating or submerged structures are involved. Poorly made joints may leak. They may provide paths for corrosion. If porous, they may provide a location for electrolytic cell formation. Poor joints are usually caused by lack of consolidation of concrete in the joint due to congested space, by shrinkage in the joint, or by lack of bond with the adjoining concrete. Structural impairment may be caused by lack of shear transfer from one face of the joint concrete to the other face. Where reinforcement must be spliced in joints, the highly congested area may cause numerous difficulties, such as spalling from welding heat, eccentric stress transfer at bar laps, inability to place concrete around the bars, and lack of proper cover where two bars overlap.

The faces of the joints of concrete segments must be properly prepared, by casting or chipping of keys, by light sandblasting or by bush-hammering. Even with perfect-fit joints (dry or epoxy joints), a light sandblasting is desirable to remove the skin of cement paste from the faces.

Use of an epoxy bonding agent, for bonding of the segment to the concrete in the joint, or from segment to segment, will prevent cracks from forming at the boundary.

If the joint is to be concreted, best results are obtained when the concrete is poured and internally vibrated. This means a joint width of 3 inches as a minimum, preferably 4 inches and the use of pea-gravel (3/8 inch = 1 cm) as the coarse aggregate in the joint concrete mix.

Shrinkage can be minimized by the use of a low water/cement ratio in the joint concrete. Since workability is essential, a water-reducing, shrinkage reducing admixture is desirable. Shrinkage-compensating cement is another solution provided adequate water curing can be assured.

The steel details at the joint require extremely careful design. An excellent method is to prepare a full-scale drawing to insure that the bars and tendons, etc., can be physically fitted into the space provided. Eccentricity should be avoided in the splices of bars carrying principal stresses; this can usually best be accomplished by butt welding or use of an angle iron splice bar. For these reasons, use of numerous small-diameter bars will generally be preferable to a few large bars. Sleeve details for connecting ducts must be provided, as well as means of sealing the splice grout-tight, as by taping.

Proper attention to detailing and proper execution in the field can assure that full continuity will be obtained across joints and that the joints will be crack-and-trouble-free.

14.2.4   Ferro-Cement

Ferro-cement techniques have been successfully applied for many years to small boat construction, starting generically with Lambot's boats in 1855, resurrected by Nervi during the 1940's, and becoming a sound industry in the 1960's. Prestressing and ferro-cement techniques are compatible. The ability through prestressing to keep the concrete under compression and to make the structure act as a whole, may be as important as the structural strength possible through prestressed ribs and beams.

In the other direction, however, prestressed concrete floating and submerged structures have much to gain from ferro-cement techniques. The use of closely spaced wire mesh near the concrete surface holds the concrete together under impact and collision and restricts and limits shrinkage cracks. In some cases, the use of the multi-layers of wire mesh may provide a needed resilience and flexibility. Finally, the advantages are shown of a high steel proportion, in the form of finely divided steel, whether in the form of mesh, stressed tendons, or small bars and rods.

14.3   Design

Design for prestressed concrete floating and submerged structures should follow good prestressed design practice with specil attention to the following:

1. Durability in sea water environment.
2. Reversals of stresses due to waves.
3. Pressure differentials due to waves and operating conditions.
4. Accident conditions: explosion, internal or external; collision; etc.
5. Detailing of joints and connections.
6. Abrasion from moving sand.
7. Marine growth.
8. Manufacturing and construction aspects; stages of construction including draft, stability, connections. Special attention must be paid to tolerances in unit weight and dimension, and their effect on draft and stability.
9. Operational aspects; draft, stability, mass.

14.4 Manufacture and Construction

Almost all structures designed for submerged or floating service have a common characteristic: they are extremely large and massive. This poses serious problems for manufacture and construction. Just as is the case with long-span bridges, the methods of manufacture and construction must be considered as an integral part of the design process. The functions are interrelated. Constructors have evolved some extremely ingenious methods for manufacturing, launching, assembly and final installation. A brief review of

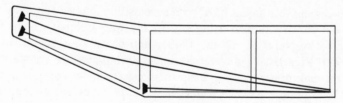

Fig. 68. Prestressing.

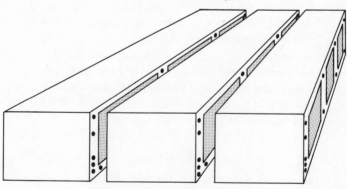

Fig. 69. Segmental construction.

a number of these methods follows, to serve as a guide to past practice and a stimulus to the development of new and better methods. Certainly this is a field where ingenuity has wide scope for further advance. (See Fig. 68 and 69)

14.4.1 Launching

Many concrete barges and caissons have been constructed on launching ways, and slid down the ways to flotation in the water. The ways must be built to withstand the loads involved and provide necessary structural support to the barge or caisson during both manufacture and construction. (See Fig. 70)

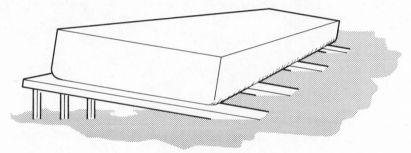

Fig. 70. Launching.

It is usually easier to construct a barge on the level rather than in an inclined position. One solution is to employ a launching cradle, which rides down the inclined slope, while keeping the barge or caisson level. Another solution is to build the barge or caisson on the level at the head of the slope and then rotate it, by jacking beams, to the inclination of the ways.

Most concrete sections have been launched by gradual lowering down the ways by hoists, or, if moving freely under gravity, with the ways extended well out into the water. There is no reason, however, that the side launching or free-launch technique, so successful with steel ship hulls, cannot be used.

14.4.2 Direct Lift

Floating derricks and shear legs with capacities from 600 to 2000 tons are now available for direct lift from barge or bulkhead, and for lowering into place.

14.4.3 Tidal Launching

A large number of concrete structures have been launched by making use of the tidal rise, either alone or in conjunction with other methods. The effect of salt water on freshly poured concrete must be considered and consideration be given to provision of impermeable coatings such as asphalt, timber lagging, polyethylene, steel, etc. (See Fig. 71)

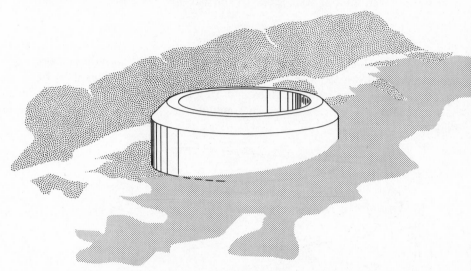

Fig. 71. Tidal launching.

14.4.4   Drydock or Graving Dock

Many large caissons and concrete ships, tubes, etc., have been constructed in a floating drydock or graving dock. These have the advantage of providing stable support and dry working conditions during construction, and controlled flooding and launching. Charges for the use of drydocks are high,

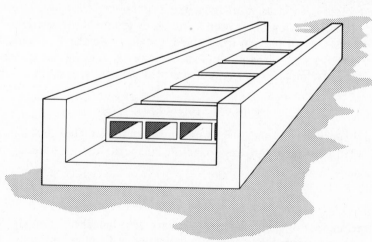

Fig. 72. Assembly in drydock or graving dock.

and the speed of construction of a concrete hull, for example, may be slow. A drydock is usually a very congested working area, and frequently presents some problems of accessibility. (See Fig. 72)

Many of these problems can be minimized by manufacturing precast segments or sections elsewhere, floating or lifting them into the dock, and making the final connections in the dock. Means must be provided for alignment and positioning as each section or segment is set on the keel blocks. Guides can be used for this purpose, assisted by jacks.

14.4.5   Basin

Many of the largest concrete structures have been cast in a basin, which is later flooded, the dike breached and the structure floated out. (See Fig. 73)

Fig. 73. Casting 200,000 bbl. Underwater oil storage vessel in basin.

In the Netherlands, for example, a bow levee is constructed behind an existing dike, the concrete structures constructed, then the basin is flooded and the dike opened.

In other areas, basins may be excavated, kept dewatered during construction, then flooded and an access channel dug to the waterway.

A steel sheet-pile cut-off wall may be necessary to control water inflow, especially if the basin is to be used more than once. Well-point systems may be needed, or French drains.

Basins should be designated with full consideration for access of materials and equipment. Because of long side slopes for deep basins, it may be necessary for the cranes, trucks, etc., to work in the bottom, or else for access trestles to be constructed.

It is very important that a good working surface be provided, such as a thick blanket of rock.

Support during construction should be well thought out. Pile supports may prove too rigid if any settlement occurs, and thus crack the concrete. For many structures, concrete or timber sleepers on a rock base are more effective; they can be ballasted to grade, and they are sufficiently flexible to distribute the load rather uniformly.

14.4.6   Lowering Down

Many caissons, large-diameter pipes, etc., have been launched by lowering down. A platform is built on or over the water, supported at the edges, or outside the final concrete structure. This platform may support the load partially by flotation. (See Fig. 74)

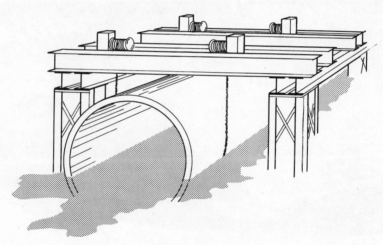

Fig. 74. Lowering down.

At time of launching, overhead beams are placed and either hoists, jack rods, or jack cables, are used to lower the structure and platform to a point where the structure can float free. The platform is then raised for a second use. For reasonable-size structures, in which many uses are contemplated, this is a very efficient system.

14.4.7   Barge Launching

Concrete barges, caissons, and similar structures are frequently constructed on a barge. Launching is then accomplished by flooding the supporting barge. (See Fig. 75)

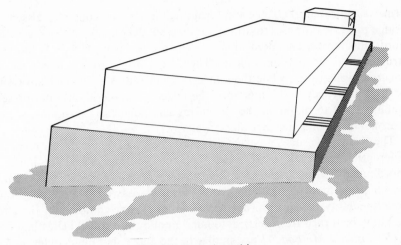

Fig. 75. Barge launching.

Once the deck of the supporting barge dips under water, stability is lost and the supporting barge will usually list heavily. The caisson then slides sidewise. Such intentional launching during sinking is often practiced and even aided by flooding the support barge so as to initiate the list in the desired direction. However, side launching does place very heavy concentrated loading at the edge of the barge. If supports are not properly designed, this can cause buckling of the support barge side or break the concrete caisson.

The support barge may be controlled by performing the launching in shallow water with a relatively flat bottom, so that it cannot tip to any degree. Or the support barge can be guided by spuds and hoists or jack cables. Additional stabilization can be accomplished with side columns; this then becomes a small floating drydock.

The support barge must be able to resist the external pressures at the maximum depth of launching without inward buckling. For safety, the maximum depth of submergence should be controlled, and this is usually attained by launching in a predetermined depth of shallow water. When launching must be done at sea, control may be effected by winching down from adjoining barges.

14.4.8   Construction Afloat

By this method, the structure is built up progressively , while it is supported by the buoyancy of the water. A protected, accessible site is chosen, with water of the proper depth.

(*a*) Slip Forms. A very intriguing system has been proposed by a Swedish firm,* as an extension of the slip-form method used on concrete caissons for bridge piers. It proposes to utilize a protected, deep body of water, such as a fjord, and use an anchored floating construction pen. A base is first constructed and floated into place. Then slip forms are installed and the concrete walls raised while the unit slowly sinks down. The unit always has positive buoyancy and water must be added to make it sink and preserve stability, and to maintain the slip forms and concreting positions within a reasonable working distance above water.

The walls of such a structure must be thick enough to resist the differential external head. The concrete-pouring schedule must be coordinated with the rate of sinking to give a sufficient setting and curing period prior to entering the water to insure it has adequate strength and impermeability, and to prevent wave erosion. After completion, the structure may be turned from vertical to horizontal by controlled ballasting.

(b "*Caisson-Method.*" Very similar to the above, but much older in practice, is that of constructing caissons while the structure is afloat. The walls are raised, using panel forms, and each lift formed, poured, and allowed to

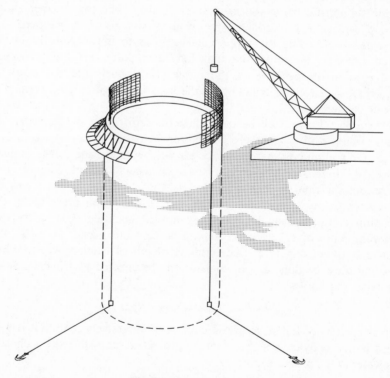

Fig. 76. Construction afloat.

*Hydrobetong, of Stockholm.

attain the necessary age for strength and durability before it enters the water. If desired, waterproofing by bitumastics, epoxy paint, or other material may be added before submergence. (See Fig. 76)

Stability and control during sinking is accomplished by an anchored or fixed enclosing pen, and anchor lines leading from the caisson down to the base, and out to the anchors.

The side walls have, on occasion, been formed by precast panels, acting compositely with poured-in-place concrete. In smaller structures, a complete lift has been precast and set on top of the previous lift, then post-tensioned to it. Use of precast segments offers many advantages of durability, speed, and economy for this form of construction.

14.4.9 Prejoined Sections

For several outfal sewer projects, the precast concrete pipe sections have been prejoined into longer units to facilitate setting and connecting in deep water. These concrete sections may be joined by bolting or, as more recently practiced, by prestressing. They may then be floated out, supported by pontoons. Internal steel tanks, like sausages, may be used to provide flotation and longitudinal strength. Temporary internal prestressing will provide longitudinal strength but not flotation. Inflated rubber sausages provide flotation, but no strength. They are easily removed by deflation. A final alternate for providing flotation is to install an internal bulkhead, sealed against the side walls by inflating rubber tires or gaskets.

External buoyancy tanks may be used to decrease draft at time of launching and to control stability during final installation. These external tanks may be rigidly connected to the structure, or hinged, so as to be articulated. Cylindrical tanks, for example, may rotate to vertical during sinking and thus stabilize the structure, especially during the stage when the upper surface first passes through the water plane.

Bottom assembly of units is basically the method used in joining subaqueous vehicular tube sections. The first section having been set in place on the bottom, the second is lowered and moved to proper position relative to the first.

This may be done by means of tapered guides, or by tensioned guide lines. By installing sheaves on the first unit, the second may be pulled towards it until juncture. Rubber bumpers may be provided to ease the shock of impact.

In heavy swell or surf conditions, the second unit may be lowered in a "horse" (a structural steel framework support). The horse and unit are set on the bottom, then hydraulic rams move the second unit into proper relation with the first. Control may be by divers or remote control assisted by underwater television. Actual longitudinal seating of pipe sections may be accomplished by a longitudinal "suck" line, which leads through the first tube and pulls the second to it.

If the second unit is seated in proper alignment with the first, then a positive seating may be accomplished by the sea pressure itself. A joint gasket of substantial dimensions is installed on one of the units. The second unit, when placed, compresses the gasket sufficiently to effect a seal. Pumping out the water contained between the two units permits the full hydrostatic force to be exerted to push the second unit into contact with the first, further compressing the gasket.

With large units, such as underwater vehicular tube sections, the mass is so large that the unit possesses great inertia. Therefore, positive restraint must be provided during lowering to prevent the massive second unit from swinging into the first and damaging it. With lighter units, such as pipe, the same problem may occur due to wave surge. Restraint can best be applied by lines leading down through sheaves on the second unit and off the anchors or deadmen.

When the jointing is of vertically disposed segments (i.e., the joint is in the horizontal plane), guide lines, spuds, and guide brackets are extremely useful in guiding one unit down into position on top of the other. Cushioning should be provided to prevent spalling on impact: e.g., neoprene, or asphalt-impregnated fiberboard.

14.4.10   Assembly Afloat

This has similarities to the previous section, except both first and second units are afloat at the time of joining. Presumably time and site will have been selected to eliminate or minimize wave surge. The units are maneuvered into correct position relative to one another and pulled into firm contact by hoist or jack lines. Since hydrostatic seating is relatively ineffective due to the shallow depth, the hoist ("suck") lines must be installed both top and bottom. They must exert a sufficient force to compress the gasket material. Horizontal guide brackets or pins are needed to insure accurate positioning of the units relative to one another. High-strength bolts or, better, prestressing tendons may be used to tie the units together into one structural element, and the joint may be made rigid by concrete grout or epoxy.

Where the joint is to provide flexibility (articulation), the joint material must have proper flexibility (rubber, neoprene, etc.) and the joining tendons must be positively protected against corrosion. While such articulated joints offer many structural advantages, the detailing to insure proper shear transfer, corrosion resistance, and water-tightness, requires extreme care and the development of a proper technical system. Consideration has to be given to expansion and contraction, and the cycles of movement under wave action, with the problems of fatigue, corrosion, shear transfer, and dynamic longitudinal forces. Failure to provide for all of these has required the replacement of one major floating bridge (Derwent Bridge at Hobart, Tasmania), and the major reconstruction of another (Hood Canal Bridge, in Washington).

Despite these past problems, the knowledge gained and the new techniques and material available make joining afloat an extremely attractive method for very large strucutures.

14.4.11 Sand Jack

An old method of launching heavy concrete structures is by use of the sand jack. The structure is built on a natural or artificial island of sand. Then by means of controlled jetting, air-lifting, and dredging, the sand is gradually removed from under the structure until it floats. This is a cheap and simple procedure provided the structure itself has sufficient strength to span across the variations in support conditions during the lowering. (See Fig. 77)

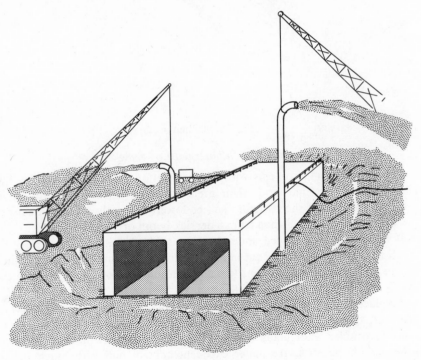

Fig. 77. Sand jack

14.4.12 Rolling In

Although seldom practiced in recent years, many cylindrical pipes, tubes, and caissons have been launched in the past by simply rolling them down a uniformly sloping beach, either on the beach itself or on launching skids.

The beach must be firm (sand or gravel) and the unit must have adequate structural strength; longitudinally, transversely, and against crushing. (See Fig. 78)

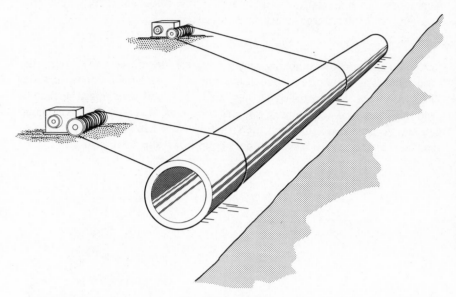

Fig. 78. Rolling in.

14.4.13   Submerged Lowering

One of the great problems with the installation of structures on the bottom of the sea or harbor is their great weight and size. The problem is greatly simplified if the structure can be lowered directly (i.e., by means of direct hoist lines) rather than from a derrick boom or A-frame.

If the unit is first placed underwater, in shallow water, then a relatively small barge or barges can float in over the top and, by means of hoists, pick up the structure for movement to final location and lowering. Because the unit is underwater, the net weight is reduced by 40%.

To place the unit under water in the first place requires controlled sinking; however, the site, depth, and bottom conditions can be pre-selected to be most favorable. Once under water, this method permits the transport and lowering of the structure under excellent control and with much lighter equipment.

14.4.14   Successive Basins

Previous sections described the use of tidal differences and basins for launching. A system used successfully for a number of very large concrete caissons has been as follows:

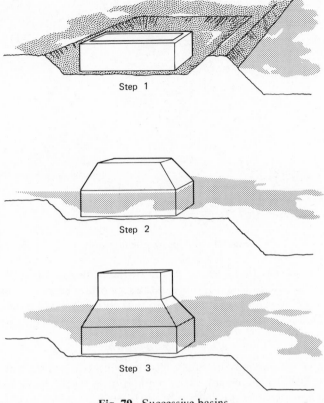

Step 1

Step 2

Step 3

Fig. 79. Successive basins.

The first lift, including bottom plate, is constructed in a tidal zone, so that it may float off as a unit at high tide. Then it is moved to a prepared underwater basin, where it is sunk onto a level bed of gravel or sand at a depth just sufficient to expose the top of the walls at low tide. A second lift is poured, and, if necessary, the unit moved again. Thus, progressively, the unit is floated at high tide, sunk in a new basin at low tide, then its walls are constructed to a higher stage. (See Fig. 79)

14.4.15   Collapsing Pile Platform

In northern countries, such as Sweden and Alaska, heavy structures have been launched by building them on a timber pile platform, then cutting the piles in sequence by dynamite, so as to launch the structure sidewise into the water.

As practiced, this is a highly dangerous method, dependent on uniform failure of non-uniform piles. The danger exists of unbroken or wedged pile stubs punching through the bottom of the structure and holing it. Also, there

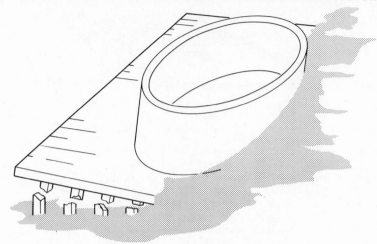

Fig. 80. Collapsing pile platform.

is always the danger that only a portion of the explosives will fire or, conversely, of premature firing or failure of the supports. Therefore, this method should be employed only where no other method will do, and then only with due recognition of the hazards and positive steps taken to minimize them.

14.4.16 Pull-Down

Floating structures may be submerged by pulling down, the hoists reacting against positive buoyancy. This method gives excellent control at all times. The hoists may be located on fixed structures, such as dolphins, at the corners, or on barges, or on the upper extensions of the structure itself.

Pull-down requires a reaction on the bottom. This may be concrete block weights or piles. In either case, positive precautions have to be taken to prevent pulling the structure down onto the reaction and holing the structure itself.

Reaction points may consist of predrilled anchors in the bottom and prestressing techniques may be used for pulling the structure down.

Pull-down gives the ability to also pull into position. Depending on the reaction points and length of line, etc., most effects of wave and surge can be readily countered, including the dynamic effects (up to 200%) when the upper surface first passes through the water plane.

14.4.17 Buoyancy Control

As an adjunct to other methods of sinking, the buoyancy may be controlled progressively to overcome different densities of the water. On large structures, a variation of 1 pound per cubic foot in water density may cause a variation of several hundred tons in net weight of the structure as a whole.

The buoyancy may be affected by differential water velocities at top and bottom. These may be sudden, as by a passing ship in a narrow channel, which on several occasions has caused large vehicular tube sections to pop back up to the surface. More commonly, the differential water velocity may be due to the Venturi effect of a steady river or sea bottom current, accelerating under the structure as it nears the bottom, causing scour and a change in effective density. While above the bottom, this change is of opposite sign to the increased density due to salt content and silt that normally also occurs near the bottom. When on the bottom, currents and wave-induced water movements tend to lift the structure.

Weight control and balance may also be exercised by the judicious combination of lightweight concrete with normal and heavyweight concrete.

When internal air pressure is used to control the submerged buoyancy, the air pressure must be adequate to keep the walls from collapsing inward during sinking. At the same time, the initial air pressure must be within the structural limits of the tank in tension. Therefore, when lowering in very deep water, a second stage of repressurization may be necessary. This takes considerable time if the size of the tanks is large. In addition, the net weight of added air may be significant.

The air pressure may be added from external sources or from internal sources, such as compressed air cylinders or gas generation. These latter eliminate the change in weight, but are rather complex to actually install and control. Especially with gas generation, positive means must be taken to prevent over-pressure and outward explosion.

14.4.18 Up-Ending

Prestressed concrete floating structures may be constructed in a basically horizontal position, on a ways, or in a dock or basin, then towed to the site in deep water, and up-ended to vertical position by flooding. In other cases, the structure may be constructed vertically, as a caisson, and up-ended to a horizontal position for towing and/or service. The rotation should be carefully checked by model tests as well as by calculations, since once such a rotation is started, it should be carried through decisively. Sometimes model tests reveal some interesting phenomena, such as a secondary rotation around its axis while being up-ended, and the speed and inertia at different phases.

Up-ending should either be controlled, as by a derrick line, in accordance with carefully determined data from calculations and model tests, or else the up-ending should be completely free and uncontrolled. The forces with large structures are very great.

Planning for up-ending must include the arrangement for air, water, and control lines during the up-ending; how to prevent them from fouling or breaking; and provision of automatic check valves in case of rupture or failure.

Planning must also include the mooring during and after up-ending, and again means of preventing fouling of lines during the operation.

14.4.19 Summary

The above sections describe some of the methods which have been used *in the past* to launch and install concrete structures. Some of the more successful major installations have utilized two or more methods in sequence. The immense importance of this phase requires the integrated efforts of engineer and constructor. Extremely careful planning is essential.

Since the actual work is done by specific human beings, the entire crew must be thoroughly indoctrinated. Practice with a model in a testing basin or swimming pool is invaluable to both workmen and engineers. A dress rehearsal should be held so that every man understands his function and the correct sequence, communications, etc. For very large and complex installations, an emergency procedure should be developed; that is, prior planning of what steps to take if actual behavior differs from planned behavior.

Adequate communications and instrumentation are essential. The recent advances in ocean technology have made available a great many sophisticated instruments to measure depth, orientation, level, etc., as well as underwater television and diver reports. These instruments should be utilized to the fullest for all deep and important installations.

The launching and installation of large floating and submerged structures of prestressed concrete can be one of the most challenging and demanding opportunities a construction engineer faces, for personal qualities of courage and decision are required as well as technical ability in the highest degree.

14.5 Research

A considerable amount of research is currently being done on the use of concrete in underwater applications. This only concerns the engineer-constructor as it highlights problem areas, suggests new ways of coping with problems, and leads back into the cycle of actual construction practices.

Underwater spherical hulls have been extensively tested for simulated depths up to 10,000 feet. Their mode of failure, implosion, and the triaxial stress conditions preceding failure reflect the great importance of shape and thickness to diameter ratio. The effect of openings in such spheres is also a subject of current research, as is the effect of depth on permeability rates.

R. G. Morgan of the University of Bristol has been accumulating data on construction and actual behavior of concrete ship hulls, from World War I on through the present day, including the most recent practice in the USSR and USA, in order to determine the most effective and efficient means of applying concrete to shipbuilding.

A number of companies (Sea Tank of France, Hydrobetong of Stockholm, the author's firm, Santa Fe-Pomeroy of USA, as well as firms in Italy and the Netherlands) have been conducting research on the use of prestressed concrete for underwater storage vessels, primarily for storage of crude oil. Their studies include effect of wave forces on stability, effect of oil on the concrete, means to minimize emulsion of the oil-water interface, etc.

Other research has been directed at the use of prestressed concrete for shipment of LNG (liquefied natural gas at temperatures as low as —259°F), and the behavior of the prestressed concrete vessel as as whole, as well as the components. Thermal gradients and differential thermal contractions must be thoroughly considered. Non-stressed reinforcing steel must be a suitable alloy if embrittlement is to be prevented. Prestressing steel itself is satisfactory in behavior even at the lower temperatures, and concrete gains in strength and modulus of elasticity. It may be appropriate, therefore, to use unstressed prestressing strand in lieu of the more conventional reinforcing steel for lateral and auxiliary reinforcement.

One of the most promising developments for undersea applications is that of polymerized concrete, which is essentially impermeable, high in strength, and extremely durable.

14.6 Utilization

Concrete has already been more widely utilized for floating and submerged structures than is generally realized. Performance has been generally excellent; economy has been somewhat erratic. A primary factor in the utilization or lack of utilization of prestressed concrete for ocean applications has been the lack of sufficiently strong and viable industry to develop, design, construct, and promote its use.

Among the applications for which concrete, both conventionally reinforced and prestressed, has been utilized to date, are the following:

14.6.1 Ship Hulls for Floating Oil Storage, Etc.

In both world wars, an extensive concrete shipbuilding program was undertaken, amounting to a total of some 800,000 deadweight tons. The ships proved their seaworthiness in hurricanes, bombing attacks, collisions, and fire. They were watertight, vibration- and condensation-free. They were, however, very heavy.

Both normal sand-and-gravel and lightweight aggregate concrete was used. Considerable difficulty was experienced during construction in fitting a double-curtain of reinforcing steel into wall thicknesses of 4-1/2 to 5 inches

(11 to 12.5 cm). Problems were also encountered with consolidation of concrete; vibration of reinforcement and form vibration was employed. Rich cement mixes were used.

Considerable variation in durability was experienced; with corrosion problems of the reinforcement mainly attributable to inadequate cover, lack of consolidation, and high water/cement ratio. On the other hand, many of the vessels survive to the present day, in excellent condition, despite little or no maintenance. The ship Selma (6340 tons) built in 1919 of expanded shale lightweight aggregate and having minimum cover over the reinforcement, was beached near Galveston, Texas, in 1921. It is still in a remarkable state of preservation despite the semi-tropical salt-water and splash environment.

14.6.2 Floating Drydocks

Floating drydocks of reinforced concrete were similarly built during the wars and some are still in service today. They were generally built with adequate cover and with walls thick enough to enable proper consolidation of concrete. Those the author is familiar with have required a minimum of maintenance, and that which has been necessary has generally been directed at the metal fittings and appurtenances.

14.6.3 Ocean-going and Inland Waterway Barges of Concrete

A considerable fleet of ocean-going barges is now operating in Southeast Asia. Pretensioning has been extensively employed to provide a crack-free, durable hull. Steel frames have been incorporated in some of these barges to span cargo hatch openings.

Despite the increased draft of these prestressed concrete barges, they reportedly tow at about the same speed due to less tendency to yaw.

Inland waterway barges of concrete, both conventionally reinforced and prestressed, have been used for a number of years in rivers and harbors around the world. Such barges have served as cargo carriers, bulk product and liquid carriers. However, their use has been scattered and generally of small extent. The greater draft of the concrete barges has been a deterrent in many cases. Construction methods have only recently begun to be developed to provide quality and economy. These include systems employing precast pretensioned segments, joined and post-tensioned transversely.

14.6.4 Floating Piers and Docks

An extensive number of concrete pontoons have been constructed, of various shapes and sizes, to provide floating docking facilities for small boats, sea planes, recreation and, in some cases, ship cargo wharves. Concrete was selected for the pontoons for seaplane docks to eliminate the electrolytic corrosion of the magnesium hulls from steel pontoons. Concrete was selected for the other docks primarily for economy and low maintenance, but in actual service has given additional benefits of durability and localization of impact damage.

14.6.5   Concrete Barges for Compressor Station, Production Facilities, Small Refineries, Etc.

The oil and gas industry has made extensive use of concrete barge hulls on which have been mounted production and processing equipment of various sorts. These barges have been towed to the site (as far as from Louisiana to Nigeria and from Antwerp to Libya) and either moored afloat or sunk in shallow water.

Many of these barges are designed so that the barge hulls will be underwater in the final installation, with only columns extending through the wave zone, and the deck with its equipment well above the storm-wave height.

The submerged hulls have been utilized to a limited extent for storage of oil and fresh water.

14.6.6   Pontoon Bridges

A number of notable pontoon bridges have been built of concrete, initially of conventionally reinforced concrete and more recently of prestressed concrete. These include the several bridges across Lake Washington at Seattle, the Hood Canal Bridge, also in the State of Washington, Lake Okanagan in British Columbia, and the Derwent River Bridge at Hobart, Tasmania. Both the Hood Canal and the Derwent River Bridges suffered extensive damage from wave action. Although attempts were made to reinforce the Derwent River Bridge by cables and new expansion joint material, damage continued to be excessive. The bridge was eventually replaced by a high-level bridge of prestressed concrete.

The Hood Canal Bridge was reinforced by longitudinal post-tensioning together of pontoons in groups of four, with epoxy joints. Apparently these modifications have been successful.

The more recent bridges in Washington have used normal sand-and-gravel for the bottom and sides of the hull, and structural lightweight concrete for the decks. They are post-tensioned longitudinally. The box-section pontoons have been constructed in shallow graving docks, and floated out at high tide.

For the Lake Okanagan Bridge, the bottom portion of the pontoon was constructed in a shallow basin, then once afloat, the walls were raised and the deck poured to complete the bridge section.

14.6.7   Caisson Gates for Graving Docks

In the USSR and in West Germany, the caisson gates for large drydocks have been constructed of prestressed concrete. This is an excellent selection, because the moldability of concrete can be fully utilized in the complex shape and varying cross section required. Weight is an advantage; steel caisson gates must usually be partially filled with concrete for stability.

The security provided from the favorable mode of failure, i.e., restricted zone of damage, provided by prestressed concrete is again of great importance. A drydock being usually located in busy shipyards, the caisson is always subject to the possibility of collision or impact, and must still protect the exposed ship and workmen behind it. Concrete gates minimize the internal stiffening required and reduce maintenance costs.

14.6.8   Caissons for Bridge Piers

A number of major bridges are founded on piers formed by concrete caissons. They caissons are usually constructed as follows: the cutting edge is formed and constructed to the height to give structural strength and rigidity, say, 20 feet. The bottom is then closed by a timber, steel, or concrete false bottom, the cutting edge section is launched and moored in position. Then the walls are progressively raised while the caisson sinks lower in the water. In some cases, after the caisson has penetrated the bottom, the false bottom is removed and, by a combination of dredging, weighting, jetting, and vibration, the caisson is sunk to its founding stratum.

In other cases, the bottom may be permanent and act as a spread footing, bearing on the soil directly. For the Duwamish River Bridge in Seattle, Washington, the bearing pressures were kept low by utilizing buoyancy from empty sections of the concrete caisson.

Piling may be driven prior to sinking of the caisson, and the juncture between caisson and piles made by tremie concrete or by cast-in-place concrete placed in a chamber maintained free from water by pneumatic pressure. Tremie concrete has been extensively used on the bell-pier caissons for such bridges as the San Mateo-Hayward and Richmond-San Rafael Bridges in San Francisco Bay, and Columbia River Bridges in Oregon. Grout-intruded concrete has been utilized on similar steel caissons for bell-piers, e.g., Narragansett River Bride, Rhode Island. Placement of the concrete in the dry, under pneumatic pressure, was the solution chosen for the Lillabelt Bridge in Denmark.

The Martinez-Benicia Bridge in San Francisco Bay used floating concrete caissons, supported by drilled-in steel caissons. The buoyancy of the concrete caisson was used to offset the greater portion of the bridge's dead load.

The raising of the caisson walls has followed a number of variant procedures. In France, precast concrete rings have been lifted on top of the floating caisson and post-tensioned successfully to it. On the second Carquinez Bridge in San Francisco Bay, the exterior walls of the caisson were made of precast concrete panels, which worked in composite action with the cast-in-place main portion. Slip forms have been used. Perhaps the most common method has been panel forms, jumped up for each pour, and supported off the floating caisson.

Prestressing has been used in recent years to provide the structural strength to the cutting-edge section during launching and the initial phases of concreting. It has also been used very effectively for structural purposes in the distribution block, in connecting segments as noted above, and to a limited extent, in the walls.

14.6.9   Caissons for Breakwaters, Quays, and Wharves

Concrete caissons have been extensively utilized for this purpose. They have been constructed on a launching way or in a graving dock or basin, launched, the walls raised while the structure is afloat, towed to the site, and sunk on a prepared foundation. Caissons have been towed as far as from Italy to Libya. Some of the more recent of such structures have utilized prestressing to provide a more efficient and durable structure and to provide the initial structural rigidity during launching.

14.6.10   Underwater Vehicular and Railroad Tunnels (Tubes)

It has been interesting to note the evolution of these structures. Initially, the tubes were formed of double walls of structural steel, in effect, a double-walled ship hull. The space between was then filled with concrete, with the concrete serving the function of weight, compressive hoop strength, and rigidity and security. The next development utilized conventionally reinforced concrete, adding to concrete the task of providing longitudinal structural strength and resistance to non-uniform circumferential loading. An external skin has sometimes been provided to give added water-tightness: this has usually consisted of steel plate or bitumastic protected by wood lagging. The most recent example of this concept, the Transbay Tube of the San Francisco Bay Area Rapid Transit System, employs the external steel skin, strengthened with steel ribs so as to provide an external support and form for the concrete construction (the steel section is launched and the concrete added afloat) and to work in composite action with the reinforced concrete tunnel section.

The Rotterdam Metro tube under Rotterdam Harbor was comprised of reinforced concrete segments with built-in joints for flexibility, joined for launching and installation by temporary prestressing inside the tubes.

A number of more recent tubes have selected a rectangular box section to provide greater roadway width and reduced depth. To resist the bending stresses under external water loads, they have been built of prestressed concrete. Tremendous quantities of prestressing steel are required; thus, the LaFontaine tunnel across the St. Lawrence River at Montreal is one of the most notable prestressed concrete structures in the world today. The external loads, due to water pressure, are not fully acting until the tube section is finally in place; thus, it has been necessary to provide countering temporary post-tensioning during construction phases.

Prestressing makes practicable the more efficient rectangular cross section. This reduced depth of section permits shallower dredging to provide the same depth of ship channel overhead, and may reduce the required length of underwater tube section. Prestressing is not only structurally efficient but provides added security against cracking, the self-closing of cracks due to temporary overload (as from collision or a ship's anchor), and greater security against catastrophic failure.

14.6.11   Subaqueous Pipelines for Outfall Sewers and Intakes

Subaqueous piplines for outfall sewers and intakes have long been constructed of precast concrete sections. Both normal sand-and-gravel and structural lightweght concrete have been used. In the latter case, the concrete immediately adjacent to the joint may be made of normal concrete to minimize the danger of spalling. Such pipelines have been laid in diameters up to 15 feet and in water depths up to 220 feet or more.

Longitudinal prestressing has been employed, both within individual pipe sections, and to join several sections, e.g., four, together to enable them to be installed as a single unit.

In New Zealand, this joining of sections has been extended so as to join the entire line, up to 1000 feet and more, with longitudinal post-tensioning. This has enabled the pipelines to be floated to position and sunk or, alternatively, to be pulled out to sea like steel pipelines. The extreme length-to-diameter ratio permits the necessary flexibility while maintaining stress across each individual joint.

14.6.12   Caissons for Lighthouses

A number of extremely interesting lighthouses have been built in exposed ocean locations by using concrete caissons, built in telescopic concentric segments. The caisson is built in a basin or series of basins in a harbor, completed afloat there, then towed to the site. It is sunk on a prepared bed and the inner segments are raised by water flotation. These are secured in position by gravel fill and concrete grout injection.(See Fig. 81)

14.6.13   Anchors

Prestressing techniques have been used to stress concrete block anchors to the underlying soil or rock. To overcome the consolidation and non-elastic compression of the soil, a number of stages of stressing (re-stressing) have been required. Prestressing has similarly been used to anchor structures down to the sea-floor. A combination of grout injection under the structure

Fig. 81. Kish Bank lighthouse caisson under construction.

and prestressing down can be effectively used to provide both shear resistance and resistance to uplift.

Moorings for floating bridges have been prestressed so as to overcome the stretch of the wire ropes and to dampen movement under wave action.

14.7 New Applications

Among the many proposed new applications which are under development, the following are of particular interest:

Very Large Specially Shaped Hulls for Offshore Storage of Oil.   These include both bottom-supported storage and floating storage. The configurations and size are selected to minimize wave forces (both lateral and uplift), promote maximum stability, facilitate operational requirements, provide structural strength, and minimize interface emulsification of oil and water. To meet these parameters and still maintain economy and practicability, cylindrical, dome, and spherical shapes have been selected for a number of the proposed systems. (See Fig. 82)

A number of new schemes provide for the storage hull to float submerged, either held down by anchor moorings or suspended from the surface, either from floats or by means of column-stabilization.

Large Breakwaters.   Several systems have been proposed for using floating assemblies of pontoons, where the width of the breakwater is related to

Fig. 82. Underwater oil storage vessels*: (a) manufacture in basin (b) towing to site (c) sinking (d) installation on ocean floor. *Concept developed by Santa Fe-Pomroy Inc.

the wave length. Various schemes for energy dissipation have been developed, including holes, baffles, etc.

Special Barges of Prestressed Lightweight Concrete. Barges transporting cryogenic materials, such as liquefied natural gas, are under development. The generally excellent performance of prestressed concrete at

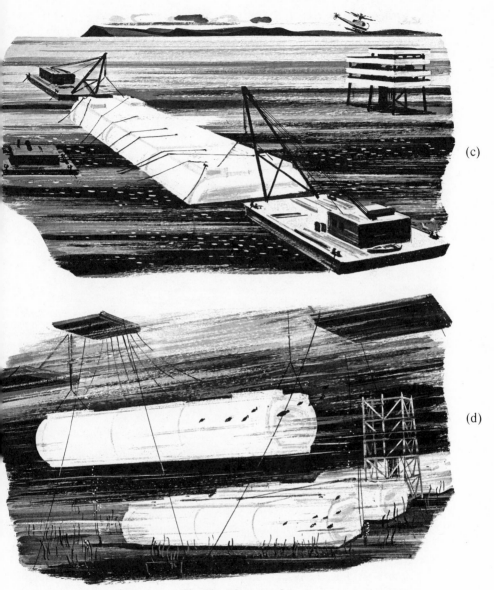

(c)

(d)

Fig. 82. *(Continued)*

low temperatures is a major reason for selection of this material. With prestressed concrete, it appears that one hull, properly insulated, can replace the double hull currently required for steel barges. The favorable behavior of prestressed concrete under collision is another factor in its favor.

Floating Caissons and Barges of Prestressed Concrete. On top and inside of these are mounted drilling and production equipment for towing to remote areas, such as the Arctic, with subsequent sinking and securing to the bottom in shallow water. Here also, the favorable impact resistance of prestressed concrete at low temperature is a determinant.(See Fig. 83)

Fig. 83. Proposed prestressed concrete caisson for deep water oil drilling facility in Arctic Ocean.

Floating Airport Structures. These structures will be assembled afloat from large pontoon structures of prestressed concrete and post-tensioned together to act as a unit of significantly greater size than the design storm-wave length. Moorings would consist of huge concrete anchor caissons sunk and jetted and/or dredged into the bottom and connected to the floating structure of prestressed anchor cables.

Column-Stabilized Ocean Platforms. The column-stabilization principle has been successfully used to stabilize deep-ocean drilling vessels and, more recently, an offshore construction derrick-pipelayer. By designing the entire structure so that a preponderance of mass is below the surface and comparatively small cross section penetrates the sea-air interface (e.g., columns), the response of the structure to the waves can be reduced and the period of movements lengthened so as to be significantly beyond the period of resonance with the waves. Mooring forces are also substantially reduced. To get large mass below the surface, prestressed concrete is obviously well adapted. In this application, its weight is an advantage. Existing and proposed structures of this sort lend themselves to the use of cylindrical, cone, and spherical surfaces to which concrete can be readily molded. Concrete is also very efficient in resisting the predominantly compressive stresses without buckling, and in absorbing impact in collision. Prestressing provides the necessary structural strength to resist tension and, in connection with mild steel, to resist torsion and shear.

Vertical Float Structures. Following the same hydrodynamic principles, a number of vertical float structures have been proposed, based on use of a large vertical cylinder, moored or otherwise fastened to the bottom, and free to oscillate with the waves. The diameter may be reduced through the wave zone to minimize wave excitation. These can be constructed horizontally and then up-ended by controlled flooding, or they may be constructed vertically in deep water, by successive additions of concrete like a floating caisson.

Undersea Habitats. A number of proposed solutions for the operation of equipment (e.g., oil production and drilling equipment) or for support of man's activities underwater are based on the use of large spheres or cylinders with domed ends moored below the surface or sunk to the ocean floor. Concrete is ideally suited to this application; it is easily molded to the desired shape and is resistant to buckling and impact. Prestressing enables the sphere to take localized bending moments as, for example, at penetrations and discontinuities. Recent research indicates its suitability up to depths of 3000 feet and more. Very-high-strength concrete is particularly suitable for such deep applications.

Underwater Mining Chambers. At the present time, such chambers are in the concept stage only. Many such concepts, however, are based on use of a large cylindrical structure, sunk to the ocean floor, and providing controlled access at both upper and lower ends. Prestressed concrete appears to be the logical answer to these structures, with its economy, weight, durability, safety, and moldability. Presumably, such structures would be sunk into the bottom by dredging and jetting, then sealed by extensive grouting, and perhaps post-tensioned to the sea floor.

Submarine Hulls. Submarine hulls have been proposed for transport of oil, with particular reference to transport under the ice of the Arctic Ocean. The large size involved, and the favorable configuration for the use of concrete, make prestressed concrete particularly well qualified for such construction. Most concepts involve cylindrical hulls with hemispherical closures, for which precast concrete, joined and reinforced by post-tensioning, offers substantial structural and economic benefits.

14.8   Conclusion

Much of the future exploitation of the ocean depends on the construction of large, fixed installations, either floating on the surface or beneath it, or sunk and anchored to the ocean floor.

This is a relatively new concept in naval architecture, since the overwhelming preponderance of man's concern with the oceans in the past has been as a means of transport. The outrigger canoe of the Polynesians, oarpowered galleys of the Phoenicians, the wide-ranging boats of the Vikings, differ only in size from the modern supertanker or ore carrier. This difference in size, however, is more significant than would appear at first thought, because of its relation to the wave length. Therefore, as indicated earlier, prestressed concrete now deserves serious consideration for vessels in the transport industry, particularly specialized vessels.

Storage, operational, and support facilities, however, are of a completely different nature from the transport function. The criteria no longer requires lightness of weight. Durability, security, economy, moldability, and rigidity become the fundamental parameters when selecting a structural material. Prestressed concrete is rapidly becoming recognized as a primary material for the emerging exploitation of our ocean resources.

15

Prestressed Concrete Tanks

One of the earliest uses of prestressed concrete was for water tanks. Since then, many thousand tanks of prestressed concrete have been built for water, oil, gas, sewage, granular and powdered dry storage (silos), process liquids and chemicals, slurries, and, more recently, cryogenics. The nuclear reactor pressure and containment vessels are a special form of tank (See Chapter 16), as are some cooling towers.

When utilized for tanks, membrane stresses are involved in addition to bending and other structural action. The objective is normally to prestress the concrete sufficiently so that it remains crack-free under normal working stresses with a suitable safety allowance selected on the basis of the probability of overload and the consequences of cracking. For gravity storage of water, oil, etc., this usually requires zero tensile stress at design load. In some areas, seismic loading of tank plus contents controls the design loadings.

Most prestressed concrete tanks have been circular in cross section, with circumferential prestressing sufficient to eliminate tensile stresses at each horizon. The circumferential tendons may be continuous, applied by wire-winding under stress, or applied hand-tight and wedged or jacked out by a sequential operation to provide the necessary stress. Circumferential tendons may also consist of overlapping tendons between buttresses. For example, there may be 6 buttresses with each tendon spanning one-third of the circumference.

Prestressed concrete tanks may also be prestressed triaxially by helical tendons crossing at 45°. This permits the use of relatively short tendons, such as bars, and minimizes friction.

303

Tanks may utilize prestressing in the vertical direction in combination with circumferential reinforcement in the form of prestressed tendons or conventional mild steel.

Square tanks may be required for industrial uses and may span either horizontally or vertically. The effect of deflections under load must be considered. Square tanks offer advantages for storage in congested urban and industrial sites because of more efficient space and land usage.

Tank walls may span vertically between a top and bottom ring or frame. The rings may be prestressed circumferentially, and the walls vertically. Thin concrete shells, such as hyperbolic paraboloids, may be used to span vertically between the rings. Once again, the matter of deflection at vertical joints and its effect on leakage, etc., must be considered.

Multiple tanks and silos have been constructed by utilizing interlocking polygons, such as hexagons, prestressed together to act as a unit with transverse and/or vertical tendons.

Elaborate double-curved shells have been built to store sewage and other active substances where both efficiency of storage and efficiency of operation can be facilitated by the shape. Similarly, double-curved shells have been built for water tanks and cooling towers, to take advantage of the efficiency of shell-action of the concrete, combined with prestressing at the edges.

Tank designs have not always taken full advantage of the possibility of internal bracing and tying. With water, oil, etc., such ties offer no functional problems. Obviously, corrosion problems must be considered in the light of the usage, and the necessary protection provided.

Prestressed concrete is essentially water-tight but it is not gas-tight. Where vapor must be retained, a thin membrane liner of steel may be used. This has proven so successful that the metal liner concept is being increasingly employed, even for water tanks. The liner may be fluted, to permit expansion and contraction under alternating filling-emptying conditions and under prestress, or it may be flat, with suitable anchors to the concrete. Such a liner may also be considered to provide a portion of the vertical reinforcement.

Such composite construction permits the ductility and membrane ability of steel to be combined with the economy, rigidity, and virtually unlimited tensile capacity of prestressed concrete. The thin liner is readily welded; thus tanks of larger diameter and height, or greater pressure resistance, can be obtained than are possible with steel alone.

This concept is especially important when special steels are required, as for cryogenics, or where localized rigidity or external loading may occur. For underground or underwater tanks, prestressed concrete is able to economically resist both the internal and the external head.

The concrete for tank walls may be cast-in-place concrete, precast concrete panels, or shotcrete.

Cast-in-place concrete walls are often cast in alternate segments to the full height, to allow shrinkage to be dissipated. Panel forms are utilized, usually so framed that no ties pass through the walls. When an internal steel liner is used, form ties using welded studs may be used, with the outside end of the tie being later removed and patched. Form ties which incorporate integral waterstops may also be used.

At construction joints, both vertical and horizontal, the joint must be cut back by water jet or sandblasting so as to expose the aggregate. Thorough soaking before the next pour, or an epoxy-bonding compound, will aid bond. The epoxy must, of course, be stable and compatible with the material to be stored.

Slip-forms are widely employed for cast-in-place concrete construction of tanks and silos.

Precast panels may consist of vertical "staves," horizontal "planks," diagonal slabs, geodesic or folded plates, or thin shells. Such precast panels, staves, etc., may be pretensioned, with post-tensioning in the crossing direction applied after erection. (See Fig. 84)

Buttress-type tanks have been constructed by utilizing precast buttresses with embedded anchorages assembled in the field with cast-in-place walls.

Shotcrete may be applied against inner forms or steel liner thus forming

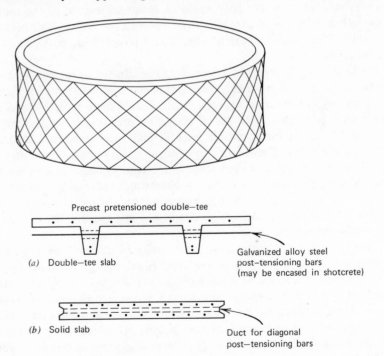

Precast pretensioned double—tee

(a) Double—tee slab

Galvanized alloy steel
post—tensioning bars
(may be encased in shotcrete)

(b) Solid slab

Duct for diagonal
post—tensioning bars

Fig. 84. Helical post-tensioned tanks constructed from precast pretensioned slabs.

the entire concrete wall, or it may be employed as the outer, protective cover over the prestressing tendons. The techniques for installation of shotcrete are discussed in a later section of this chapter.

A tank usually consists of floor, walls, and roof. A great deal of attention has been given to the design and construction of the walls, as compared to the roof and floor slabs.

The floor slab is actually a composite system of soil, sub-base, and slab. The floor slab may be structural or may be a membrane supported by the sub-base. In this latter case, footings are required under the walls.

Floor slabs have been shaped concave to provide drainage, or convex, to provide shell loading on the soil; that is, a hyperbolic paraboloid foundation shell.

In most sizeable tanks the floor slab and walls are provided with a non-rigid connection, so as to reduce bending stresses in the walls and permit movements of the walls under prestressing and loading variations. These connections may allow limited or full translation and rotation.

Roof slabs have consisted of cast-in-place or precast slabs, supported on internal posts, or of thin shell domes, reacting against a prestressed edge ring. Wall-roof connections may be either monolithic or separated so as to allow translation and/or rotation.

Relatively little use has been made in design of the use of the roof and floor slab as a means of providing transverse ties at the top and bottom of the walls. This would appear to offer useful possibilities, particularly with square tanks.

The design of prestressed concrete tanks is affected by internal and external loads, by conditions of edge restraint at the wall junction with floor and roof, and by construction aspects, such as shrinkage, moisture gradients, creep, steel relaxation, modulus of elasticity, prestressing stages, and time intervals. External loads include backfill and surcharge, roof loads, and aerodynamic loads. These latter may occur during construction and must therefore be considered at each stage in the case of high, thin-walled structures. The design must also consider ground water uplift on floor slabs.

Construction tolerances, both in wall thickness and out-of-round (or other shape), affect both the final design conditions and the stability of the structure during prestressing.

Shrinkage is particularly severe with tanks because of the thin sections and large exposed areas. Adequate curing is extremely important for durability and the prevention of surface cracks and crazing. Water cure may be provided by soakers or spray, with burlap. However, high walls exposed to strong winds may require more positive protection, e.g. membrane curing.

Non-prestressed reinforcement may consist of welded wire mesh or conventional mild-steel bars. A liner plate may also serve to provide a portion of the requirements.

When buttresses are employed, they must be detailed so as to provide adequate clearance for the anchorages and the stressing equipment. The buttress is usually a highly congested area and details should be laid out full scale to ensure adequate room for concreting. Particular attention should be paid to vibration in these areas and to selection of a mix that can be properly placed. Anchorages must be rigidly held to prevent displacement during concreting.

Roof joint details must not only provide the desired condition of restraint or freedom but must also provide a positive seal to prevent moisture penetration between steel and concrete, or between concrete and shotcrete, etc., which would lead to corrosion of the prestressing tendons. If the roof is constructed before the tank is post-tensioned, the roof-joint detail must permit free movement inward of the tank walls as they are stressed.

Joints between precast panels must be designed to transmit the shear and local bending stresses and deflections. This is of particular significance with shell-type units such as geodesic plates, folded plates, hyperbolic-paraboloids, etc. Welding details must consider the effect of heat and the possibility of spalling of adjacent concrete. Concreted or grouted joints must be detailed to ensure that the joint will possess the required strength. Fully concreted joints, 3 inches or more in thickness, are preferable to thin grouted joints. Epoxy-bonding compound and/or use of expansive cement in the joints will help to prevent shrinkage cracks. Joints should be well-cured.

The installation of the prestressing wires and the stressing of the tendons will produce temporary wall bending stresses. The sequence and stages of stressing must be set forth so as to maintain wall-bending stresses within allowable limits.

Openings are generally accommodated by deflecting the strands above and below, in bands. Individual tendons may have to be spaced so as to prevent excessive concentration of force such as might occur with bundling. Additional mild steel may be placed at right angles to the prestressing tendons to contain bursting and radial forces.

Brackets introduce localized structural loads which must be distributed. Any change in concrete cross-sectional area should be gradual (tapered) to avoid stress concentrations.

A long delay in filling of a tank may permit excessive shrinkage and creep.

Mild-steel anchor bars from floor-to-wall and wall-to-roof must have adequate embedment and anchorage to prevent pull-out under dynamic loading.

Prestressing tendons may consist of one of the following:

(a) High-tensile wire, in which the stress is obtained by drawing through a die or on a series of brakes.

(b) Bars of high-tensile alloy steel, joined with couplers where necessary.

(c) High-tensile strand, black, galvanized, or plastic-encased. This may be stressed at buttresses, or continuously wrapped and stressed by appropriate machines, or by wedging.

(d) High-tensile cold-drawn steel strip.

Tendons may be external, as noted, or internal, enclosed within ducts. They are subsequently injected with grout. (See Chapter 5, Subsection 5.1, Grouting.)

Earthquake cables, consisting of unstressed galvanized strands, are frequently used to anchor the walls to the floor slab when the joint details permit rotation. Sleeves of rubber or similar material may be placed around the strands so as to permit radial movement of the wall. The portion of cable to be encased in the sleeve should be given a protective coat against corrosion.

With dome roofs, two layers of steel should be used near the edge, in the meridional direction, to resist edge-bending movements. The dome may require thickening in this edge region. These movements can be minimized by proper detailing of the joint.

To prevent shrinkage and temperature cracking in the dome roof prior to prestressing, mild steel, such as wire mesh, should be placed and water curing continued until the prestressing is completed.

Positive keys or stops or anchor cables must be provided with unrestrained joints so as to prevent displacement of the roof relative to the walls.

Floors must be designed to resist hydrostatic uplift when empty. They may be anchored to the subsoil by stressed or unstressed rods. With proper gravel underdrains, floor pressure relief valves may be a solution. The gravel should be protected by a sheet of polyethylene from contamination and clogging by soil.

If sand is used under the floor slab, the edges should be protected from erosion or wash out. A low concrete wall or bitumastic seal coat may be used to retain the sand.

For steel diaphragms (liners), vertical ribbing or fluting is desirable. Sheets should be seal-welded, brazed, or sealed with elastomeric caulking and sealing compound to form a permanent, flexible seal. Polysulfide liquid polymers are used, applied with a caulking gun. Epoxies are also used, applied by pumping up the joints between the sheets and the wall.

Waterproofing of the exterior is occasionally required, especially when the tanks will be backfilled. The thin concrete cover over highly stressed tendons makes such additional steps desirable to ensure durability despite any latent errors in concreting or shotcreting, or any porosity, etc. External paint may be bitumastic or one of the paints listed in the next paragraph.

Internal paints are employed when the tanks will contain acids, sewages, gasoline, etc. Rubber-base paint, polychloride vinyl-latex, polymeric vinyl-acrylic, and epoxy paints are used, selected for their durability in continuous contact with the stored material.

Waterstops are usually of rubber, or plastic such as extruded polyvinyl chloride. Butted joints should be fused together to ensure water-tightness. Waterstops should be placed in joints in floors and wall footings. Waterstops must be used in wall joints unless other positive means are taken, such as liners, or especially prepared joints (expansive cement, epoxy bonding compound, etc.).

Bearing pads are normally of neoprene or natural rubber. A combined bearing pad waterstop of extruded virgin polyvinyl chloride may be used.

Sponge filler should be closed-cell neoprene or rubber.

Concreting of a dome roof should be performed in circumferential strips, working from the exterior edge towards the center. Circumferential and tangential screeds should be placed to ensure uniform thickness. The thickness can also be checked by probing the fresh concrete. The dome roof should be supported by scaffolding or posts until prestressed.

15.1 Shotcrete

Shotcrete should conform generally to the provisions of ACI 506. Calcium chloride must *never* be used in the mix. Shotcrete should be applied with a very dry mix, so that frequent slightly dry spots appear on the regular glossy surface. Joints should be cleaned with an air-and-water blast. They should be as nearly at right angles to the surface as possible.

Walls should be built up of individual layers of 2 inches (5 cm) or less in thickness, using tensioned vertical ground wires not more than 3 feet apart to control thickness.

Protective coatings over prestressing tendons may consist of shotcrete or cast-in-place concrete. Shotcrete is most commonly employed. It must be emphasized that the placement and control of shotcrete is a skilled art. Failure to properly apply shotcrete may result in trapping rebound behind the tendons, leading to porosity and susceptibility to electrolytic corrosion.

Each layer should be protected either with a cement slurry coat or a flash coat. The cement slurry is desirable but difficult to apply; the flash coat usually achieves satisfactory results due to loss of sand by rebound. The outer layer should be protected by a flash coat, body coat, and finish coat. A suitable paint sealer may be provided over the final coat to ensure water-tightness, and to prevent shrinkage cracks. This author believes such painting is justified because the cost is small in relation to the assurance achieved. Painting is strongly recommended in coastal and arid regions, and near industrial areas.

The flash coat is especially important in providing protection to the steel. Care must be exercised to ensure absence of voids, trapped rebound, etc. The flash coat should be a 1:3 mix, wet but not dripping, and should provide 1/8 inch (3 mm) cover over inner wire layers, 3/8 inch (1 cm) over the outside layer.

The nozzle should be held at a 5° upward angle, moved constantly and regularly, pointing always at right angles to the surface (i.e, towards the center of a circular tank). The nozzle should be held at such a distance that the shotcrete does not build up or cover the front face of the wires until the spaces between the wires are filled. Immediately after placing the flash coat, a visual inspection is made. If wire patterns show up as continuous horizontal ridges, the shotcrete has not been driven behind the wire properly. If the surface is substantially flat, with no particular showing of the wire pattern, then the space behind the wire is essentially filled.

The flash coat should be damp-cured.

The body coat is usually 3/8 inch (1 cm) thick, of 1:4 mix. It should be screeded prior to final set, and damp-cured.

A finish coat of 1/4 inch (6 mm) is frequently applied. Use a 1:4 mix.

Total coating thickness over the tendons should be not less than 1 inch (2.5 cm). The completed shotcrete coating should be continuously damp-cured for 7 days.

Where practicable, the tank should be filled with water prior to the application of the body coat.

15.2   Wire Wrapping

Wire wrapping under tension is extensively employed for circular tanks. Specifications commonly required a tolerance in tension of plus or minus 7%. As the wire is wrapped, temporary anchoring clamps should be attached to prevent loss of prestress if a wire should break. Anchoring clamps may be removed as the cover coat is applied. Terminals of individual wire coils may be spliced with splices capable of developing the full tensile strength.

A calibrated stress recording device should be employed continuously to record the tension levels during the wrapping process. In smaller tanks, intermittent readings are customarily taken; however, continuous readings are obviously preferable. Readings should be identified with height and layer of wrap.

If wire stress falls below the specified tolerances, additional wraps can be placed to compensate.

Wire spacing in a layer should normally be 5/6 inch (8 mm).

When strand or bar tendons are used externally, small saddles of steel are usually set at frequent intervals to minimize friction. These saddles should be staggered, or placed on a diagonal, so as to distribute the concentrated radial stress. Such saddles may be part of a wedge stressing device.

A ratchet-type jack has been developed that can stress bars at a coupler, pulling them together and anchoring. Other jacks permit tensioning on a slight outward curve, with minimum deviation from the circular path.

When tendons are placed, it is essential that corrosion protection be applied within as short a period as possible, e.g., 24 hours or less. It is especially important that no salt crystals be deposited by salt spray or salt fog. Prior to coating, the tendons should be thoroughly washed with fresh water under pressure.

Tendons likewise should not be exposed to a combination of moisture and sulfides, such as can occur near a refinery, especially during a fog. Tendons inside ducts can be protected by VPI powder, dusted in and sealed. External tendons should be washed as above and coated as soon as possible.

Under conditions of severe exposure, underlying concrete (or flash coats) should be jetted with fresh water both before wrapping and before coating.

Vertical tendons which are to be grouted after stressing should have the ducts thoroughly flushed with fresh water after each lift of concrete is placed, to prevent blockage through in-leakage.

Wire-wrapping machines impose a vertical bending movement on the walls. The walls should be suitably reinforced for this condition, since wire winding usually is performed prior to vertical prestressing. When wire is extruded through a die, means must be provided to cool the wire so that the heat generated does not affect the wire properties. In any event, temperatures should be kept well below the 300°F level.

15.3 Tolerances

A maximum out-of-round tolerance of plus or minus 3 inches (7.5 cm) per 100 feet (30 m) of diameter should be maintained for circular tanks. Wall thickness should be kept to a tolerance of + 1/4 inch (6 mm). All transitions should be gradual. Walls should be vertical with a tolerance of +3/8 inch (1 cm) per 10 feet (3 m) of height.

Elastomeric bearing pads should be attached to the concrete with adhesive to prevent displacement during concreting. Any voids or cavities occurring between butted ends of pads, pads and waterstops, pads and sleeves, or sleeves and waterstops should be filled with a soft non-petroleum-base-mastic compatible with the pad, sleeve, and waterstop materials, and with the material to be stored in the tank.

Sponge-rubber fillers should be ordered 1/2 inch (1.25 cm) wider than the gap, in order to facilitate placing and reduce possibility of voids between sponge rubber, bearing pads, and waterstops. Sponge rubber should be secured to the concrete with an adhesive.

When wire mesh is employed in thin shells, "dobe" blocks or plastic spacers should be employed to keep the mesh at the proper distance from one face. For such thin shell applications, galvanized mesh is desirable.

15.4 Evaluation

Prestressed concrete tanks have a long history of use for water storage and to a lesser extent for storage of other materials. Unfortunately, a few cases of corrosion and failure have occurred. Most of these can be traced to the following causes:

1. Sewage or other aggressive materials penetrating to the tendons through large, unclosed shrinkage cracks in the primary wall.
2. Salt-cell electrolytic corrosion in coastal regions due to porosity of shotcrete.
3. Use of calcium-chloride in the shotcrete mix.
4. Leaving tendons stressed for long periods prior to encasement, in an industrial area exposed to fog.
5. Separation of the joint between wall and cover coat, allowing moisture to penetrate to the tendons from the top.
6. Corrosion of sleeves and valves at penetrations through tanks. Inadequate sealing and electrolytic corrosion of fittings.

Any tank used to store salts, particularly those containing chlorides, should be waterproofed on the inside to prevent salt-cell electrolysis.

The use of a steel liner, external painting, and sealing of the top of the wall (juncture between roof and wall) will solve practically all durability problems.

Prestressed concrete tanks can be reliably and economically built to perform a wide range of services, from underwater oil storage, to buried fuel tanks, to elevated water storage, to silos for granular materials. Both hot liquids and cryogenic materials are safely and practically stored in prestressed concrete tanks. As with pressure vessels, the failure properties of prestressed concrete are inherently safe; they fail by cracking and relief of pressure rather than by sudden or catastrophic ripping. They are durable. They can be built to extreme diameters and heights, and to accommodate high pressures. Some of the elevated water tanks of prestressed concrete in Western Europe are among the most attractive and beautiful of all architectural structures. The future of prestressed concrete tanks, therefore, depends primarily on increased reliability and dependability, especially in regards to corrosion, by following the provisions outlined in Chapter 4 (Durability) and this chapter.

16

Prestressed Concrete Pressure Vessels

16.1 Introduction

The construction of prestressed concrete pressure vessels is both a challenge to the constructor and a major consideration in the design of the vessel. A pressure vessel is a subsystem within the overall system, and as a subsystem, may involve a complex integration of concrete, prestressing tendons, anchorages, ducts, mild reinforcing steel, penetrations, liner, and cooling system. The successful integration of all of these into a completed vessel requires that the best and most sophisticated techniques of construction engineering, planning, scheduling, coordination, prefabrication, materials handling, and installation, all be brought into full utilization.

We have present in pressure vessels three aspects not normally present in other types of engineering construction:

1. The quality control, accuracy, tolerance limitations are extremely severe.
2. The work site is very compact and, therefore, highly congested.
3. The time available is very limited and very critical.

The solution, of course, is adequate construction preplanning, using the best planning tools available. Thus, methods can be specified in detail that will assure adequate quality control; forms, supports and concreting methods can be chosen to achieve accuracy; prefabrication can move much of the labor away from the immediate focal point; and proper scheduling can assure completion on time.

However, it must not be overlooked that we are dealing with real, tangible

313

materials and equipment, and with groups of individual workmen and site supervisors. We are subject to all the multitude of practical limitations and variations that occur on all construction; the difference is that with pressure vessels, we cannot tolerate anywhere near the range of variances and mistakes that we accept elsewhere.

It must be again emphasized that all of the activities, as well as others relating to the main plant itself, are going on simultaneously at one focal point—a cube perhaps 100 feet in each dimension. After discussion of the individual activities and the techniques required, further examination will be made of the scheduling, coordination, and management of construction.

Prestressed concrete pressure vessels have been primarily connected with nuclear reactor power plants. This development took place as a result of the demand for larger vessels, capable of withstanding higher pressures. The present generation of light-water reactors has utilized a heavy steel primary reactor vessel and a secondary containment structure. This containment structure serves to contain fission products in case of accident, and thus must also serve as a radiation barrier. Prestressed concrete is thus excellently suited to utilization as a containment structure and is being so employed on a number of nuclear power plants in the United States, France, and Canada. A containment structure is a rather unique structure in that the only times it functions as a pressure vessel is during acceptance and annual tests, and for a short period following an accident.

For the CO_2 gas reactors a single pressure vessel of prestressed concrete is utilized. Prestressed concrete is ideal for this use because of its behavior during progressive stages of loading and failure. The several stages may be as follows:

Stage 1—Elastic behavior.
Stage 2—Cracks form and widen, liner keeps vessel tight. Upon reduction of pressure, cracks close.
Stage 3—Cracks widen, liner ruptures, gas escapes locally, reducing pressure.
Stage 4—Wide cracks, spalling, wide leaks, but *no* catastrophic ripping or bursting.

These reactor pressure vessels must withstand large thermal gradients as well as pressure. There are numerous transient stages which have to be thoroughly investigated in design.

Future generations of nuclear power plants will probably involve fast breeders, cooled by sodium, gas, or steam. In the meantime, there is an existing development of advanced reactors such as the high-temperature gas reactor utilizing helium, and the light-water breeder reactors. Far in the future are the deuterium reactors, with their almost unlimited available energy from the oceans and with no radioactive waste. Dual purpose plants, combining desalination and power, and "energy centers" are presently undergoing feasibility engineering studies. Thus, a tremendous challenge is being presented to the engineer and constructor, both now and in the future.

While attention today is concentrated on pressure vessels for nuclear reactors, prestressed concrete offers similar significant advantages to pressure vessels for chemical plants, refineries, etc. The non-catastrophic mode of failure, the ability to assemble and construct on site, the lack of limitation on size and pressure, make prestressed concrete an ideal material for such use. The problems of penetrations are no more severe than for nuclear reactor pressure vessels and may even be simpler of solution.

While "prestressed concrete pressure vessels (PCPV)" has become almost a generic term, actually a pressure vessel may embody a large amount of conventional reinforcing, especially high-yield steel, and may act as a partially prestressed concrete structure as it passes through its several "limit states" or behavior stages.

The prestressed concrete pressure vessel requires a "systems engineering" approach to construction. The typical system includes the following:

1. Concrete, including forms, batching, mixing, placing, curing.
2. Reinforcing steel.
3. Liners.
4. Cooling tubes.
5. Penetrations.
6. Ducts (sheathing and protective caps).
7. Sheathing couplers, trumpets, transition cones.
8. Air vents and vent piping.
9. Tendons.
10. Anchorage assemblies.
11. Bearing plates.
12. Corrosion protection.
13. Duct filler retaining caps.
14. Placement drawings.
15. Stressing record forms.
16. Equipment selection.
17. Quality-control procedure.

16.2 Concrete

16.2.1 Production and Delivery of Concrete to Site

While the design engineer will undoubtedly have selected the aggregate sources, the type of cement, mix proportion, and water/cement ratio, it is the province of the contractor to establish methods to produce concrete of consistent, uniform quality. Many properties of the concrete have an important effect on its performance for pressure vessels. These properties include thermal expansion, conductivity, and stability, creep, and shrinkage at normal and elevated temperatures, strength, modulus of elasticity, Poisson's ratio, tensile strength, and heat of hydration.

The aggregate is normally crushed, washed, and screened at the crushing plant and delivered by cars or trucks to the site, dumped in piles or hoppers at the site and stored pending elevating to the batching plant. It is obvious that the chances are manifold for contamination, chipping, and dusting, etc.

First concern should be uniformity of production at the quarry to be sure weak or unsatisfactory material from seams cannot be blended in. Depending on the source, a careful geological inspection should be made, a particular block of the quarry selected for this production and continuously inspected to prevent accidental blending from other blocks or sources.

At the site, aggregate for the vessel should be stored on concrete slabs, or in bins or metal hoppers, not on the ground. For a reactor vessel, the aggregates should be kept separate from all other concrete supplies at the plant.

Aggregates should be protected from excessive heat (by covering or the water soaking-evaporation method) and from ice and snow. Rescreening just above the batch bins will remove chips. The batch bins should be large enough to assure uniformity of moisture content.

For the cement, it is important that it be properly aged and that its temperature be within established limits.

It is important to limit the contamination in the mixing water and in the concrete to prevent the possibility of electrolytic corrosion. Recommended limits (by weight) are:

Mixing Water
Chlorides as Cl   250 ppm
Sulfates as SO_4 250 ppm
In Wet Concrete
Chlorides as Cl   1000 ppm
Sulfates as SO_4 1000 ppm

The ambient temperature of freshly mixed concrete should be between 50° F and 60°F (10°C and 15°C). This can be accomplished in summer by using crushed ice in the mixing water and by cooling by evaporation in the aggregate storage. Vacuum cooling may be utilized effectively in extremely hot climates. In winter, the mixing water can be heated, and the aggregates kept in warm storage or heated.

For the low-slump concrete involved here turbine mixers are preferable to ready-mix trucks. Mixing time should be carefully regulated; variations will produce wide divergence in quality and workability.

Transportation to the vessel site should then be in hoppers, buckets, truck-mounted hoppers, etc., which are specially designed for low-slump concrete. These usually have vibrators or screws for discharging. Agitating vibrators or paddles may be employed to prevent segregation and false set.

To prevent segregation, and to increase workability while maintaining the required water/cement ratio, an admixture may be effectively utilized. This should be one containing no calcium chloride and should, of course, be subject to the approval of the engineer.

This matter of workability is of prime importance to the contractor. The vessel is full of highly congested spots and it is often just at these spots that the best consolidation is needed. It is recommended that prior to any concreting of the vessel itself, a full-scale mock up of one area of typical congestion be made and actually poured with the design mix. Then it can be stripped, cut into, and the thoroughness of consolidation verified. This insures that there will not have to be last minute changes in the mix, placement methods or vibrators, as concreting is underway. In addition, it is the best possible means of indoctrinating the workmen as to the why and how of the concreting procedure. This writer has used this mock-up approach very successfully on other special and congested prestressed construction, such as the ends of long-span girders.

For the Fort St. Vrain, Colorado, HTGR pressure vessel, the following mock-up procedures were followed:

1. A penetration liner mock up simulating an actual penetration liner, to demonstrate the adequacy of the concreting procedure in producing sound concrete in the area of penetrations, cooling tubes, and shear anchors.
2. A full-scale mock up of a 60-degree sector of the bottom region, extending 5 feet below the bottom head liner and involving 80 cubic yards of concrete.
3. Two mock-up demonstrations for evaluating the technique for applying concrete by special means to the bottom head liner, to demonstrate production of void-free concrete of specified strength in the bottom head.
4. A full-scale mock up of a 120-degree sector of the PCPV head to establish the sequence for installing rebar and other embedments.

16.2.2 Placement

The concrete pours will have been carefully delineated by the contractor and engineer. Generally, lifts of 5 or 6 feet are scheduled, and pours in a size range of 100 to 300 cubic yards per day. Form pressures and deflections, heat and shrinkage are all related to the size and height of lift.

For containment structures slip forms have been used, requiring continuous pours over a period of up to 10 days.

Because of protruding ducts and reinforcement, it will probably not be possible to get the bucket closer than about 6 feet vertically to the surface. A rubber elephant trunk on the bucket bottom can be used to prevent segregation during fall.

Thorough internal vibration is essential. In addition, form vibration on the liner plate can be used to insure intimate contact between liner and concrete. The frequency of vibration required depends on the mix; for some very dense harsh mixes, higher frequency and more powerful vibrators may be required. The effect of powerful vibration on form pressures must not be overlooked.

Care must be taken to prevent dislodging or mis-location of ducts or reinforcement during vibration. This may be helped if position of ducts and critical areas are clearly marked on the side forms.

Curing procedure should be selected so as to insure adequate maintenance of moisture for concrete maturing and may be used to absorb some of the heat of hydration. Circulation of water through the ducts and cooling tubes may be employed for heat reduction. Whether water curing or membrane curing is employed, the important thing is to apply it early enough to prevent surface drying. In very hot, dry areas, especially with a wind, this surface drying may present a major problem. Enclosures and scheduling of concrete pours for night may be required. Night pouring also aids in reducing the ambient temperature. Some reactor vessel construction has been completely enclosed and relative humidity and temperature control maintained.

Construction joints present a major concern. Horizontal surfaces may be jetted so as to expose the aggregate, while the concrete is green (say, 12 hours old), with air and water. Side forms may be stripped at age 12 to 24 hours, and jetted. High-pressure hydraulic jets can clean off laitance and mortar even after a few days of age.

Retarders have been used to facilitate this; they are painted on the fresh surface or on the forms. This author recommends against their use if at all feasible; the chances exist for retarders to spill or get onto reinforcement, etc. Their use is not necessary if the jetting is performed at the proper time with the proper equipment.

In starting a new pour, the old concrete should have been kept thoroughly wet for 24 hours. The new concrete lift should then be well vibrated near the old surfaces. Use of a rich grout coating, as practiced in other engineering construction, is not believed necessary nor desirable for the mixes and requirements of pressure vessels. Here again, a mock-up pour, including a horizontal construction joint, with all the congestion of ducts and bars across the joint, will prove the proper technique of placement and vibration.

Sand-blasting can also be used to prepare construction joints, but produces a lot of sand rebound which is difficult to keep out of ducts, etc. Therefore, its use is not recommended.

Bush-hammering of vertical joints appears to be a very practical solution if jetting is delayed too long. This is being used to prepare vertical joints on segmental bridge construction and, for the relatively small congested areas involved, appears practicable. It should be followed by wire-brushing.

One of the main problems in all of these operations is keeping mortar and debris from contaminating other surfaces. Ducts must be tightly covered, as must cooling-pipe openings and anchorages. Mortar splash onto reinforcement must be kept to a minimum. Before starting a new pour, the cleanliness of adjacent areas must be restored. Curing water must be kept out of ducts, or well drained.

To ensure complete consolidation of concrete around anchorages and penetrations and the underside of the liner base, grouting has been employed. Holes near the surface are formed by inflatable tubes; after concreting, the tubes are deflated and withdrawn, and grout is injected under pressure.

At the base of the liner it seems best to set the liner on supports a few inches above the screeded concrete surface, and then inject grout in one or two courses.

Systematic testing procedures must be carried out on all concrete pours. A fully equipped concrete laboratory should be maintained at the site.

Forms should be designed to limit deflection and to be rugged against accidental impact from buckets and vibrators. It is particularly important that anchorages and penetrations be rigidly and ruggedly held. Form ties must be considered by both contractor and designer, as they represent a heat transfer path. Thus, it may be necessary to use the type that is subsequently removed and, in this case, they must be later filled with grout. For this reason, ties within a pour must be kept to an absolute minimum; ties across above the top of the lift and anchors in a previous lift may eliminate the need for internal ties in a pour.

If, in spite of careful planning and supervision, rock pockets or honeycomb do occur in the concrete, repairs must be instituted. For example, on one containment structure, the first pour was too large, slump too low, and consolidation by vibration proved inadequate. For this containment vessel, a specially formulated epoxy, designed to be stable at 180°F, was injected. In other pressure vessels, grout injection may be needed, or specific spots may have to be cut out, with keys and a well-prepared surface, and new concrete placed. In such cases, shrinkage of the repair grout or concrete must be reduced by careful control and curing, and by judicious use of non-corrosion-producing admixtures, or shrinkage-compensated cement.

16.3   Conventional Steel Reinforcement

This may be either mild-steel or high-strength deformed bar reinforcement. There is a strong tendency in design to use greater amounts of such reinforcement, particularly with unbonded tendons. This presents a special problem in pressure vessel construction in congestion and support. Reinforcement must be thoroughly tied. Where feasible, prefabricated cages may be installed, as for example, around penetrations.

Splicing of small bars is generally by lapping. For bars of #11 size or larger, splicing should be by welding or mechanical connectors. If butt welding is required, care must be taken not to accidentally burn adjoining ducts, and any tendons must be positively protected.

This reinforcement may be used effectively to support the ducts; in fact, a careful design may produce the necessary framework for duct support. Once again, the contractor must work with the designer.

16.4   Liners and Cooling Tubes

General practice has been to prefabricate steel liners to the side of reactor vessels, then skid and jack them into position on the concrete base. In other cases, prefabricated sectors are made and lifted into position, requiring jointing at the site. It is important to note that the erection sequence chosen may influence the behavior of the liner.

In most reactor construction there is a huge gantry or similar crane available to lift the reactor core; this can be often used to lift in the liner assembly.

Liners must be tight; therefore, a high-quality welding technique must be employed with radiographic and other testing. Both the steel and the welds must have ductile (non-brittle) behavior at the lowest temperatures to which the liner will be subjected, both while in service and during construction.

It seems most desirable that the cooling tubes be installed on the liner plate in the prefabrication phase. Care must be taken in welding not to distort or warp the liner excessively. Similarly, stud anchors for the liner should also be installed during prefabrication. Gun-welded studs may prove very practicable. Care must be exercised not to create any notch effects in either liner or anchors.

Once the liner is set in general position, it must be jacked and aligned into exact position and held there during concreting. It may be possible to use the cooling tubes as structural reinforcement, or to use the liner rib reinforcement as temporary support. Local deformation due to concreting pressures must be severely limited by adequate bracing, supports, or ties.

When welding on the liner after concrete has been placed (as for example, sealing holes in the base through which grout was placed) extreme care must be taken to minimize heat which may warp the liner away from the concrete. This also applies to welding of liner plate joints where concreting has been carried to or past the joint. This joint must be detailed properly to prevent warping or pulling away from the concrete.

In the case of containment structures such seams may be purposely designed for flexibility, with anchorages in the center panels only.

Cooling water tubes must be protected at all stages from internal corrosion, accidental filling with debris, grout, etc. They are generally filled with nitrogen during welding. They are tested with nitrogen for leaks after welding and after concreting. Silica gel may be injected after welding. Cooling pipes must be capped at all stages to keep debris from falling in.

Corrosion protection must be provided to the liner assembly (liner, cooling tubes, anchors) during fabrication, assembly, and installation, as well as during service. When organic compounds are specified, surface preparation must be carefully carried out in accordance with specifications, and coating thicknesses maintained with no "holidays." Additional corrosion protection may be required at local points of concentration, such as low spots, etc.

During concreting, sufficient supports, both ties and internal struts, should be provided to insure that the roundness of the liner is maintained, with no local flattening or bulging.

By applying shotcrete to the liner, after attachment of anchors and cooling tubes, the liner may be stiffened and protected from corrosion. Shotcrete requires the specification and enforcement of procedures that ensure against entrapment of rebound and non-uniformity. Rebound must be removed from the site. Only skilled and experienced operators should be employed. Careful visual and sonic inspection should be carried out on the completed shotcrete layer to verify uniformity and non-porosity. Shotcrete should not be used to fill pockets or corners.

Plastic liners are under serious consideration, including those applied by spray and those put on in the form of sheets with adhesives or anchorages. Extreme care must be taken in protecting plastic sheets during storage and installation, including protection from sun's radiation, to insure no change in color or shade, cracking, blistering, or swelling, since these defects are usually progressive.

The adequacy of plastics is often dependent on the substrata, that is, the concrete surface and the adhesive. Surface preparation must be carried out with agents and techniques that are compatible with the plastic, for example, use of hydrochloric acid to clean the concrete may cause corrosion of stainless steel sheets or disintegration of adhesives, etc. This care applies to the form oils and form parting agents used for the concrete, to the materials and techniques for correction of defects in the concrete surface, and to the chemical and mechanical means for preparing the surface.

During application of the plastic liner, strict adherence must be had to the specifications and limitations on the mixing time, temperatures during application, time intervals between successive applications, thickness of coats (or adhesives), qualification testing of applicators, and local tests on completed liner to check final dry film thickness.

Procedures must be set up for repairs of defects.

Laps of plastic to steel and of plastic to adjacent sheets or previously applied layers must be given particular care. This also applies to the juncture at penetrations.

16.5   Penetrations

These are almost always prefabricated. They may be attached to the liner during prefabrication, but usually are installed after the liner is in position. They require especially careful support to prevent dislocation due to dead weight or concreting. The welds are then made to the liner. Most penetrations are surrounded by mild reinforcement and the prestressing ducts are deflected past them. This results in extreme congestion, yet it is precisely here that the best concrete is required. Thus, prefabrication of a complete assembly, including encasement in a precast block, seems to be a good solution for large penetrations.

16.6   Ducts for Prestressing Tendons

The selection of the proper duct material is of prime importance in determining the quality of final construction, friction losses in prestressing, and even concrete quality. Duct tubes must be able to withstand the external wet concrete pressure (approximately 4 psi) without collapse.

Sheathing, trumpets, and transition cones should preferably be of ferrous metal to prevent electrolytic corrosion with the tendons. If other than ferrous metal is used, then electrolytic couples must be prevented. Copper and aluminum are dangerous in this regard and should not be used. If organic materials, such as plastics, are used for ducts, or if organic coatings are applied, they must not introduce sulfides, nitrates, or chlorides due to decomposition during the life of vessel.

In some vessels, thin flexible tubing has been used. Such tubing is not absolutely water-tight, and is liable to local deformation and "wobble." When used, an internal mandrel should be inserted to give support. Inflatable ductubes and light gauge metal conduit have also been used as mandrels. These will follow most curvatures required.

Thin-gauge rigid metal ducts appear to offer the most satisfactory solution. Rigid ducts must be preformed, either in the factory or at the site. Factory preforming is practicable only for reasonable lengths; the longer lengths are prefabricated at the site.

Joints and seams must be sufficiently tight that they do not leak grout or laitance. All joints and connections should be taped.

Standard water pipe, pre-formed to the correct curvature, with screwed or rubber sleeve connectors, has been utilized as tendon ducts.

Ducts are installed progressively with concreting. This may prove exceedingly tedious and detailed and requires the utmost care to insure proper positioning. The usual tolerance on positioning is plus or minus 3/4 inch at any point.

Ducts should be protected from corrosion in storage, during installation, and after placement. In particular, water must not be allowed to collect in low spots, and the ends must be sealed against moisture entry. Vapor-phase-inhibitor crystals may be dusted in prior to sealing.

Ducts must also be protected from dents and deformations during handling, installation, and concreting. No dent should exceed 3/8 inch.

Extreme care must be taken to prevent debris and mortar from accidentally entering the ducts. The only sure way to do this is to provide and install caps.

Air vents and vent piping must be installed at specified points to permit air to escape during injection of duct filler (grease or grout).

16.7 Prestressing Tendons

16.7.1 Selection of System

The selection of the tendon system will generally have been made by the designer in collaboration with the tendon and anchorage manufacturers. Construction considerations include size of tendon, duct, anchorages and bearing plates; stressing, seating and anchorage procedure; and protection of the tendon system during storage, assembly, and installation.

There is a definite trend toward ever-larger concentrations of force in a single tendon. Several manufacturers offer large tendons assembled from 7-wire strands, others offer large cables formed from individual wires. Anchorage systems are wedged, swaged, or button-head. A new French development uses high-tensile steel strip.

Tendons are subjected to static tests for strength and elongation at rupture, and to dynamic tests of the entire system, i.e., tendon plus anchorages.

Whatever system is selected, its degree of success will be, to a large extent, determined by the accuracy of manufacture, and the attention given to the details of installation.

16.7.2 Fabrication

Tendon material is usually fabricated into large cables on the jobsite. During fabrication and storage, it is essential that the material be protected from corrosion. The storage warehouse, etc., should be heated and dehumidified. Usually this can be accomplished by heaters alone. The intent is to keep the tendon steel clean and bright, with no more atmospheric rust than that which can be removed with a soft, dry cloth. Temporary coatings for corrosion protection must be compatible with the permanent protection system. For example, a wax permanent system is compatible with an oil-base temporary system, but a Portland-cement grout system is not. Temporary coatings must be designed for a specific time and exposure, and this must be monitored.

During fabrication and again in handling and threading, the tendon cable must be protected from nicks and abrasion. During shipment, the tendon shall be protected from a permanent set or kink, notches, and abrasion, and contamination from the outside (e.g., dirt, smoke, effluents).

16.7.3 Installation

A fabricated tendon is usually wound on a drum. This is hoisted to one duct end. The duct is blown clean with compressed air. A messenger wire is sent through the duct, a nose piece (grip) attached, and the tendon pulled into place. As the tendon is drawn in, vapor-phase inhibitor may be dusted on. As soon as the tendons are in place, the ends should be sealed by polyethelene bags.

Anchors are installed, jacks lowered to position, and stressing performed. A field telephone should be provided for instant communication. Usually stressing from both ends is required. An exception is straight or nearly straight tendons. Strands may be over-stressed by up to 5%, with the engineer's approval, to compensate for seating and friction losses. Complete records must be maintained.

Both elongation and pressure on the calibrated gauge must be recorded; they should check to an accuracy of about 5%. Greater variation than this indicates excessive friction or duct blockage. There are remedies available for emergencies; a water-soluble grease may be injected to reduce friction.

Generally, at some specified time after initial stressing, all or a selected number of tendons are re-stressed and shims placed.

Tendon stressing and re-stressing is performed in accordance with a specified sequence. Usually this requires jacks to be so located as to permit stressing in sequence at multiple points.

For tendons which will be permanently protected by grease, a coating may be put on during threading; this, of course, reduces friction and provides corrosion protection.

After stressing, it is important to seal the anchorages with either grease or vapor-phase inhibitor inside.

Prestressing steel should not be subjected to excessive temperatures (over 450°F), welding sparks, grounding currents, etc. Tendons must be protected when any welding or burning is to be performed in the vicinity.

A great deal of discussion has taken place concerning the problem of individual strands or wires twisting over one another when installed in a curved duct. As far as this author can determine, this is not a serious problem in actual practice and can be ignored except for very short tendons of high curvature. If no twisting or overlap is to be permitted, then the cable must be bound during fabrication as by wrapping continuously with soft-iron wire.

16.8 Anchorages

The anchorages are, of course, an integral part of the tendon system. Experience shows that troubles and breakage, if they occur, are most often associated with the anchorages. This requires extremely close and detailed inspection at the shop, identification of shipments, protection during storage in a dehumidified atmosphere, care in placing to avoid nicks, etc., or damage from welding, and corrosion protection after.

Bearing plates should have clean surfaces, free from rust and grease, before installation and from concrete grout afterwards. They must be set to the tolerances specified; usually plus or minus 1/2 inch in length and plus or minus 2° to 5° in angle.

The spiral reinforcement at the anchorage serves extremely important functions in confining the bursting zone and in restraining spalling. It must, therefore, be set accurately and held rigidly during concreting.

Button-heads, if used, should be formed so as to prevent any serious indentation in the wires, with no seams, fractures, or visible flaws in the heads.

Wedge-type anchorages are highly dependent upon accurate fabrication of the cone to very close inside tolerances.

Because of their exposed position, anchorages are more liable to accidental damage after installation than other elements of the vessel system; therefore, they should be protected, either by being recessed in the concrete or by temporary guards, etc.

Anchorages and bearing plates must remain ductile, i.e., not subject to brittle fracture, at the lowest temperatures anticipated during installation and the life of the structure.

16.9   Encasement and Protection of Tendons

The use of very large tendons consisting of multiple wires or strands presents a high ratio of steel surface area to total steel cross-sectional area. This means that special attention must be paid to corrosion protection. The continuous curvature of the tendons produces radial components, forcing the tendon against the sides of the duct, and restricting the penetration of filling materials.

Special corrosion-inhibiting greases are extensively used for filling ducts and protecting the tendons. It is important that the temperature of the grease and duct be relatively uniform and high enough to maintain fluidity. Tests should be run to determine the permissible temperature variations. The duct can be heated if necessary, by blowing hot air through it.

Likewise, the grease may be heated. Care should be taken against accidental fire.

Grease should generally be injected from one end until a small quantity of grease of the same consistency emerges from the other end. The grease injection should be shut off under pressure.

Other systems in Europe use an electrostatically deposited wax, or bituminous compounds.

Permanent coatings should:

(a) Remain free from cracks and not become brittle or excessively fluid over the entire range of temperatures, including those during transportation and storage.

(b) Chemically and physically stable for the life of the structure. Stable under total radiation anticipated during life of structure.

(c) Non-reactive with concrete, tendons, ducts, etc.

(d) Non-corrosive, and corrosion-inhibiting.

(e) A barrier to moisture.

It is important that the coating be continuous and of the thickness specified after installation (pulling in) of the tendons. Pumps, hoses, etc., shall be suitable and adequate over the entire range of temperatures and viscosities expected during installation.

Provisions shall be made for removal of spilled material from concrete surfaces.

All greases or other injection materials must be shipped in tight containers that will prevent contamination.

In some recent French reactor vessels and at least one United States containment vessel, the stressed tendons are grouted. In such a case, grouting practice should follow the recommendations of Chapter 5, Section 5.1, Grouting. Grout should have a low water/cement ratio, be thoroughly mixed by machine, injected under low pressure until grout of the same consistency emerges from the other end, and shut off under pressure. The following limits on contaminants (by weight) are recommended:

$$\text{Chlorides as Cl} \quad 650 \, \text{ppm}$$
$$\text{Sulfates as SO}_4 \quad 800 \, \text{ppm}$$

Grouting practice should follow strictly the recommendations of Chapter 5, Section 5.1, Grouting. Extensive tests have shown that with multiple strand tendons, a path exists for grout flow regardless of radial force and bearing. With multiple parallel wires, grout flow is impeded to the inner wires and where wires are forced against the sides of the ducts. In these cases, special procedures, such as the use of soft-iron wire spacers, are necessary.

Recent tests on vertical and semi-vertical tendons consisting of bars and wires indicate sedimentation may be reduced by inclusion of a gelling agent plus expansion additive plus a water reducer (but not retarder). Free expansion is permitted through a vertical extension above the highest point of the tendon. With vertical strand tendons, bleeding of 10% to as much as 20% may occur. A vertical standpipe should be provided and the gelling agent should be omitted.

Grout, properly injected, offers both corrosion protection due to alkalinity, heat protection, additional safety in the anchorage zone, and possible reduction in size of cracking under accident conditions.

The protection of the anchorages is a very important matter. Not only is corrosion more likely here but the steel is under greater stress and more susceptible to brittle fracture. Therefore, it is important that a mechanical seal be installed around the anchor and the specified material be injected (grout, grease, or vapor phase inhibitor).

16.10 Precast Concrete Segmental Construction

There are many construction problems in reactor vessel construction which can be best met by judicious use of precast segments, joined by cast-in-place concrete.

Penetrations, especially large penetrations, are critical areas. The tendons must be deflected around them, thus being spaced more closely. Mild reinforcement is usually required to resist local secondary cracking. Concrete quality and consolidation must meet the highest standards.

Therefore, it seems logical to precast these penetrations in blocks of a size that can be handled by the available crane capacity. Segments would be concreted in their most favorable position, cured, and set in place. The joints would be treated like other construction joints.

Already it is standard practice for many reactors to cast the anchorages into precast beams. These can be concreted in their most favorable position, with anchorages held in exact alignment. These segments are then set in the forms. Thus, they provide the rigidity and resistance to displacement, as well as solving the difficulty of concrete consolidation.

Where tendons are anchored in buttresses, these could be precast in large segments, and vertically prestressed. Here, either concreted or, better yet, dry joint techniques borrowed from segmental bridge construction can be applied.

The liner presents many problems in its relation to concrete. It is thin and flexible. Segments could be made, cast with the pre-formed liner face down. All cooling pipes and anchor lugs would have been pre-welded. Such segments would, presumably, be of large plan size but only a foot or two in thickness. Since this is the area requiring greatest compressive strength and the most heat resistance, special mixes and consolidation could be applied.

Joints would be recessed to permit welding and later grout injection.

Designs have been prepared and models tested in which the entire reactor vessel is made of precast blocks. These have socketed joints to insure biological shielding, and may be joined with non-shrink or expansive grout.

It is the opinion of this author that a combination of precast and cast-in-place concrete offers the best solution.

16.11 Prepacked Concrete

Preplaced aggregate, with grout intruded, is being used on the Fort St. Vrain reactor in Colorado. The specially selected aggregate is placed in lifts containing all the elements (ducts, reinforcement, etc.) and, in addition, grouting tubes and instrumentation tubes. A battery of grout pumps is used to pump grout into the aggregate from the base. This method was selected to minimize displacement of inserts and to substantially eliminate shrinkage. The grout also has the ability to penetrate highly congested areas.

Extreme care has been taken to prevent rock dust, chippings, dirt, or other contamination in the preplaced aggregate. The grout, being of relatively small total volume, could be carefully controlled as to consistency, viscosity, temperature, entrapped air, etc. Spare pumps and mainfolds were provided to ensure that the pour could continue without interruption due to failure of equipment or power supply.

The use of prepacked methods, with carefully controlled rates of placement and pressures, reduced the magnitude of head acting to distort the liner and forms during concreting.

Actual pressures, as well as full development of the techniques, were determined in mock-up tests at the site.

16.12   Insulation

The insulation on the inside of the liner, if required, may be pumice concrete blocks, stainless steel foil, or fibrous ceramic blocks. Attachment is usually very complex. The constructor can minimize the cost and time required by providing adequate lighting, ventilation, scaffolding, and material handling.

16.13   Scaffolding, Hoists, Cranes, Access Towers

These are required for many phases of the work and with many activities going on simultaneously. The available space is extremely limited and congested by other activities of plant construction. The selection of equipment, its scheduling, and most effective utilization deserve the closest possible attention. Practically every item of material and equipment has to be raised by power, and supported during installation. Tendon coils are particularly cumbersome and awkward. The jacks are heavy and must be held at odd angles, and men require staging from which to affix anchorages, stress tendons, and inject grout or grease.

Such hoists must also be designed for safety, with controls and limits built in to prevent accidents, and with maximum visibility, including night lighting, for the operators.

16.14   Inspection and Records

Inspection includes maintenance of the Quality Control and Assurance Program, for all elements entering into the complete system. Complete and accurate records must be maintained for the engineer, owner, inspection agency, and the governmental regulatory bodies. This requires a carefully thought-out system of inspection, of identification, of quality standards, permissible tolerances, procedure in case of deviation, and records. Records must be maintained of all shop drawings and details, welding procedures, welder qualifications, weld inspection results, radiographs, ultrasonic magnetic particle reports, liquid penetrant test results, concrete batch records, concrete pour check-off lists, vibration times, temperatures, stressing sequences, forces, and elongations, grouting or greasing pressures, temperatures, etc.

One of the most critical of all inspection functions is the thorough check of each pour prior to concreting, to insure proper placement of all elements.

Sampling and testing of tendons and anchorages must be carried out in accordance with the prescribed procedures and recorded in detail.

16.15   Scheduling and Coordination

The prestressed concrete pressure vessel is reportedly on the critical path for two-thirds of the total construction time. The total work force at a nuclear plant has reached 2,250 men. While the vessel is being constructed, the construction of the reactor core, boilers, boiler shield wall, etc., must also proceed on an integrated schedule.

The "systems design" approach means that construction must also proceed on a "systems" basis. All possible activities should be dispersed from the immediate site, and prefabrication used to the highest extent possible.

At the vessel site itself, every effort should be expended to assure optimum working conditions, that is, dry, light, clean working conditions, and adequate hoisting and access equipment (see Section 16.13 above).

Scheduling obviously lends itself to the critical-path approach. There are numerous activities, all are interrelated, all are essential to the timely and efficient completion of the vessel. Programs are set up and monitored by a computerized technique. The critical path must include working drawings and their approval, computations submissions, and tests, as well as procurement and construction activities.

However, the practical limitations of critical-path analyses are well known. One of the largest and most successful nuclear reactor power plant builders in England controls the work primarily from a detailed bar chart. A detailed critical-path analysis is made before the start of the job, from which the bar chart (operations vs. dates) is prepared. Thereafter, if any activity falls behind, he augments the work force or lengthens the hours of work as necessary to get back on schedule.

It is believed that the critical path is the better scheduling method and can be effectively utilized, provided it is not made in such minute detail, i.e., too many activities, as to prevent its useful application in the field.

A typical schedule for a prestressed concrete pressure vessel for a gas-cooled reactor may show the accompanying Table 16.1 steps and approximate time requirements.

After the scheduling is prepared, a detailed layout must be made for each stage. Details of duct and penetration placement are prepared for each pour. Often these can be integrated with the forms insofar as layout is concerned. Equipment and access and material flow must be pre-planned. Spare parts and spare equipment must be provided for tensioning, injecting, grouting, and instrumentation, so that the schedule will not be delayed by breakdown. Proper tagging and identification, including shipping instructions and description, will obviate delays and confusion.

	Steps	Duration
1.	Preliminary work at site	12 weeks
2.	Excavation	18 weeks
3.	Base and Foundation Slab	29 weeks
4.	Foundation Steelwork Erection (liner supports)	4 weeks
5.	Base slab of vessel	5 weeks
6.	Prefabricated liner moved in and placed	4 weeks
7.	Final positioning of liner	3 weeks
8.	Fit and weld cooling water pipes Grout under liner floor	4 weeks
9.	Walls of pressure vessel	19 weeks
10.	Weld upper penetrations	4 weeks
11.	Complete walls	3 weeks
12.	Top slab supports, lift in top disk	4 weeks
13.	Set and weld top disk	9 weeks
14.	Concrete top slab	6 weeks
15.	Cure top slab	2 weeks
16.	Remove supports	3 weeks
17.	Reactor construction inside	19 weeks
18.	Thread tendons	8 weeks
19.	Stress tendons	14 weeks
20.	Interior completion	8 weeks
21.	Pressure test	
22.	Completion of plant ready for full operation	1 year

As one leading designer-builder recently stated: "The largest savings in the future are likely to be made in the field of construction improvements such as simplification of ducts and stressing arrangements and streamlining of erection procedures."

16.16   Notes on Construction Techniques and Practices from Recent Experience

For a number of economic and technological reasons, the art and practice of prestressed concrete nuclear reactor pressure vessels started early in France and England, and has progressed much more extensively than in the USA. Prestressed concrete reactor vessels are now standard for all nuclear power plants in England and France. One high-temperature gas-reactor pressure vessel is currently under construction in the United States. Also, there are a substantial number of prestressed concrete containment vessels completed or under construction in the USA. Because of their huge size (of the order of 160 feet high and 160 feet diameter), they present many similar construction aspects to pressure vessels.

In England, there are two completed and tested reactor vessels just commencing operation at Oldbury (600 MWe); two vessels at Wylfa nearing completion (1180 MWe); and two under construction at Dungeness B (AGR) (1200 MWe); with construction starting on similar AGR plants at Hinkley Point, Hunterston, and Hartlepool.

In France, there are the two pioneer prestressed concrete pressure vessels operating at Marcoule; EDF3 operating at Chinon; St. Laurent I (formerly known as EDF4) now nearing completion. Bugey 1 and St. Laurent II starting construction; and Fessenheim I about to commence.

In Spain, the Vandellos plant is under construction.

All of these are reactor pressure vessels for gas-cooled reactors; all but EDF3 and Marcoule are of "integral" design with the boilers and boiler shields contained within the pressure vessel as well as the reactor core.

In addition, there is the completed prestressed concrete containment structure at Brennilis, France, and Gentilly, Quebec.

The reactor vessels of England and France are parallel in concept to the Ft. St. Vrain plant in Colorado; the Brennilis containment structure is parallel in concept to the containment structures in Florida, Michigan, North Carolina, New York, and California.

16.16.1   Oldbury, England

The shape chosen was a vertical cylinder, 77 feet interior diameter, and 60 feet internal height. Wall thickness is 15 feet. It is interesting to note that the wall thickness was determined by the pattern of tendons and physical room required for anchorages and jacking.

The unique thing about Oldbury was the invention and use of a double helical pattern for the tendons, which gave simultaneous hoop and vertical prestress, reduced friction, and required anchorages only on top and bottom. This reduced the number of total anchorages required, and made them accessible. There are 11 clockwise layers and 11 counterclockwise layers, with 160 tendons in each layer. The tendon path is a helix at a 45° slope. Each tendon consists of 12 x 0.6-inch strands, with a Freyssinet-type anchor, for a total of 273 tons ultimate strength per tendon.

Construction proceeded as follows:

A 1-2-inch thick neoprene mat was placed on the foundation plug, then the 22-foot thick vessel base was poured in 5 lifts. The pours were broken by vertical construction joints so as to give individual pours of 230 cubic yards.

The top lift contained embedded structural tees, set with great accuracy, and the concrete was screeded to these with precision. The steel liner base plates were then set on the slab and welded to the tees. This welding caused shrinking of the base plate and curling of the concrete at the edge. Eventually, after the remainder of the vessel was nearing completion, the cracks under the base plate, which varied from 0 to 1/8 inch, were filled with a pressure-injected resin.

Precast anchor ring beams had been prefabricated, with anchors accurately set and aligned. These were now set on the base slab.

Meanwhile, at a site some 200 feet away, the steel liner was prefabricated, including cooling water pipes and anchors. This was now jacked up onto small rail cars, rolled on rails sidewise to location, then jacked down into position. It weighed 1000 tons.

The concrete walls were now constructed in ten 6-foot high lifts. Ducts were 3-1/2 inches outside diameter, 11 gauge milk-steel seam-welded tubing. Rubber joints were used to join the extension for each lift. A total of 1,210,000 lineal feet of duct was required. After each lift, the concrete, ducts, etc., were thoroughly cleaned. Considerable time and effort was required in positioning the duct extensions as their angle and position was constantly changing.

Vertical construction joints were formed with wire mesh, backed by timber and scaffolding. They were stripped while the concrete was still green and were jetted with compressed air and water to expose the aggregate.

The top slab (22 feet thick) was similarly poured in 5 horizontal lifts, each lift broken into reasonably sized pours. Two concrete mixes were used; a 6000-psi (cube) mix for the general vessel construction, and an 8000-psi (cube) mix for use in anchorage bearing zones. Slump was kept about 2 to 3 inches for all pours.

Meanwhile, the tendons were made up in an on-site shop, and wound on a small drum. This drum was taken to the top, a messenger line sent through, and a nose attached to the tendon; then, it was pulled through by air winch. Actually, the helical tendons had to be held back during threading. Base and top slabs were prestressed by tendons arranged horizontally in layers, forming a square mesh pattern around the vessel axis near the center, diverging near the perimeter.

Stressing was by a 250-ton jack, working from both ends. Considerable difficulty was encountered in fitting the jack into the congested anchorage area, and, in fact, a special jack had to be designed. Tendons were stressed, checked, restressed, and shimmed. Then the ends of the duct tubes were sealed with a polyethelene bag containing silica gel and a moisture color indicator. This job required 17,576 anchors and considerable time was expended in perfecting these to eliminate slip, notching of tendons, etc.

Some 350 sonic strain gauges, as well as numerous moisture content gauges, temperature sensing gauges, and thermocouples, were installed during construction.

It is interesting to note that only 5 out of 8788 ducts became blocked during construction. Fortunately, provision had been made for spare ducts.

16.16.2   Wylfa, Wales

These consist of two gigantic concrete spheres, 96 feet interior diameter, with a minimum wall thickness of 11 feet. The general tendon arrangement consists in the vertical planes of 528 tendons inside the walls forming great circles; in the horizontal planes, inside the top and bottom slabs, of 426 tendons forming small circles, and of 384 hoop tendons circumferentially wound outside the vessel walls, through exterior concrete ribs.

Each tendon consists of 36 strands of 0.6 inch diameter that are spread near the ends into three groups which are anchored separately by Freyssinet anchors.

Internal ducts are made of 5 1/2-inch flexible metal ducting, whereas ducts through the exterior ribs are formed from accurately curved 1/4 inch steel tubing.

The vessel liner was prefabricated several hundred feet away, in the form of two hemispheres, then lifted into position by the 400-ton capacity Goliath crane which served the reactor site.

Much of the extensive cooling-water system was welded to the liner after it had been placed. This caused considerable congestion of m and materials at the site.

Concrete lifts wre kept to 5 feet deep, 250 cubic yards per pour. A 7-day interval was required between adjacent pours to minimize the effect of drying shrinkage and heat.

The vessel was supported on the foundation slab by 147 Freyssinet laminated rubber bearing pads, surrounded temporarily by sand. The sand was jetted out after 5 lifts had been poured, leaving the vessel supported on the rubber bearing pads. During construction, the exterior ribs were supported by Freyssinet flat jacks, enabling construction stresses to be controlled. After stressing, these were removed.

After lift number 5 of the base slab was poured, the liner was set in position 4 feet above its final position on 16 short steel columns. It was then jacked down to within 4 inches of the base slab, and the large penetrations were fitted.

This 4-inch space was then grouted in two stages. In the first stage, 1 1/2-inch duct tubes were attached to the liner; after grouting, these were then deflated and withdrawn and the resulting voids were grouted under a pressure of 30 psi. The liner was anchored down by the first stage engaging the ribs of the liner.

Where large circular penetrations intersect the liner, the complexity of cooling water pipes and reinforcement necessitated special care in concrete placement. Concrete of a special mix was placed in small bays, working out from the liner. Ductubes were used to form holes which were later pressure-grouted.

The strands were coated by the manufacturers with electrostatically deposited wax. During on-site shop fabrication of the tendons, a grease having lubrication and corrosion-inhibiting qualities was applied and each tendon was bound in a continuous helix of soft-iron wire.

Fifty-six days after all concrete was placed, the main prestressing was commenced. It was carried out in a predetermined sequence to equalize stresses. After final stressing, the ducts were injected with a thixotropic carrier dosed with corrosion inhibitor. This was pumped in at 75°C, but remained soft and flexible at ambient and operating temperatures. The exposed hoop tendons were coated with bitumastic and wrapped with polyvinylchloride fabric. A protective sheathing, containing vapor phase inhibitor, was placed over all anchorages and exposed tendon ends.

16.16.3   Dungeness B, England

This is an AGR design, which requires a very close standpipe arrangement in the cap. Standpipes are at 15-inch centers, and are 9 1/2 inches in diameter. The standpipes are then insulated, so the net concreting space is between 2 and 3 inches. There are 465 standpipe channels within a region 31 feet in diameter.

A vertical cylindrical shape was chosen for the vessel, with only hoop and vertical tendons. Hoop tendons extend over 180°, and are anchored in buttresses. Successive tendon anchorages are offset 90°. Each tendon consists of 163 wires of 0.27 inch diameter, giving a design load at transfer of 700 tons force. Anchorages are BBRV button-headed type.

The liner is 1/2 inch thick mild steel, with cooling pipes on the outside and stainless-steel foil insulation on the inside.

The liner was prefabricated off-site, and was jacked into position just above the concrete base slab. Lugs had previously been welded to the bottom of the liner. Grout was then placed in two stages, the first stage anchoring the lugs and holding the liner down while the second stage was injected. This entire procedure was developed by a series of mock-up tests at the site.

For concreting the top slab between the standpipes, a rather unique solution was found. Instead of dividing this into horizontal lifts, it was formed and poured in small full-height lifts, with vertical construction joints. Standpipes were set up and welded in rows and concreted successively. This minimized standpipe installation problems and also solved the concrete placement problem.

Particular care was taken in construction planning to ensure orderly procedures and minimum congestion. During construction stages, care was taken to avoid corrosion of cooling water pipes, etc., and accidental blocking.

Anchorages were protected after final stressing by a spray-applied plastic.

16.16.4   Hinkley Point and Hunterston B, England

These were constructed as cylindrical vessels with a helical system of tendons which were continued on an extension of the cylinder wall beyond the flat slab ends. Precast anchorage beams were used as at Oldbury. Tendons with an ultimate load of 250 tons, using the CCL system of 7 x 0.7-inch strands, were stressed individually. This permitted use of a small jack which was easily handled and aligned. Headrooms and clearances in the stressing galleries were reduced and construction time shortened. The erection of the vessel liner and installation of plant were simplified by early erection of the building superstructure and a 600-ton capacity gantry crane.

Ducts were mild-steel pipe, 1/4 to 3/16 inch thick, bent on site, and joined with screwed, watertight fittings.

16.16.5   Hartlepool, England

Barrel walls were constructed, free from major penetrations. The walls and caps were then prestressed by external wire-winding with 0.2 inch diameter wire, with the tendons falling in preformed circumferential channels. These channels consisted of mild steel channels, set in precast concrete units, which then served as the permanent exterior forms for the vessel. After the tendons were in place, a cover plate was placed over the open face of each channel. At the caps, the greater prestress requirements were met by closer spacing of the channels. This procedure permitted the development of tendon capacities in excess of 27,000 Kips, using as many as 48 layers of wire of up to 120 turns each, in a single channel.

The high-tensile wire was wound under tension by means of a winding machine which was "locked-in" during winding operations to a predetermined tension. Traction was obtained from a continuous chain encircling the vessel. Tension in the wire was maintained by clamping it between a pair of caterpillar tracks whose rotation and resistance was controlled by the hydraulic system. Tension was continuously monitored and recorded. The wire used was of a "low-relaxation" type and was treated in manufacture and during the winding with corrosion-protective compounds.

Anchorages for the wire were provided by anchorage pegs which slid into sockets in the channel wall.

Wire winding offers a number of advantages for circumferential stressing of large prestressed concrete reactor vessels, including elimination of friction during stressing, reduction and distribution of anchorage forces, more effective location of tendons because of elimination of buttresses, higher concentration of tendon forces, and reduction in over-all diameter of pressure vessel.

16.16.6 French Research and Development Reactors

These include the Melusine Pool Reactor Containment and the "Triton" Reactor at the Nuclear Research Center, Fontenay-aux-Roses. These are small boiling-water reactors. Efforts were made to achieve maximum density in the concrete. At Melusine, some barite concrete was used with a unit weight of 224 pounds per cubic foot. Concrete placement led to difficulties in consolidation which were overcome by using very strong vibrators at a frequency of 20,000 rpm.

The Triton is chiefly of interest because the concrete was placed in segments with 6-inch (15-cm) wide joints. These joints were later filled with vibrated fine concrete. This is essentially the same technique that would be required if precast segments were used and jointed by cast-in-place concrete.

16.16.7 Marcoule (G2 and G3), France

These were the first prestressed concrete pressure vessels for commercial nuclear reactor power plants. Each is a horizontal vessel, interior diameter 45 feet (14 m), internal length 65 feet (20 m), with walls of 10 feet (3 m). The ends consist of domes convexed inward; thus, the radial component is resisted by concentrated hoop tendons, and the longitudinal component by longitudinal tendons. Each tendon consists of 715 5-mm wires.

Circumferential (hoop) tendons are external to the concrete and slide on shoes and sliding plates, lubricated by molybdenum bisulphide grease, giving a friction coefficient of only 2%.

After concreting the shell around the liner, grout was injected through holes in the liner, the holes being later welded closed.

The only corrosion protection was by humidity control. However, after several years, corrosion was found, especially in those cables wrapped by jute. Tendons in question have been replaced, and more stringent steps of humidity and temperature control and monitoring are being enforced.

16.16.8 Chinon (EDF3), France

This is an operating plant, in which the prestressed vessel encloses only the reactor. One point is worthy of note. First was the decision to grout the tendons after stressing. This reflects growing confidence about the long-term behavior of prestressing steel under the effects of irradiation, combined with concern over corrosion protection and concern about the long-term stability of grease. Concurrently, it was found that the grout did transfer the full anchoring forces to the concrete even if the mechanical anchorage was removed.

16.16.9 St. Laurent I (formerly EDF4), France

This brilliantly conceived and executed pressure vessel is being adopted as a standard concept for further nuclear power plants in France. This is an integral design, with boilers and boiler shield being contained within the vessel as well as the reactor core.

The vertical cylinder has an interior diameter of 62 feet (19 m), an internal height of 118 feet (36.2 m), with a minimum wall thickness of 15 1/2 feet (4.75 m).

The prestressing arrangement involves curved cables at right angles under the base slab and over the top slab, vertical tendons in the walls, and circumferential tendons which overlap as they span an arc of 120°, and which are staggered in succeeding layers.

The exterior walls are formed by large precast concrete slabs containing the anchorages fitted in at proper angles. This not only positions these anchorages, but assures dense, well-compacted concrete under the bearing plates, and prevents dislocation, deflection, and misalignment. (See Fig. 85)

Fig. 85. Use of precast panels for external skin and anchorage blocks of nuclear reactor pressure vessel.

The liner was prefabricated in sections about 24 feet high with cooling-water pipes attached. It was then lifted into place with the 400-ton gantry crane. Prestressed concrete columns of 20 feet (6 m) length were set in the pressure vessel in the prolongation of the openings for the blowers, and constitute a rigid support for the turbines and drive shaft, anchored to the pressure vessel.

Particularly notable during the construction of EDF4 was the careful planning and sequencing, the organization of the labor force, and the maximum use of prefabrication.

The tendons used were of the SEEE type, using 19 strands, fabricated in an on-site shop and coated with protective wax. Some 2500 cables totaling 118 miles of tendons were required. After installation and stressing, they were grouted.

An indication of the scope of the work is given by the fact that 32,300 cubic yards of concrete were placed in the vessel, 800 tons of mild-steel reinforcement, and 1800 tons of liner, including cooling system and structural framework.

16.16.10   Bugey I, St. Laurent II, and Fessenheim, France

These followed the general pattern set by St. Laurent I (EDF4).

St. Laurent II is almost identical to St. Laurent I, except for minor improvements in layout of liner and tubes to permit better gas-tight welding of the liner.

Bugey I is somewhat greater in height (125 feet) than St. Laurent I, and is in the form of a vertical cylinder with three ribs. Blowers are supported on corbels cantilevered out from the vessel. The BBRV system of prestressing is used, 54 wires of 0.275-inch diameter for 200 tons force per tendon. The horizontal (hoop) prestressing of cylinder and upper slab required 2200 tendons, each traversing two-thirds of a circle.

The top slab has 935 standpipes and control-rod holes, necessitating the use of peripheral hoop tendons.

The liner is made of 1-inch (2.5 cm) thick mild steel, and the cooling-water pipes, 1 inch diameter (2.5 cm), are welded both sides to the liner.

Fessenheim I is similar to St. Laurent I. Extensive testing of concrete aggregates for creep and thermal properties were undertaken and the results were utilized in the design calculations.

16.16.11   Monts d'Aree Nuclear Power Station at Brennilis (EL4), France

Unlike the gas-cooled reactor vessels described above, this prestressed concrete vessel is a containment structure designed for low pressure (60 psi) and large volume under accident conditions. It is 160 feet (48 m) high and 150 feet (46 m) in diameter.

The prestressed concrete pressure vessel is a cylinder 156 feet (48 m) high, with a diameter of 150 feet (46 m), and is very similar to the containment structures under construction in the USA, for which Brennilis served as a forerunner.

A thin Freyssinet hinge was built in at the junction of floor and shell to reduce the vertical bending moments due to radial pressure. The cylinder walls were constructed by sliding forms (slip forms) at a rate of rise of 6 1/2 feet (2 m) per day. Use of slip forms required highly detailed planning. The upper dome was supported on a rotating steel bridge. Prestressing was by the Freyssinet, 12 x 0.5-inch strand system.

16.16.12   Siemens Developmental Reactor Vessel, Germany

This vessel is designed for a pressurized heavy-water (D_2O) reactor, with an inside diameter of 33 feet (10 m) and internal height of 66 feet (20m), and a design pressure of 1850 psi.

The construction scheme employed the use of precast concrete segments, assembled ring by ring. The blocks were jacked out by multiple hydraulic jacks, thus stressing external hoop rods. The spaces between the precast blocks were then filled with grout.

The top closure is a concrete cone, held by the top ring by means of wedge blocks and corrugations.

The insulation is stainless-steel foils in a nitrogen atmosphere, with a separate steel membrane separating the nitrogen from the heavy water.

16.16.13   Fort St. Vrain, Colorado

This is the first US gas-cooled reactor. It is a high-temperature gas reactor, employing helium, rather than CO2, and operating at a relatively high pressure. As described earlier, specially selected aggregate was preplaced, and then grout was injected through grouting tubes by a battery of pumps.

Prior to the construction of this vessel, extensive tests were conducted to verify performance of all aspects, including tendon anchorages. Thorough controls were instituted to insure uniform properties of concrete aggregates and grout.

16.16.14   US Containment Vessels

These relatively large, low-pressure vessels serve as a containment structure in event of accident. Their large diameter and height has required careful regulation of pour size and control of shrinkage. Most are similar in design to the Brennilis Containment Structure previously described in Subsection 16.16.11 of this chapter. Several different prestressing systems have been utilized for the internal hoop stressing between buttresses and for the vertical stressing. These include button-headed wires, swaged strands, and alloy bars.

The containment vessel at Rochester, New York, was anchored to the bedrock by prestressed bars, which continued up into the walls to resist the vertical bending moments. The tendons were left unbonded (grease-injected ducts) through the transition zone between rock and concrete to reduce seismic forces.

Concreting has been carried out both by slip-forming and by panel forms. Difficulty was reportedly experienced in the first of the containment structures in consolidating the concrete in the base ring, which is highly congested. Rock pockets had to be pumped full of epoxy mortar.

In a number of these structures, conventional reinforcing steel has been used in conjunction with prestressing; for example, hoop steel has been conventional and vertical steel has been stressed. It is the belief of a number of designers that a combination of stressed and unstressed steel in each direction (partial prestressing) may prove most advantageous for the particular criteria imposed on containment structures.

In construction, the main problems appear to have been associated with the handling of tendon coils to the various elevations and their feeding into the ducts. This is complicated by the congested activity at the site, engaged in the construction of the reactor and building as well as of the containment vessel itself. The solution adopted was detailed planning of space and time activities, including crane service, so as to minimize interference.

Most of the US containment structures have employed unbonded tendons with grease injection. However, a few have utilized cement grout injection.

16.17 Implications for Future Construction of Prestressed Concrete Pressure Vessels

As operating pressures increase, there will be a demand for higher tendon capacities and higher concrete strengths. Already a number of manufacturers have developed tendons and anchorages with capacities of 600 to 1000 tons force per tendon.

These highly concentrated forces mean higher concentrations of bearing pressures. Concrete is somewhat vulnerable to sustained uniaxial stress. This may then again raise the question of grouting in the transfer zone to help distribute the load to the concrete. Alternatively, means of ensuring a triaxial stress condition (as by spiral hoops) must be further explored.

Higher strength concrete is essential if prestressed concrete is to be effectively and economically utilized as the primary vessel for pressurized water and boiling-water reactors. These will require cylindrical vessels with an interior diameter of some 20 to 35 feet, an internal height of 40 to 60 feet, and a wall thickness of 16 feet, to resist working pressures of 1000 to 2000 psi. Concrete strengths of 10,000 psi and more can be utilized most effectively. It is the author's opinion that this can best be obtained by precast segments, joined by cast-in-place concrete. The precast segments can be used in the regions of highest compression. The joining concrete can be located largely in zones of less critical requirements.

Precasting of anchorage segments, penetration segments, and liner segments appears practicable and economical.

Efforts continue toward the development of a concrete possessing the special qualities desired: low coefficient of thermal expansion, low modulus of elasticity, stability under high temperature and sustained loading, and high compressive strength. Structural lightweight aggregates already possess some of these qualities. The future may lead to the production of ceramic aggregates having the special properties desired.

Prestressed concrete with its favorable behavior through all criteria up to failure, and its inherent economy and efficiency, has a major role to play in the future development of pressure vessels for nuclear power. There should be a parallel development in the utilization of prestressed concrete for pressure vessels in the refining and chemical industries.

17

Prestressed Concrete Poles

Prestressed concrete has been widely adopted for poles throughout Europe and, to a lesser extent, elsewhere. The requirements for poles are rather unique structurally and economically. Properly utilized, however, prestressed concrete is a useful material for this application and may be the most advantageous for a wide variety of situations.

Historically, prestressed poles were first designed and constructed by French engineers in Algeria. The specific environment, blowing desert sand, destroyed both wood and steel. Prestressed concrete poles have seen more than 25 years' successful service in Algeria, despite what might be considered minimum concrete cover over the tendons.

Prestressed concrete poles are extensively employed in Europe, even in the Black Forest of Germany, where suitable trees grow alongside. There, a careful and proper decision has been made of relative values: the trees are to be utilized for the interior of housing, and prestressed concrete is to be used for long-term service poles. It is interesting to note that one of the few regions of the United States in which prestressed concrete poles are utilized is Oregon. There, also, the true value of a tree is recognized, as is the true value of a prestressed concrete pole.

Prestressed concrete offers durability from corrosion and erosion in desert areas, and ductility and freeze-thaw durability in cold temperatures and in mountainous regions. Prestressed poles are clean and neat in their appearance, both at time of installation and for succeeding years; thus, they are highly suited to urban installations. They are fire-resistant, particularly to grass and brush fires near the ground line. They can be set directly in drilled holes, in the ground, with or without concrete fill.

Prestressed concrete poles have been utilized for:

Railway power and signal lines.
Lighting.
Antenna masts.
Overhead pipeline supports.
Telephone transmission.
Low-voltage electric power transmission.
Substations.
High-voltage electric power transmission.

As compared with conventionally reinforced concrete, prestressing increases the crack resistance, rigidity, and resistance to dynamic loads.

17.1 Design

The design criteria for poles varies widely. For those carrying electric current, grounding must be provided. For those supporting wires under tension, the effect of line breakage, including torsion and impact, must be considered. Wind load is a major factor and, in some regions, seismic behavior must be considered.

A recent and extremely important parameter is the effect of collision and the behavior of the pole afterwards; does it fall on the automobile? Does the pole snap off suddenly when overloaded, or does it fail gradually, with large deformations and energy absorption?

Most poles to date have been hollow and tapered. The hollow core has been usefil in reducing weight and in providing a raceway for electric wires, etc. The taper is a reflection of the reducing bending moments and the desire to reduce wind (and seismic) loads over the top portion.

Cross-arms and diagonal bracing have been constructed both of prestressed concrete and steel. It would appear that, in most cases, use of galvanized steel or wood for this purpose would be proper because of light weight and ease of assembly by bolting. The one exception may be in TT or H-frame structures, where the main cross-arm may be advantageously made of prestressed concrete.

The greatest moment resistance is required at the base, and here both elastic resistance and a satisfactory plastic "toughness" is required. This can best be met by the inclusion of high-yield steel reinforcement or by additional unstressed strand.

A great deal of debate has occurred over the cross section of the pole. Should it be round or square? Wind is not normally the controlling factor; rather, torsional strength and bending strength on line breakage. Thus, square and rectangular sections and H-sections are suitable for poles and are widely employed. (See Fig. 86)

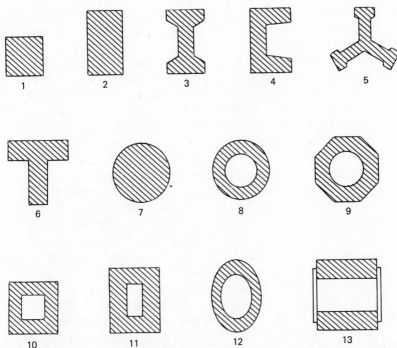

Fig. 86. Typical prestressed concrete pole cross sections. Note: (a) in 13 the lattice members are galvanized steel; (b) in 2, 3, and 4, the web may be perforated to give a trussed action, usually Vierendeel-type; (c) most piles in service today are tapered throughout their length.

Tapering has presented major problems to the application of prestressing. With full length tendons, the effect of taper is to increase the effective prestress at the top. By conventional design theory, a reduction in bending strength occurs both because of reduced section depth and because of high initial compression.

Manufacturing systems have been developed to reduce the effective prestress in the upper portions of tapered poles, e.g., by preventing bond, or by dead-ending or looping some of the tendons at mid-height. Another more common method is to add mild-steel bars in the lower portion. This author feels that "taper" is a concept that is inherited from timber poles (trees) and that a constant cross section may, in many cases, prove a better solution. Naturally, there will be an increase in seismic and wind stresses but, for torsional resistance, the top must be as strong as the base. Adoption of a constant cross section permits more effective use of prestressing and easier connections.

Wherever the effective section is changed, either in concrete cross section or in embedded steel, the transition should be gradual. Steel reinforcing bars in the lower end, for example, should be terminated at different levels, rather than all at one elevation.

The collision parameter requires toughness and ductile behavior at the base. This can be obtained by a solid concrete section, reinforced so as to have a large plastic deformation after cracking. This reinforcement can consist of the stressed tendons, augmented by additional unstressed tendons, or by mild- or high-yield reinforcing steel, or by a structural steel core.

The weight of poles should be kept as low as possible to ease transportation, erection, and to reduce seismic loads. Lightweight concrete is ideally adapted to this use, provided it is suitably bound by spirals.

For many uses, the poles require a duct-way leading up the pole. This can be formed by a hollow core, an embedded duct, or an external pipe conduit.

If copper wires are used for grounding, they should be insulated from the concrete to prevent electrolytic corrosion of the prestressing tendons since wet concrete can act as an electrolyte. Perhaps a soft-iron ground wire of larger cross section would be a safer solution.

Since maximum bending moment occurs at the base, the tendons should either be mechanically anchored (as by post-tensioning) or with as short a transfer length as possible (e.g., by use of strand and gradual release).

Inserts or embedded plates for attaching cross-arms and bracing are usually of galvanized steel. Epoxies might be used to advantage, especially where the attachments are affixed prior to erection.

Lighting poles present special cases, in which the means of supporting the light must be aesthetically blended into the support. For this reason, many are made with curved upper portions of conventionally reinforced concrete or steel or aluminum. However, there is no reason why a central post-tensioning tendon should not run the curve and, in fact, many are so constructed.

17.2 Manufacture

Prestressed concrete poles are produced either in fixed forms or by centrifugal spinning. Fixed forms may be tapered on the sides to facilitate stripping. With fixed forms, compaction is accomplished by internal vibration, use of a vibratory table, or vibration of the core.

The deflection of prestressing strands to match the taper is readily accomplished on a long bed by merely seizing them as they pass through the end gate. When several poles are to be manufactured in a long-line pretensioning bed, they should be arranged alternately, so that small ends adjoin; this permits the prestressing tendons to be seized and spread in the gaps between poles.

Tapered cores are designed for easy removal by such means as:

(*a*) Wrapping with paper.
(*b*) Lubricating.
(*c*) Rotating at time of initial set to break bond.
(*d*) Jacking out.
(*e*) Longitudinal vibration during pouring and until set.

In one particular manufacturing plant, the exterior and core forms remained fixed, with the completed pole being jacked off longitudinally.

Poles can be manufactured by using precast segments. The segments can be assembled, jointed, and post-tensioned, giving the ability to use standard elements for a wide variety of pole heights.

17.3   Erection

Poles are installed in one of the following ways:

(*a*) Bolted to previously constructed bases.
(*b*) Set in drilled holes in the ground.
(*c*) Concreted in drilled holes in the ground.
(*d*) Concreted in sockets in previously constructed bases.

Under (*b*), the hole is usually packed with tamped gravel. Under (*c*), the concrete may be poured or pumped. Grout may be advantageously used where the annulus is small. This ability to set directly in a drilled hole and be assured of durability is a major advantage of prestressed concrete.

Erection may be by a derrick on a truck, a truck crane, a guy derrick, or a helicopter. The main problem is that of insuring lateral stability, as by use of guys, and to prevent accidental impact of the pole with the crane boom.

Segmental poles have been erected by telescopic action; this appears to be a special case, applicable to very large or unusual poles. Segments have also been joined by welding during erection.

Guy wires and stays are often employed, particularly on very high poles and at angles in the alignment of the line. The vertical component of guy stress should be investigated, particularly for high poles, to prevent buckling. The behavior of the guyed pole after breakage of a line or guy must also be considered.

17.4   Service Behavior

The prescribed cover differs in the various countries. However, most agree that 0.65 to 0.75 inch (15 to 20 mm) is adequate.

Longitudinal cracking has plagued highly stressed hollow-core poles having thin walls. The answer seems to be more closely spaced spiral, along with thicker walls.

17.5 Economy

Most objective studies indicate that the first cost of prestressed concrete poles is greater than timber but less than steel. The economic justification, therefore, is usually based on the following:

(*a*) Durability against erosion, insects, corrosion, etc. This is of special importance in tropical (jungle conditions), desert, and shore-side environments.

(*b*) Fire resistance. This is of special importance in areas of grass, low brush, and low timber. Ground fires should not normally cause the failure of a prestressed concrete pole.

(*c*) Service life.

(*d*) Aesthetics, especially long-term appearance.

Wider acceptance in the United States must await a re-evaluation of true economy, service life, maintenance, installation costs, etc. Nevertheless, the use is growing and, if we can draw valid conclusions from other countries, environments, and economies, prestressed concrete poles will enjoy a major utilization in the future.

Trends of development should include the following:

1. Re-evaluation of taper vs. constant cross section.
2. Re-evaluation of the need for a hollow-core as opposed to a simple duct.
3. Selection of best cross section: square, circular, retangular, channel, etc.
4. Development of positive means of insuring against brittle behavior under impact.
5. Use of lightweight concrete.
6. Lower costs through mass production.

18

Prestressed Concrete Pipes, Penstocks, and Aqueducts

Prestressed concrete is widely utilized for the conveyance of water in pipes, penstocks, siphons, and aqueducts in sizes up to 15 feet diameter and more. It has been utilized for large-diameter, high-pressure penstocks for hydroelectric power plants in New Zealand and Scotland. Today's prestressed concrete aqueducts in Italy carry on the tradition of their Roman ancestry.

The use of prestressed concrete pipes for outfall discharges and intakes in the ocean and lakes is discussed in Chapter 14, Prestressed Concrete Floating and Submerged Structures. Prestressed concrete pipes are excellently adapted to siphons and underwater lines because they combine circumferential strength with longitudinal beam rigidity and strength; durability with weight for stability.

Prestressed concrete pipes lend themselves to factory production and many proprietary techniques are utilized. In the following sections a brief description will be presented of a number of the manufacturing techniques.

Concrete pipes are cast by the vertical-cast process, the centrifugal spinning process, or horizontally, with a removeable mandrel.

The vertical-cast method is used for the larger diameter pipes, usually those above 72 inches, in lengths of 16 to 24 feet. A machined cast-steel ring is used for the base. Forms must be very rigid, since the concrete is placed very dry with heavy external vibration. After the central core ring is placed, a welded reinforcing cage is set, containing ducts for prestressing if so designed. The outer form is then placed, concrete placed and consolidated,

349

followed by steam curing and stripping. Supplemental water curing improves impermeability.

If lightweight concrete is to be used, then a small amount of conventional hard-rock concrete may be placed at the top and bottom to provide greater resistance to spalling. Internal vibration blends the transition zones effectively.

Joint details may be flexible, with machined steel rings and rubber O-ring gaskets; or rigid, for mortar joints; or semi-rigid (lead caulked).

Centrifugally spun pipes utilize the contrifugal acceleration to consolidate the concrete and to drain water from the inside. One process uses steel balls which run around the inside and provide additional compaction of the concrete.

Longitudinal duct forms must be stressed to prevent outward deflection under spinning, which might leave voids behind the ducts. Use of removable formers, tensioned, and later filling the ducts with grout, eliminates this problem.

Horizontal casting utilizes a moving mandrel, or a withdrawn mandrel, in a manner similar to that described for hollow-core concrete piles (see Chapter 11).

Prestressed concrete cylinder pipe uses a steel liner, with a mortar lining placed centrifugally on the inside. Then the composite section is wrapped with wire under constant tension, the pipe being turned against a wire or rod feed-off device. The wire-wound pipe is then given a coat of cement slurry and a mortar coat is placed by spinning brushes as the pipe slowly rotates.

This same process may be used with an all-concrete core, wrapped under tension as described above, and coated.

As with all prestressed concrete, calcium chloride must never be incorporated in the mortar coating. One of the most disastrous cases of corrosion occurred with concrete pipe where calcium chloride had been used to accelerate set of the mortar to prevent freezing.

Helical prestress has been applied in Czechoslovakia, the wire feed-off device being moved longitudinally as the pipe is rotated. This gives a triaxial stress condition.

Self-stressing cement techniques have been successfully applied to the manufacture of small-diameter pipe in the USSR. A pre-formed spiral of high-strength steel is placed in the forms, the concrete made with expanding cement is placed and consolidated, and the completed product taken through a carefully controlled curing process to ensure the correct degree of expansion after set.

Longitudinal prestress improves crack resistance, gives greater beam strength, and may be used to join several sections or even a complete pipeline together. In the latter case, tendon couplers are used to join convenient lengths of prestressed pipe. Joints may be concreted (3 inches or more in thickness) or epoxy. A thin coat of mortar is generally not satisfactory. Joints may also be flexible, with the tendons passing through the joints. In these cases, the prestressing is considered to be of temporary benefit only, for aid in setting and adjustment of bearing, backfill, etc., unless positive means are taken to provide corrosion resistance at the joints. Simple coating with bitumastic has proven to be insufficient.

Aqueduct sections are made by horizontal casting in standard forms. Such units are usually semi-circular or trapezoidal. They are conventionally reinforced transversely and pre- or post-tensioned longitudinally. They are usually set as simple spans on flexible bearings, with a flexible seal. Box sections, manufactured in a similar manner to box-beam bridge girders, have been used for over-chutes and for underwater sewage mains. They are prestressed longitudinally and possess high beam strength. They are particularly useful where the supports must be widely spaced, as with pile-supported piers. Prestressed concrete cylinder piles, with increased spiral reinforcement, have been used in a similar manner for these over-chutes and underwater pipelines.

By proper balancing of weight and buoyancy, river crossings and siphons may be readily sunk into position. If water is admitted to the pipeline to sink the pipe, the longitudinal movement of the water must be restricted by end and intermediate bulkheads. These may consist of inflatable tires or balls.

Large-diameter concrete penstocks have utilized cast-in-place concrete or precast segments. The circumferential prestressing has been applied by high-capacity tendons, curved over 180° to 270° sectors. While frictional losses due to curvature are high, they may be kept within reasonable limits by the use of smoothly curved rigid ducts and by stressing from both ends of the tendon.

Such penstocks may also be prestressed longitudinally by internal tendons to provide beam action and longitudinal resistance to thrust and gravity forces.

Prestressed tendons are frequently used to anchor penstocks, both concrete and steel, by inserting tendons in holes drilled into rock, anchoring them with grout, and stressing them against an anchor block around the penstock. See "Prestressed Concrete Soil and Rock Anchors" (Chapter 22).

Canal linings have been built in Italy by using thin, prestressed concrete planks. These are usually 1 inch (2.5 cm) or so thick and are pretensioned. They are so flexible that they readily adapt to a semicircular canal cross section, with high compressive stress (and thus impermeability) on the inner face. They must, however, be set on an accurately prepared bed, and the joints between planks must be water-tight.

While prestressed concrete has not so far been used for the transmission of petroleum products, as far as is known, it would seem to have a potential use in swamps, stream crossings, and off-shore lines where weight is required, and where beam strength may be desirable. The use of a steel liner or diaphragm is indicated in the same manner as for prestressed concrete fuel tanks.

Joint details would, of course, have to be developed to suit the pressures involved and flexibility desired. The ratchet-type jack developed for tensioning bars at a coupler provides one means of constructing a longitudinally prestressed joint.

Prestressed concrete is also suited to the conveyance of certain chemicals, and to salt-water at desalination plants. The very highest-quality concrete must be used, with a hard-finished surface. A high degree of prestress is desirable. Internal coatings, such as epoxy, may be warranted. Extreme care must be taken in detailing and executing joints and anchorages; thus the longest possible pretensioned sections will usually be found to provide the best solution.

The use of prestressed concrete for pipes continues on a major and expanding scale, particularly as growing utilization is made of its structural capabilities.

19

Prestressed Concrete Railroad Ties (Sleepers)

Prestressed concrete railroad ties (sleepers) have been extensively utilized in Europe for many years. In the United States, their application has been limited to a few main line extensions, industrial tracks, and test sections. Rapid transit has opened a new opportunity: prestressed concrete ties are being utilized for rapid transit in San Francisco, Chicago, and Boston.

Prestressed concrete offers the advantages of anchoring and stability for welded rail, an estimated life approximately twice that of timber, elimination of under-rail cutting under heavy wheel loads, and greatly reduced ballast and track maintenance. The trend to heavier axle loadings and higher speeds is an important impetus for the more extensive use of prestressed concrete ties.

The retarding factor to their more widespread use in the United States has been one of first cost, compounded by the fact that many railroads own their own forests, mills, and treating plants; thus, the comparative costs are sometimes based on unrealistically low timber tie prices. The exclusion of freight charges may falsely benefit either timber or concrete in particular cases. However, when a comparison is based on market prices, the prestressed concrete tie will generally be found to be competitive in first costs.

Another adverse factor for the wider utilization of prestressed concrete ties has been the inability to date of satisfactorily replacing individual ties in an existing line. The greater stiffness of the concrete tie leads to its carrying an excessive portion of the track load.

Where prestressed concrete ties have been installed in a length of track, some difficulties have arisen at the transition at the ends to timber ties.

Perhaps a more resilient under-rail pad on the end ties will prove the answer.

Having acknowledged the problems facing the greater utilization of prestressed concrete ties, the projections on a structural performance and economic basis clearly indicate an imminent widespread change to prestressed concrete.

The heavier axle loads, higher speeds, high labor cost for track maintenance crews, demand for greater safety, need to anchor welded rail, and the growing scarcity, diminishing quality, and higher costs of timber ties make the switch to prestressed concrete a matter only of time.

Prestressed concrete ties are one of the smallest individual items manufactured. They are, however, used in great quantities under extremely severe conditions. Thus, they deserve the closest attention in development, design, and manufacture.

The exposure conditions include poorly drained sub-grades, freeze-thaw, excessive drying, and salt drippings from refrigerator cars. They are subjected to dynamic loadings up to 10^8 cycles. They must also perform in cases of derailment to insure no loss of gauge, so that following cars will not also derail.

Fastenings must take accelerations up to 60 g and more. Insulation must be provided between rails to permit electric signal operation.

Finally, the consequences of failure are serious to catastrophic; the ties must perform their function as part of the total load-carrying system which comprises wheel, rail, sleeper, ballast, and road bed.

Initial developments of prestressed concrete ties were based on analogy to timber ties, and attempts were made to duplicate timber tie properties, fasteners, spacing, etc. This proved to be an inadequate approach. More recent studies have been properly based on a systems analysis of the actual materials and performance required.

19.1 Design

19.1.1 Spacing

Timber ties are usually spaced at 20 to 24 inches (50 to 60 cm). With the heavier rails now in use, and a wider base for the prestressed concrete tie [12 inches (30 cm) as against 8 inches (20 cm)], it has been found satisfactory to use 28-, 30-, and in some cases even 36-inch spacings (70 cm, 75 cm, and even 90 cm). At 30 inches, tests show the rail stress is increased only 10% and the ballast bearing pressure is the same.

19.1.2 Shape

Since the most serious structural problem for timber ties is the center-bound condition, attempts have been made to relieve the load by several means, including:

Wedge-shaped center.
Narrower width at center.
Under-cut center.
Ditching of the ballast at the center.
Creation of a partial hinge at center of tie.

While all these methods have worked in test sections, in actual field experience the results have been somewhat unsatisfactory, due, primarily, to inadequate ballast maintenance, and to longitudinal creep of the whole track system. The alternative has been to make the tie stronger, usually with more prestress.

19.1.3 Length

Under the rail, a high positive moment condition exists. Since this is near the end of the tie, means must be taken to ensure that full prestress is available even after 2×10^6 or more cycles of loading. This means that the tendons must be anchored outside the rail. Anchoring may be by mechanical anchors, as in post-tensioning, or by bond, as in pretensioning. In the latter case, the tie must extend beyond the rail by the transfer length.

In pretensioning, a number of steps can be taken to ensure a short transfer length. (This is the transfer length under repetitive loading.)

1. Gradual release reduces the transfer length by 40 to 50% as compared with shock release.
2. Slightly rusted steel has a shorter transfer length than clean bright steel. The difficulty is how to obtain a uniform "slightly rusted" condition.
3. The surface of the steel, as determined by the wire drawing process, has a great deal to do with the effective adhesion.
4. Strand has a much shorter transfer length than wire.
5. "Dimpled" and deformed wire gives better performance. Ribbed bars have been used in East Germany.
6. Strand of slight irregularity in lay appears to perform better than absolutely perfect lay strand.
7. Greater steel tendon perimeter (e.g., more strands or wires) with no increase in total effective prestress on the concrete section reduces transfer length.

In actual manufacturing practice and tests with pretensioning to date, use of an adequate area of strands, gradual release, and a slight increase in tie length has provided fully satisfactory results.

19.1.4 Prestress Tendon Profile

This is usually deflected to provide greater positive moment capacity under rail, and approximate concentric prestress at the center. "Deflected," as used in the preceeding sentence, is in relation to the center of gravity of the concrete, since the tendons are usually straight and it is the concrete cross section which is varied.

19.1.5   Rail Seat

Current American practice is to use direct fastening of rail to concrete; thus, the 1:40 inward cant must be cast into the concrete. The seat is recessed so that the lips will provide lateral gauge support against forces in and out.

19.1.6   Fastenings

These must resist several million cycles of up and down loads with acceleration up to 60 g. Various types of wedge and spring clips have been developed to hold the track to the tie under positive stress. It has been found that a fastening which is neither excessively rigid nor excessively flexible is most satisfactory.

19.1.7   Insulation

For electrical signal operation, the rails must be insulated. Attempts have been made to develop concrete that will provide adequate electrical insulation even while wet; this has not been successful.

Therefore, insulation pads of various materials, such as nylon, rubber, or plastic are placed under the rail. Also various means, such as pads, are adopted to insulate the rail-fastening assembly. In one case, an insulating pad is placed between the clip and the rail. In another, the bolt and nut are insulated from the concrete. In a third, the insert is insulated from the concrete by coating it with an epoxy.

19.2 Manufacture

19.2.1   Pretensioning by Machine

A machine method has been developed by the American Concrete Crosstie Co. in which the ties are pretensioned against individual rigid forms. The dry concrete is heavily compacted on a vibrating table, and subjected to high-temperature steam curing. Very high concrete strengths are obtained, but some lack of uniformity in prestress may occur.

Equipment and form costs are high and this method has had difficulty in economically adapting to different lengths and cross sections.

19.2.2   Pretensioning, Long-Line

This method has been highly developed for the production of both rapid-transit and main-line ties, using multiple forms, gang-handling, and special machines and tie designs to facilitate production.*

*Patent pending.

Prestress is extremely uniform; excellent concrete surfaces are obtained, and accurate location of inserts assured. To date, the lowest man-hour labor consumption has been achieved by this method of manufacture.

19.2.3 Post-tensioning

A number of clever schemes of post-tensioning have been developed, primarily in Germany, in which the ties are cast with very dry concrete in individual molds, with formed ducts. They are then stripped, subjected to autoclave treatment, and the post-tensioned tendons inserted, stressed, and grouted. This readily lends itself to mass-production techniques. End anchorages provide positive assurance of maximum moment under the rail; thus, the tie length can be reduced as compared with the pretensioned tie. However, the end anchorages must be protected against corrosion.

19.2.4 Mild Steel

Stirrups are generally not required for shear. They are, however, used in some designs to contain the tendons to reduce transfer length, and next to inserts to prevent local cracking.

Other forms of ties and under-rail supports have been developed and utilized.

The duo-block tie consists of two blocks connected by a steel tie. In France, the connection has been made by a steel angle. In a number of countries, a light-gauge section of rail has been used. Such ties are, of course, not prestressed.

In Sweden, a unique tie has been developed and widely used. It is now being used in other countries, including the United States. The two concrete blocks are made by machine, and connected with a galvanized steel tube. The assembly is then post-tensioned with a prestressing bar, and the tube filled with grout or bitumastic.

Duo-block ties have the advantage of great flexibility and eliminate center-binding as a problem. However, the slant of the rails may be distorted in service and this may raise problems for high-speed, heavily loaded traffic.

In Czechoslovakia, longitudinal slabs have been manufactured, prestressed transversely, and set as a unit on prepared and graded ballast. These, of course, give very low bearing pressures, but present serious problems of drainage and ballast maintenance.

In the USSR, rectangular frames and slabs have been produced, using continuous wire-winding machines, such as the turntable or the DN-7. These are pretensioned and the wires are essentially self-anchoring, so that transfer length is not a problem. Their use does, however, reportedly present some track and ballast maintenance porblems, and this solution appears to be more expensive than the standard monoblock ties.

A number of longitudinal track support beams have been proposed and some have been installed on short sections of rocket-launching tracks. Gauge must be maintained by struts and ties.* There is a problem of concentrated bending in the rail and excessive ballast pressures at the ends of segments under high speed traffic.

19.3   Problems and Experience in Use

Concrete ties have proven in service to maintain better alignment and grade than timber ties and have performed very well in derailment accidents.

Wedge-shaped ties, so shaped to relieve center-binding, have developed some cracks at the top center due to inadequate ballast maintenance.

Torsional cracks have occasionally occurred on a 45° line at top center. This appears to have been only with the wedge-shaped tie with its reduced center cross-section. It points up the need for adequate torsional strength in the design.

Under-rail cracking, extending up from the bottom to the bolt or insert holes has occurred with a few pretensioned ties and is apparently due to bond slippage, i.e., excessive transfer length.

Rail joints have presented problems due to over-loading of the adjacent ties. Better rail-joint connections may be the answer, or use of wider ties with more resilient pads at either side of the joint. All-welded rail will, of course, largely eliminate this problem.

Change to timber ties at switches or ends causes overloading of the adjacent concrete ties. Here again, it may be desirable to use more resilient pads or to develop a more flexible tie for this location.

Switch ties can be made of prestressed concrete but require special location of inserts. Prestressed switch ties are being used on major rapid transit systems.

19.4   Future Developments

Proposed high-speed freight trains will have axle loads up to 80,000 pounds and speeds of 100 mph. It is probable that wider ties will be used to reduce ballast pressure, and that rail fastenings will be improved to match the greater forces involved.

Fully bonded behavior for pretensioned ties under repeated loadings under rail, may be further assured by such steps as:

(a) Dimpled strand or "bundling" (if this proves to accomplish mechanical anchorage).

*Japan National Railroads is conducting extensive tests on this system in an effort to further reduce labor for ballast maintenance.

(*b*) Slightly increased tie length which, in turn, will also reduce ballast pressure.

More "flexible" ties may be developed for insertion on an individual basis in existing timber tie tracks.

At the present time, some 50,000,000 prestressed concrete ties have been manufactured and installed, principally in West Germany, the USSR, and England.

However, the potential market in the USA has scarcely been touched, with less than 1,000,000 prestressed concrete ties in service.

The demands of rapid transit, higher speed and higher loaded trains, and changing economic factors will, undoubtedly, require that a complete and nation-wide industry be set up to supply high quality, mass-produced, prestressed concrete ties.

20

Prestressed Concrete Road and Airfield Pavements

Prestressing theoretically offers many advantages for pavements. These have attracted many designers and a considerable number of experimental roads, pavements, and airfield runways have been constructed. Yet so far, only the airfield runways could be said to have justified the cost and efforts. The basic problem stems from the fact that road and pavement construction by conventional means has been highly developed, with mechanized construction processes, and established sources of material supply. Prestressed concrete offers a substantial reduction in material cost, but an increase in the cost of skilled labor and supervision and, to date, a lack of mechanization in construction. The potential cost advantages are thus offset or exceeded by the actual inefficiency in construction.

Comparatively short sections of roads have been built in France, Italy, Belgium, Germany, and the Netherlands. Extensive research has been conducted in Czechoslovakia, Great Britain, and Switzerland.

Airfield runways and taxiways, on the other hand, with their much heavier loads, more readily respond to the reduced pavement thicknesses and greater flexibility offered by prestressed concrete. The advantages of a smoother surface, a watertight covering for the subgrade, a longer life, and a substantial reduction in construction joints show major benefits. As a result, many major airports are utilizing prestressed concrete extensively; these include the airports of Schiphol (Amsterdam), Orly (Paris), Algiers,

Kuwait, and LaGuardia (New York). Test strips have been installed in Great Britain, Belgium, Biggs Air Force Base and Sharonville (United States), and Germany.

Pavements are but one part of the system of load-carrying ability. The other part is the subgrade. The interrelationship of subgrade and pavement has been well established by numerous tests and experience for asphaltic concrete pavements, plain, and reinforced concrete pavements. As yet, the data for prestressed pavement and subgrade interaction is relatively unknown; we can postulate theories but cannot yet maximize economy in design.

Because of its thinner section, prestressed concrete pavements are more flexible; thus, they distribute their load more efficiently over the subgrade, resulting in reduced pavement stresses under wheel loads.

Three systems are considered for prestressed pavements:

20.1 Systems

Mobile System. The pavement slides on the subgrade, expanding and contracting due to temperature and mositure changes. The prestess is imparted by internal tendons, either pretensioned or post-tensioned. The mobility of the slab reduces the amount of prestress required.

Mobile System—External Prestress. In this case, the prestress is furnished by jacks, springs, etc. reacting against abutments and between construction joints, keeping a constant force on the pavement as it contracts and expands.

Fixed System. This may be cast in segments, but the joints are filled so that the entire slab acts as a unit. It is usually prestressed externally between fixed abutments although, theoretically, it could be internally prestressed yet remain fixed due to friction of the pavement on the subgrade. A fixed system is not able to contract or expand. The major criteria for which a pavement must be designed are temperature stresses, shrinkage and creep, and wheel loads.

With mobile systems, a major adverse factor is the friction between subgrade and pavement. Techniques to reduce this friction are primary considerations in construction methods and procedures.

In most such installations, a sand layer is placed first, on which the prestressed pavement rides. Contrary to expectations, a friction factor of about 1.0 is usually developed. To reduce this to 0.7 or so, a number of means have been employed between the sand and pavement slab, such as:

1. A single sheet of polyethylene.
2. A bitumastic layer.

3. Double sheets of plastic between sheets of paper.
4. Double sheets of plastic with grease between.

Prestressed concrete overlays have also been considered for strengthening existing runways. In this case, the existing runway is treated as a subbase, and a friction-reducing layer, such as sand, placed between it and the prestressed pavement.

Internal prestressing can be by means of post- or pretensioning. With post-tensioning, the ducts are inserted in the slab, usually with tendons installed in order to minimize "wobble." Both longitudinal and transverse ducts are installed. The transverse ducts are most easily supported by the side forms and "dobe" blocks; the longitudinal tendons can then be tied to them.

A major concern is to keep the duct size as small as possible. Rigid ducts are superior to flexible ducts because of less wobble and less friction. Obviously, in as thin a section as a pavement, the ducts must be accurately positioned and maintained in position during concreting.

With pretensioning, usually the longitudinal tendons are the only ones pretensioned. They are tensioned against abutments as much as 3000 feet apart or more; the abutment being concrete piers or steel H-piles. Transverse ducts are then usually supported by the longitudinal tendons.

After pouring, shrinkage cracks must be prevented. Thus, it may be desirable to pour at night. Curing must be instituted and carefully maintained. It has often been found advisable to use saturated cotton or burlap mats which are, in turn, covered by a polyethylene sheet to prevent drying out.

A little prestress should be released into the slab as soon as possible to offset shrinkage stresses. This can be done at ages of 1 or 3 days, with final prestress released at, say, 7 days.

With overlay pavements, two-way pretensioning can sometimes be employed, with the tendons anchored against the original slab.

In the Netherlands, an 1100-meter long roadway pavement was pretensioned around a curve, with the strands being deflected around sheaves affixed to an anchored horizontal frame.

An effective and economical means of providing transverse prestress, which has been developed and used in Germany, is the use of "tensile element" bars; small bars which are given a high prestress and then placed in the slab. Through the mechanism of creep, they induce a degree of prestress into the pavement slab. (See Chapter 3, Section 3.6.)

Another interesting solution to roadway pavements has been a network of diagonal post-tensioning, with the angle and spacing varied to fit the degree of longitudinal prestressing required.

Internally stressed pavements cannot buckle, and have usually shown greater elastic and ultimate strength than the design required. They do require fairly sizeable quantities of prestressing steel. Repairs to small areas are readily made; major repairs may be extremely difficult.

Externally stressed pavements offer the possibility of substantial savings in steel. The external prestress permits variable application, so that a little stress can be induced shortly after initial set and, during service, the stress can be varied to meet the change from summer to winter. This external prestress can be induced by permanent hydraulic jacks, by coiled steel springs, by flat jacks, or by pneumatic tubes. These tubes can be incorporated in the construction joints and are very efficient. The only difficulty is a tendency of the tubes to be pinched; in fact, all forms of applying external prestress except, perhaps, flat jacks, are subject to possible damage in service and must be protected.

Externally stressed pavements can buckle or "blow-up" under a combination of high stress and high temperature. This can be prevented by proper design, accurate construction, and careful adjustment of the external prestress forces as required. The incorporation of some steel in the pavement will enhance both its ultimate strength and its buckling resistance.

These pavements are easily repaired; the external prestress is temporarily released and the repair accomplished by normal methods.

The major problem lies in the joints. Much work is currently underway to develop reliable, effective joints.

One means to prevent buckling is to deepen the slab as, for example, by making a hollow-core or box slab, with the hollow cores being utilized for utilities, etc. On the Mont Blanc tunnel approach (Italy to France), the box sections were used for ventilation ducts, as well as roadway slabs.

20.2 Artificial Aggregates

Since a major portion of the stresses in prestressed slabs is due to temperature, it is obvious that a concrete with lower thermal response would be desirable. Silicious aggregates appear more favorable than limestone aggregates. Expanded shale, slate, and clay aggregates (lightweight aggregates) have a reduced thermal response and, also, provide better insulation, so that the lower surface of the slab, in contact with the subgrade, is not subjected to as great a variance, particularly, the short-term variances which are most troublesome.

Artificial aggregates have been manufactured with extremely low coefficients of thermal expansion: while, as yet, they are not commercially practicable, they offer possibilities for the future.

Strains due to temperature are transformed to stress in direct proportion to the modulus of elasticity. A concrete, such as lightweight concrete with its low E, can reduce temperature stresses by up to 30%.

Stresses from wheel loads are distributed to the subgrade by the flexibility of the pavement. Once again, a low modulus of elasticity is desirable.

Prestressed lightweight-aggregate concrete offers, therefore, the following advantages for pavements:

1. Lower modulus of elasticity.
2. Better insulating qualities.
3. Reduced thermal response.
4. Better skid resistance.
5. Improved durability under de-icing salts.

An experimental roadway employing lightweight aggregate has been built in Belgium.

20.3 Problem areas

The two major problem areas for prestressed pavements are the construction joints and edge warping.

Construction joints must permit movement, must transfer shear, must be water-resistant and, most difficult of all, must respond to wheel loads about the same as the adjoining pavement, i.e., they must not be too rigid. Combinations of steel, rubber (neoprene), and concrete are utilized. The joints must be designed so that dirt, spalled concrete fragments, etc., cannot become lodged and prevent movement.

Edge warping is a problem due to a combination of shrinkage, temperature, and prestress. Thickening of the edge helps by increasing the dead weight. Properly proportioned, it may effectively balance the warping effect.

20.4 Precasting

Precast slabs have been used for both roadway and airfield pavements. The largest such installation is at Kuwait, where triangular precast slabs were employed. Post-tensioning was in the spaces between adjoining slabs, thus permitting three-dimensional prestressing. Use of precast slabs minimizes shrinkage and creep, but requires extreme care in alignment and bedding.

The over-water runways at LaGuardia Airport, New York, are a marine structure as well as a runway. This project is discussed in Chapter 12. A combination of precast beams with a composite cast-in-place slab was employed. This, in turn, was post-tensioned in two directions. The use of the cast-in-place slab permitted an accurate and smooth surface to be achieved while enjoying the structural and economical advantages of precasting.

20.5 Mechanization

There is a definite need for constructors to develop mechanized construction means which will make prestressed pavements competitive in cost with conventional pavements. This will enable the many advantages of prestressed pavements to be realized and thus increase the volume of use, justifying further cost-saving mechanization. The technique of tensile element bars, mentioned earlier, is one such step. Longitudinal pretensioning appears to offer significant cost advantages. Lightweight aggregate concrete appears particularly applicable.

Whichever system and design is adopted, construction planning, equipment, and techniques have a proportionally great influence on the economic and structural success of the pavements.

21

Prestressed Concrete Machinery Structures

Prestressing and prestressed concrete techniques offer a potential major breakthrough in the design and construction of machinery structures, including foundations, beds, frames, and, in some cases, even the working parts. This is a new concept and one generally not familiar to machinery designers and constructors. At the same time, the quality control and tolerances required are substantially more rigid than those usually required for civil engineering structures.

Prestressing enables the achievement of results which are otherwise unobtainable. Foremost among these is reduced deformation in service, a property of increasing importance for automated machinery.

Another major property obtainable with prestressed concrete is thermal insulation and reduced response to thermal changes through the use of specially selected and manufactured aggregates.

A machinery structure is a dynamic system that includes soil, foundation, bed, frame, machine, and product. By means of prestressing, the foundation, bed, and frame can be made to act as one.

Fatigue is presented in a new dimension in machinery structures. The civil engineering structure must be designed for perhaps 2×10^6 cycles; a machine may require design for 10 to 30×10^6 cycles. Prestressing reduces the range of cyclic variation, thus greatly increasing the fatigue resistance.

Dynamic response is determined by the relation between the frequency of the machine and the natural frequency of the system. This latter can be modified by changing the mass and/or the rigidity. Prestressing enables blocks (masses) to be added in such a way as to work integrally with the structure. Prestressing is also a means of changing the rigidity through its external trussing effect.

Dynamic response is also determined by the modulus of elasticity. Selection of the proper aggregates (e.g., granite, expanded shale, etc.) gives a means of controlling this modulus through a range of 100% or so.

Prestressed concrete concepts enable the designer to preset the behavior of the structure so that it will have fully elastic behavior up to cracking; thence a plastic-elastic behavior (with micro-cracks) giving damping and ductility with reduced stiffness, from which the concrete will substantially recover upon reduction of load; and, finally, a plastic (ductile) failure mode. The achievement of such a spectrum requires the careful proportioning of stressed and unstressed reinforcement. Combinations of stressed tendons with wire mesh have proven useful in this regard.

Prestressed concrete machinery structures are usually designed to be crack-free under normal operation, as cracking reduces stiffness and fatigue strength. A structure which is in resonant frequency with the exciting force up to cracking will undergo a substantial frequency change on cracking; thus, it possesses a built-in-safe-guard against progressive dynamic failure.

Prestressing makes it possible to stress together segments to form a foundation structure of almost any size or shape which will still act mono-lithically.

Prestressed concrete machinery structures can be designed with elastic deflections as low as 1/7000 of the span compared with the figure of 1/1250 which is normal for steel.

Thus, in summary, prestressed concrete offers these advantages:

1. High energy absorption before rupture.
2. High internal damping.
3. High fatigue strength.
4. Safety through the automatic pre-testing of materials, and in inherent mode of failure (non-catastrophic).
5. Stiffness.
6. Precision.
7. Ability to design and modify dynamic response.
8. Unlimited shape and size.
9. Low deformability under repeated short-term loadings.
10. Chemically resistant.
11. Noise reduction.
12. No condensation.
13. Not notch-brittle at low temperatures.
14. Economy.

Prestressed concrete does possess some disadvantages as compared with steel. These are mainly associated with strains. Strains include those due to volume change, such as shrinkage, temperature, and creep. They also include elastic changes under prestress and under dead and live loads. These all can be controlled as noted later.

Another disadvantage of prestressed concrete may be its greater size as compared with cast steel. This may mean that the machine itself exerts a greater bending force, or that the foundation must be larger. Proper rational design makes beneficial use of this larger size and mass to reduce foundation pressures and vibration, and overcomes the bending moments by prestressing.

On occasion, the machinery structures may be tied to rock or very firm soils by prestressing, so as to make the mass of the rock an integral part of the structure. Even in soft soils, with piling, the machinery structure may be prestressed to the piles, which, in turn, "lock" to the mass of soil through friction.

Prestressing is an excellent means of securing inserts and of bolting the machine to the frame or bed or foundation in such a way as to prevent differential movement and fatigue.

21.1 Manufacture

Prestressed machinery structures are inherently more important and higher-priced structures per unit of volume than most civil engineering structures. Thus, much more sophisticated manufacturing techniques are appropriate and justified.

First, aggregates, cement, and mix can be selected to meet the specific properties required (compressive and tensile strength, modulus of elasticity, minimum volume change, low thermal response, self-insulating qualities, etc.).

Because of the generally small size of structure or segments, mixes can be carefully vibrated and compacted, with a very low water/cement ratio.

Proper curing can readily be provided and controlled. For example, a steam-curing cycle can be set with a controlled rise and fall of temperature followed by water cure and protection from sun in storage.

Forms (moulds) can be made of machined cast steel, adjusted by machine screws to exact position, within a tolerance of 3 to 5 mils. Such forms must be protected from temperature change.

By means of such procedures, concrete segments can be cast to an accuracy of 1/64 inch (0.4 mm).

An alternate means of concreting is the use of prepacked aggregate concrete. In this case all reinforcement, ducts, anchorages, and inserts, are accurately set. Crushed rock is carefully screened to remove all fines, then is placed in the forms, and the inserts re-checked. Grout is then intruded through preplaced tubes. This is a special grout, possessing great fluidity, either through the addition of admixtures to reduce surface tension, or by mixing to a colloidal state.

With prepacked concrete, the form pressures may be quite high, especially in confined and localized areas, so forms must be very rigid and unyielding. Concrete-backed steel or fiberglass makes an excellent form.

To reduce heat of hydration in large blocks, the aggregates may be pre-chilled, as may the mix. Ice may be used to keep the ambient temperature of the mix near 60°F. Cement may be low-heat or large-grain size (low Blaine fineness number).

Water may be circulated through the duct tubes or through special cooling tubes. If the duct tubes are used for this purpose, they should be of rigid tubing.

Shrinkage may be minimized by a low water/cement ratio and a relatively low cement content, using some pozzolan to replace cement. Shrinkage-compensated cement offers possibilities; however, its performance must be determined for the mass and thickness involved, as most research and tests to date have been on slabs. Shrinkage may also be minimized by sealing the surface to prevent moisture loss, as by a complete epoxy coating.

21.2   Tendons

Prestressing tendons and their anchorages must be selected for proper dynamic behavior. This requires that the system be proven under dynamic testing and that there be adequate elongation (2% minimum) at rupture.

Both bonded and unbonded tendons are used. The use of unbonded tendons permits the re-stressing and/or replacement of tendons at a later date. Techniques for grouting of bonded tendons are discussed in Chapter 5, Section 5.1; techniques for protecting unbonded tendons, in Chapter 5, Section 5.2, and Chapter 16. In any event, the anchorages must also be protected from corrosion, fire, and impact. For these reasons, recessed anchorages are preferable.

21.3   Triaxial Prestress

The behavior of concrete under multi-axial loading is radically transformed. Bearing stresses can be increased up to perhaps 3 times the compressive strength of the concrete by use of spiral reinforcement. Proper shaping of the concrete can permit greatly increased stress and strain, as is shown by the plastic hinges of concrete which develop apparent stress levels of 20,000 psi and rotational strains up to 4%.

Triaxially stressed concrete has a high elastic limit, high stiffness, and a high fatigue limit.

One means of triaxial prestressing is by using tendons in three or more planes. Stressing in two of the planes may be obtained by curving one set of tendons with overlapping anchorages, or else by using continuous wire-winding. With heavy spiral reinforcement of high-strength wire, longitudinal prestressing to a high degree produces radial prestressing also, through Poisson's effect.

21.4   Ferro-Cement

For certain structures, frames, etc., requiring great ductility, the combination of ferro-cement techniques with prestressing offers interesting possibilities. Layers of mesh are placed in the concrete so as to give a steel proportion of approximately 800 pounds per cubic yard. To properly place the concrete, it is usually applied in layers or coats, by hand, alternating with the mesh.

Approximations of this phenomenon can be achieved in more conventionally placed concrete by placing one or two layers of mesh close to the tensile surface, or by the use of finely-divided wire in the mix.*

21.5   Cold Working

One of the most interesting aspects of prestressed concrete for machinery structures is the cold-working phenomena. Repeated loading and unloading cycles show a gradual stabilization against creep. This may be achieved more readily by a combination of alternating heating and stressing treatments. Obviously, any such process has to be applied in the direction or axis of final stress and is most practicably applied to segments.

21.6   Segmental Construction

The joining of segments can be best achieved by the dry-joint or epoxy-joint techniques described in Chapter 3, Subsection 3.1.4. Dry (exact fit) joints may be obtained either by casting in sequence against the previous segment and match-marking, or by casting in machined cast-steel forms as described earlier in this present chapter. A 1/64 inch tolerance has proven adequate for dry-joint techniques of joining segments. This is *more* accurate than can generally be obtained by grinding.

If grinding is employed, controls must be set up to compensate for wear of the grinding wheels, and means established for determining a plane of reference. The difficulties attendant usually limit tolerances obtainable to about 1/32 inch.

*See Postscript vii.

21.7 Utilization

Prestressed concrete has been utilized for forging hammer blocks. One of these has had over 300 million cycles in service, with accelerations up to 100 g, and is still uncracked.

Structures subject to shock loading such as rocket test stands and explosion test chambers are particularly suited to prestressed concrete because of hair-crack resistance, chemical inertness, high thermal resistance, safety in explosion, and noise reduction.

Large foundations can be held to very rigid tolerances of accuracy, which is especially important for extrusion beds, and automated machinery.

Large-diameter hydraulic presses have been made, with internal pressures up to 6000 psi. The prestressed concrete cylinder head is manufactured, then a machined steel liner placed and grouted. The piston is likewise of triaxial prestressed concrete, with a machined liner placed over it and joined by grouting. The steel serves as wearing surface and seal, but the prestressed concrete resists the pressures and shock loading.

Turbine blocks may require adjustment in order to change from a resonant frequency. Additional masses (blocks) of concrete may be added to the block and prestressed to it, thus "tuning" it. This is especially valuable when employing a turbine block of lower frequency than the turbine itself, since the lower frequencies are harder to accurately predict and may require field adjustment.

To readily permit such adjustment, it may be desirable to install additional, unfilled ducts in the original block, into which tendons may later be inserted.

Frames for presses appear to be an excellent application for prestressed concrete because of rigidity and freedom from distortion.

Boring machines have had their pillars prestressed to the table or bedplate to prevent distortion.

A 10,000-ton testing frame, a steel strip mill, a torsion-stretcher bed, and a 600-ton hydraulic press are among the many successful applications of prestressed concrete.

There is an obvious relation between the use and techniques of prestressed concrete for machinery structures and for prestressed concrete pressure vessels (Chapter 16) and prestressed concrete floating and submerged structures (Chapter 14).

21.8 Summary

Prestressing enables low deformability and controllable dynamic response to be achieved in machinery structures. As such, it is moved into an application of high sophistication requiring the greatest of care and control in design and manufacture. It is probable that the increased demands for automated machinery and ever larger machines will force the greater utilization of prestressed concrete for machinery structures. It is important, therefore, that the design and construction be undertaken by those thoroughly versed in the properties, techniques, and art of prestressing.

22

Prestressed Concrete Soil and Rock Anchorages

Prestressed tendons have been increasingly applied as anchors and tiebacks. They have been utilized to tie down the floor of graving docks against hydrostatic uplift; to tie back bulkheads and retaining walls, both temporary and permanent; to stabilize rock cuts and roofs; to tie down structures to the sea floor and towers to their bases; to anchor penstocks and thrust blocks; to anchor suspension bridge cables; to anchor piles against uplift, and to increase the stability of dams.

Ideally, the inner anchorage is in rock, and the structure, e.g., concrete slab, is stressed against the soil. Thus, the inner anchorage is grouted to the rock, and the outer anchorage secured to the structure. In the space between, presumably soil, the tendon should be unbonded until after stressing, and may be left permanently unbonded provided steps are taken to insure corrosion protection.

Weak rock may yield through time, in an accentuated form of creep. Thus it may be necessary to re-stress the external anchorage at intervals. The anchorage selected should be such as to permit re-stressing and re-anchoring or shimming.

In many cases, rock is not available for the inner anchorage and the anchorage must be provided by friction against the soil. Obviously, the soil must be relatively stable against creep; compact sand is far superior to clay. The soil engineer must perform a thorough analysis of the creep behavior of the soil. Friction anchorage can often be obtained in a drilled hole in soils by pressure injection of grout.

One of the several schemes that have been developed for soil anchorages consists of an augur drill that has a hollow stem. The tendon is inserted in the hole with a grout pipe attached and the anchorage attached at the bottom end. Then the hole is drilled. Grout is now injected under pressure and the drill and grout pipe slowly withdrawn, while continuing to slowly rotate the augur in the same (drilling) direction.

For pile anchors a hole is formed in the pile before driving (or drilled afterwards, if necessary). After driving, the hole is then extended into the underlying soil. The tendon is inserted with a grout pipe attached, and grout is injected to achieve the necessary bond with the rock. Then the tendon is stressed from the top and anchored. Grout or bitumastic or grease is injected to fill the void in the intervening soil and the hole in the pile. This same technique has been utilized to anchor a concrete nuclear reactor containment vessel.

Consideration in design must be given to shear and concentrated bending at the juncture of the structure (or pile tip) and the soil, so as to prevent localized damage. At the same time, a positive seal must be provided against moisture penetration and corrosion.

Suspension bridges have been anchored by prestressing tendons embedded in rock. The connection between the prestressing tendons and the suspension cables may be an elaborate mechanical device of steel, with swaged fittings on the suspension cables, or it may be achieved in a large concrete anchor block.

Other forms of anchors for use in soil have been developed for low-capacity anchors, of the order of 25 tons or so. These are mechanically or hydraulically expanding anchorages, to which tendons are attached. Grout may be injected in addition.

Prestressing anchors are widely employed for temporary bulkheads; in fact, the "tied-back" bulkhead has revolutionized foundation design for large buildings because it frees the interior for excavation and construction. Being soil-retaining structures, the tendons are usually left ungrouted in the soil zone, except for the anchorage zone itself. Corrosion and its consequences must be considered, depending on the time of exposure, soil and ground water conditions, and type of tendon. Bars, being of thicker section, are more durable than wires or strand. Strand may be galvanized, plastic-coated, or bitumastic coated. Grease may be injected in the intermediate zone.

The anchorages of such bulkheads are very vulnerable to corrosion, accidental impact, and fire. Insufficient attention has been paid to this problem in many cases. The anchorage should preferably be protected by steel channels or boxes, or timber. Proper details must be used to insure against standing water and corrosive drippings. A sheet metal sleeve and liberal use of grease may suffice.

In event of fire inside the excavation, as, for example, in the formwork, the anchorages and their adjoining tendons are more sensitive to damage than the remaining steel. Each case must be analyzed to determine the exposure. In some cases, blown insulation, such as vermiculite, etc., or concrete mortar, may be packed around each anchorage.

With all retaining structures, temporary or permanent, care must be taken in both design and construction to avoid a "stack-of-cards" type of failure; that is, the failure of one tendon, whether in itself or in the soil, leads to an overload of the adjoining tendons, causing them to fail, leading to catastrophic consequences. It is customary, therefore, to design each tendon for a substantial overload, for example, 50% to 100% or more.

Each tendon should be jacked to an overload as specified by the engineer (frequently 50% overload) and then released and anchored at the specified stress. Occasional tendons should be restressed to determine creep, and slippage, etc. Tests may be run to twice design load or to failure, to verify the design procedure.

Since the usual failure is in the soil, and is a gradual, time-dependent yielding, such tests must be held for an extended period as directed by the soil engineer. Some soils are very sensitive to repeated loadings (a fatigue-type failure), and others to vibration, as from surcharge traffic effects. Dry, granular soils may suffer reduction in friction-holding power when wetted during rains.

One of the most successful types of friction anchorages for bulkheads in soft materials has been the driven H-pile, with its large surface area. Friction around the tips of these H-piles may be improved in certain soils by injecting grout at the inner end. Pipe piles may be hydraulically or mechanically expanded at the inner end. These piles may then be connected to the bulkhead walls by short prestressing rods, or may be jacked directly, enabling the pile to be anchored under stress and permitting each anchor to be pretested by jacking.

Whenever any anchor extends through soil into rock, the inner anchorage must lie well beyond the failure plane. This failure plane is determined by soil mechanics and geological engineers. Since the extra cost of extending holes and tendons a few feet deeper is small, and the consequence of too short tie-backs catastrophic, this author strongly recommends that an ample margin be provided (5 to 10 feet extra in many cases) so that any error in length is on the safe side.

When anchoring rock, consideration must be given to local failure near the anchorage and uneven bearing. Provision of a "washer" of steel or concrete, set in grout under the tendon, will serve to distribute the bearing stress and overlap local anomalies. If the rock surface has been pre-covered with shotcrete to prevent air-slaking, the anchorage block or "washer" may still be necessary, because the thin, unreinforced shotcrete coat may be inadequate.

Many soils, particularly fractured rock interbedded with clay, exert greatly increased pressures once they start to move. Prestressing of these prevents even small movements and tends to keep the pressures within the design limits.

Strain gages may be attached to tendons to monitor the change in stress through time. A sufficient number must be used to average out local conditions and distorted readings.

Prestressed anchorages employ very conventional post-tensioning techniques. Since they are working in soil, rather than concrete, the potential variations in ultimate strength must be considered by all involved, including designer and constructor. Concrete is a relatively uniform product, manufactured, placed, and controlled with care, with a coefficient of variation of perhaps plus or minus 15%. Yet current design practices usually include a factor of safety of 2 or more.

Soils, on the other hand, vary widely within small distances, particularly at the shallow levels in which most foundation construction takes place. When soils yield, they tend to proceed continuously and increasingly toward failure. Typical soil design criteria may provide a safety factor of only 40 or 50%.

This is the reason that recommendations are made herein for testing of each tendon, for extending tendons to greater penetrations than the minimum shown by calculations, and for evaluation of the consequences of failure of a tendon.

Prestressing as a concept is revolutionizing soil mechanics. It explains why a driven pile may develop more capacity and lesser settlement under load than a drilled cast-in-place caisson; the soil resistance is pre-mobilized; the soil is "prestressed." "Rammed" piles, and piles in which a bulb is formed at the tip under pressure, similarly provide increased performance beyond that which can be explained solely by bearing area. Grout injection under pressure increases frictional interlock and also "prestresses" the adjoining soil, increasing its shear resistance.

One aim of soil mechanics is to find means by which the soil itself can attain improved structural properties. A Western European development, known as "Reinforced Earth" utilizes thin friction strips of galvanized steel placed in layers of compacted earth. The weight of the fill deflects the strips and stresses them which, in turn, exerts lateral (confining) prestress against the fill.

Foundation slabs and drydock floors may be anchored by vertical pre-stressing to prevent upward displacement due to hydrostatic uplift and soil heave. In very deep excavations in unstable soils, the construction of the slab and the stressing and anchoing may have to be carried out underwater. An underwater jack, controlled by a diver, was used in stressing down the anchor slabs for the LaFontaine subaqueous tunnel at Montreal. Alterna-tively, stressing may be carried out from the surface, with the tendons splayed, and anchored by injected grout in the foundation slab.

For the controlled sinking of caissons, weight or thrust is required. Ducts may be formed in the walls, then holes drilled on down to rock, tendons inserted, grouted, and stressed. This is a very economical means of applying thrust and has the major added advantage of furnishing a means of control. Tendons will have to be progressively extended by splicing as the walls are built up.

Prestressing concepts therefore offer great promise in the field of soil mechanics, and the field is just beginning to be explored. As with other materials, prestressing offers to soils the opportunity to control behavior, i.e., deflection, as well as to provide ultimate strength.

23

Prestressed Concrete; Special Structures, Repairs, and Modifications of Existing Structures

There are, of course, a great many special applications of prestressing and prestressed concrete. New uses will constantly continue to arise and be exploited by this technique and concept.

Monumental statues present difficult and unique structural problems, exposed as they are to wind loading and seismic forces while at the same time not necessarily being designed for the most effective structural behavior and position. Fortunately, as Freyssinet pointed out many years ago, nature utilizes bones and tendons in living forms; in a statue the concrete, bronze, etc., may replace the bones, and the tendons may be of high-strength steel. Nevertheless, nature provides dynamic response to maintain balance, while a statue must remain fixed. Hence the design of a prestressing system for such a statue may present a challenging and intriguing exercise.

Prestressing techniques are similarily used to restore and preserve ancient monuments and structures, many of which were toppled by earthquakes such as the great one that shook the Middle East in the 6th century. Techniques include drilling through the existing segments, and post-tensioning, with particular emphasis on protection of the tendons from corrosion. The interstices of the existing structure should be first filled with grout under carefully controlled minimum pressures. Deflections etc., should be carefully monitored during all stages of grouting and stressing.

Towers are particularly well adapted to prestressing and most of the recent monumental towers in Western Europe and the USSR are of pre-stressed concrete. Vertical post-tensioning tendons are installed to stress the walls and, in some cases, to anchor the tower to the rock. Circumferential tendons have usually been in sectors, stressed around 180° or 270° to external buttresses.

Prestressed concrete girders and beams have recently been proposed as structural skid beams to support oil production equipment in the Arctic, because of favorable behavior at low temperature and rigidity. Prestressed concrete offers substantial economies over steel girders, fabricated from low-temperature steel, even after consideration of freight costs.

Prestressing techniques have also been applied to stone. The problem has been the non-uniformity of stone and planes of weakness and lamination. The actual prestressing consists of drilled holes, with post-tensioned tendons and grouted joints.

Many small mass-produced items lend themselves to prestressing. Fence posts are typical of a small linear item suitable for pretensioning in mass production. Multiple-form techniques are used. Manhole covers are an illustration of the application of circumferential stressing, with wires applied under tension in a groove in a precast plate. The groove is later filled with mortar. Such items become economical only when produced in large numbers by mechanized processes.

An interesting application of prestressing is to horticultural (greenhouse) planks. Such planks must be resistant to rot and constant humidity; they must be non-toxic and non-corrosive. Narrow planks of prestressed light-weight concrete can be lifted by two men, and answer the needs of the environment.

A major item of concern in large structures, such as bridges, are hinges. Concrete hinges have been developed which permit rotation across a narrow joint. Heavy amounts of steel are used in a crossing pattern; this may be mild-steel or high-tensile (unstressed) tendons. The extremely high compressive stress across the throat and the high percentage of steel prevent cracking as the joint rotates. The Cement and Concrete Association of Great Britain has conducted extensive research on these, leading to their use in such large concrete bridges as the Gladesville Arch, in Sydney, Australia.

Tunnel liners are an important application of precast concrete. Segments are mass-produced in very rigid forms. These forms are either machined thick steel plates or ground and polished concrete. If the form requires disassembly to permit stripping, then machine-thread bolts are used to insure accurate re-positioning. To prevent warping and shrinkage after stripping, extreme care must be taken in mix design, curing procedure, and in uniform, protected storage. With proper care, segments may be produced with tolerances of 1/32 and even 1/64 inch.

Similar to tunnel liners are mine supports, in which prestressed beams replace the traditional timber supports. Not only is adequate elastic strength required, but also ultimate strength. This large ductility range can best be obtained by incorporation of unstressed strand or mild steel in substantial quantities, in addition to the stressed tendons. Lightweight concrete is well adapted to this application because of its greater deflection under load, fire resistance, and lower handling weight. Such mine support timbers and tunnel liners are durable, fire-resistant, and economical.

Prestressing techniques have been very successfully applied in underpinnings. Rather than going beneath the footing to jack it, inclined tendons to new piers can be stressed so as to prevent deflection. If these tendons are left unbonded temporarily, the prestress force may be adjusted to exactly balance the deflection. As in all underground uses of prestressing, corrosion must be considered if tendons are to be exposed for any length of time. Bars are less susceptible than strand but are not as flexible around saddles, etc. Tendons may be protected with grease, bitumastic or galvanizing. Permanent underpinning tendons should be encased in grout.

Prestressing techniques have been employed with great success to correct excessive and undersirable deflections in existing structures. Tendons are usually applied externally, and may or may not be bonded to the structural member in question. Such tendons must be protected against corrosion by means such as galvanizing, greasing (with provision for maintenance), or concrete encasement.

Where the excessive deflection was due to creep, care must be taken not to over-correct, that is, not to produce excessive camber and creep in the other direction. Fortunately, this is usually obviated by the deck or floor slabs which have been poured on the sagging girders.

Steel members may be strengthened against buckling by judicious prestressing and especially if concrete encasement or fill is used to provide increased compressive area.

The tensile flange of existing steel girders was augmented by a plate of high-strength steel, jacked for elongation, and secured to the existing flange by high-strength bolts. The increased tensile capacity, coupled with the composite action of the concrete deck slab, enabled the bridge to carry substantially heavier loads.

Prestressed and ferro-cement techniques have been applied to the construction of both stressing beds and the forms or moulds for production of prestressed concrete. Consideration must be given in form design to the degree of flexibility or rigidity required for operations. Surface effects must also be considered, and concrete forms may require a surfacing of epoxy or plastic to permit easy stripping and removal of the product.

When prestressed forms can be incorporated in a permanent cast-in-place structure, additional benefits result. A very durable skin is provided. If highly stressed, some of the tensile resistance is transferred to the adjoining cast-in-place concrete, permitting greater strains before cracking.

In cable-supported structures, tents, pavilions, etc., tendons are prestressed and anchored so as to work against one another, and to support the membrane covering. Particular care must be taken to prevent corrosion, as by the use of galvanized or plastic-encased tendons, and to prevent abrasion where tendons cross. Such structures have been of great interest at recent World Fairs in Brussels and Montreal.

Additional applications will continue to be developed as engineers, architects and contractors become more familiar with prestressed concrete. As each new application is tried, the composite experience of the profession is enhanced and the horizon ever widens.

SPECIAL CONSIDERATIONS

24

Prestressed Concrete in Remote Areas

Prestressed concrete is obviously the same material and involves the same techniques and construction methods, whether constructed in urban centers or the most remote corners of emerging countries. However, the advantages of prestressed concrete in such remote areas are often even more dramatic than in highly developed centers. These advantages include:

Maximum use of local materials.
Maximum use of indigenous labor.
Minimum import of special materials (tendons) and equipment (jacks, etc.).
Ability to set up in remote and difficult places.
Development of a local industry that can be continued and expanded as a basis for housing, local industry, and public works.
Low cost.
Durability with minimum maintenance.

Many of these advantages are typical of all concrete construction, whether prestressed or conventionally reinforced. Prestressing reduces the steel consumption in weight to about one-sixth and in cost to about one-half. Prestressed concrete members themselves are lighter, have longer span potential, and are more durable.

The problems in prestressing in remote areas are the lack of local industry and material support, the lack of skilled and semi-skilled labor, the lack of technical skills, the lack of transport, the difficulty of quality control and, most of all, the lack of construction management. Many remote areas are exposed to extremes of temperature and humidity, increasing the problems of durability, or are desert areas, with problems of obtaining sound aggregates and fresh water, or Arctic environments, with problems of winter construction.

The greatest single problem has been the production of high-quality concrete. Coarse and fine aggregates must be tested for soundness and for alkali-aggregate reactivity. Fine aggregates must, in addition, be tested for chlorides, fluorides, and excessive silt or organic material. In desert areas, despite the high cost of fresh water, aggregates may require washing with fresh water to remove salt.

In extremely hot areas, such as those of the Middle East, aggregates and concrete must be cooled. Water soaking (with fresh water), vacuum cooling, shielding of stockpiles from the sun with aluminum or galvanized sheathing, and mixing with ice are techniques used to bring down the temperature of the fresh mix. Steel forms may become excessively hot; these can be precooled by a water spray. After pouring, a fog spray should be applied almost immediately, followed by moist steam curing and/or water cure. Water cure should be for a minimum of 7 days (3 days if supplemental to steam cure). Use of a retarding admixture may be found beneficial in preventing flash set. Many of these problems may be minimized if the concrete pour takes place in the early hours of the morning (about daybreak).

Prestressing steel may suffer serious corrosion in transport to remote sites, particularly in the tropics. It should be wrapped in heavy export wrapping, with VPI crystals sealed inside. Packaging and handling must be such as to prevent rupture of the package.

In the tropics it may be difficult to find water that is free from excessive organic materials and dissolved chemicals. Frequently, however, water which is discolored is found satisfactory upon tests and in actual experience. The emphasis thus is on proper testing of all materials.

Cement must be export-packaged and preferably palletized and wrapped so as to insure against moisture. The age should always be determined and be within manufacturer's limitations.

In Arctic environments, aggregates must be selected for freeze-thaw durability, a property for which tests may not be conclusive. Experience in previous use for concrete structures, whether prestressed or not, is the best indicator. Air entrainment should always be employed.

Production in Arctic environments is typical of all cool and cold-weather concreting activities. Steam curing is definitely indicated. Care must be taken to cool gradually to prevent excessive thermal strains upon removal from the steam chambers. Particular care should be taken to prevent drying-shrinkage cracks from cold dry winds immediately after removal from curing. Since drying-shrinkage and thermal strains are additive, continuation of water curing, as by soaking blankets or sealed thermal blankets, may be appropriate.

Management, training, and direction of unskilled indigenous labor is generally easier than anticipated if properly planned and organized. At least one skilled prestressing ironworker foreman and one skilled concrete foreman (expatriate supervisors) are required. A skilled crane operator and a mechanic are essential, and may be either indigenous or expatriate. These men can direct the actual crews, train them, and supervise their work. Invariably, the indigenous personnel are eager to learn. They can't learn from engineers; they can only learn from highly skilled expatriate working supervisors, who can actually show them how to do each of the required tasks.

There is one fundamental rule to apply: train each unskilled indigenous worker for one task only. Tasks should be subdivided into detailed categories, such as, for example:

Prestressing Ironworker

(a) Spiral making.
(b) Spiral tying.
(c) Strand laying.
(d) Strand anchoring.
(e) Stressing and releasing.
(f) Mild-steel fabrication.
(g) Mild-steel placing.
(h) Placing of lifting loops.
(i) Burning of strands.

In a skilled labor market in a highly developed country, one prestressing ironworker may do all of these tasks in a given day; in an emerging country, one man should do only one of these tasks. It will be found that within a reasonable period he will be very efficient in the one task. Because of lack of flexibility in assignment, manpower requirements may run about 3 to 5 times that of a United States yard, plus any special labor requirements such as watchmen, clerks, security, etc. As time goes on and men become more experienced, particularly if they have some education, their flexibility will improve.

Care must be taken to respect local craft divisions. Traditionally, certain tasks are performed in a certain way, by a specific group. As long as the end product is of adequate quality, there is no reason to upset these patterns. When this is recognized, new tasks or higher-quality production can be fully mechanized, with the most modern equipment, and this will usually be accepted. On one job in Southeast Asia, for example, rock was hand-"pitched" from a quarry, trucked to the river in 1/2-ton lorries, loaded into canoes by basket, taken across the river and unloaded by basket to a modern crushing-screening washing plant. Remarkably, the finished aggregate cost about the same as in the United States. This result is quite typical; despite much lower labor costs, prices of finished products are often in the same general range in widely varied economies.

At the engineering-management level, most developing countries have trained engineers who can serve as assistants. One expatriate prestressing engineer and one manager are usually sufficient. Accounting and purchasing roles may be handled by expatriates with indigenous assistants. If all expatriate personnel take the time and patience to train as well as to supervise, the whole job will become easier.

The engineer must serve as quality-control man and must be firm and rigid in his acceptance-rejection, but take pains to show why, and wherever practicable, to demonstrate the reason.

Equipment sent to a foreign location should be new, of 10 to 25% larger capacity than required, in order to minimize wear, and equipped with properly marked and protected spare parts. If used equipment is to be considered, it must be thoroughly overhauled and placed in first-class condition. Additional spares may be indicated with used equipment. Instruction booklets and parts catalogues must accompany the equipment, with a duplicate set in the home office to enable ordering by number. Contact should be established with local distributors to determine the spares and parts carried by them and to be sure they match the particular models.

Small tools and miscellaneous equipment and supplies are troublesome due to loss, theft, and careless misuse. Adequate supplies must be available, properly marked, and warehoused. Particular care has to be taken with such special items as strand vises.

When procuring materials from other countries, such as prestressing steel and cement, a thorough analysis has to be made to ensure the properties are known to the designer and constructor. Engineers become accustomed to trade practice, and tend to identify products by one property only, which may work in their home country but may lead to serious difficulty in other countries. Fortunately, most exporting countries are able to supply to international standards, such as ASTM or RILEM, when so required.

Steel, for example, may be customarily classified by yield point. In some countries, however, impact strength, ductility, chemistry, and rolling practices may have to be investigated. Prestressing tendons are usually classified by ultimate strength. However, elongation, ductility, and surface characteristics may vary.

Cements require investigation of their chemistry, fineness, setting characteristics, behavior in steam curing, etc. Admixtures are small in quantity and several recognized ones are distributed internationally. It is best to pay for and use these.

Mild-steel reinforcement may be deformed or smooth, and of widely varying yield strength, ultimate strength, and ductility, depending on whether it is rolled locally as mild (soft) steel or re-rolled from rails (hard steel).

Wire rope varies widely, and only well-known, well-identified brands should be employed in picking. Lifting eyes should preferably be of strand loops. Local steels may be susceptible to brittle fracture.

Steel forms may be obtained in many foreign countries. The plates and sheets may vary widely. If they have laminations, they will cause the concrete to stick in the forms. Fabricating practices vary also; tolerances and grinding of joints must be clearly directed. Finally, to ensure compliance with specifications, an international testing laboratory should be engaged to verify compliance.

Many excellent products (materials, forms, equipment, etc.) are ruined in shipping. Export packaging must be specified, with special precautions to minimize damage in handling. The moral is: "If it is possible to mis-handle a package, it will be mis-handled."

Prestressed concrete is a basic structural material upon which the emerging countries may grow. It offers wide application, it supports and stimulates the local economy, it trains indigenous workers in many skills, and it provides the highest quality of completed structures. Engineers and constructors from developed countries can take great pride in participating in this challenging extension of technology to the developing countries and in seeing the immediate and dramatic evidence of benefits to the country and its people.

25

Prestressed Concrete—
Implications and
Prospects

Prestressed concrete has been a revolutionary development, first as a material, but more importantly as a concept. As a concept it is valid and applicable to practically all materials and structures and machines. It seeks to understand the actual behavior during the lifetime of the structure; it has forced a consideration of several "limit states" in the performance of structures under loads, and an evaluation of the mechanism of failure under ultimate load.

Prestressed concrete has important implications in the economic field, causing changes in the constitution and importance of whole industries. It is a means by which more structures, housing, utilities, and facilities can be made available to more people.

Prestressed concrete has important social implications. It has a decentralizing effect in that numerous relatively small plants can compete efficiently with large centralized production. It develops a widespread technological and engineering ability, and encourages innovation and continued new development.

World-wide, prestressed concrete offers a major opportunity for the developing nations to meet their physical needs and to develop an indigenous skilled labor force.

Internationally, prestressed concrete has jumped the boundaries of nation, race, and political system. Engineers and constructors world-wide are working together on common problems. This is, in the author's opinion, the only true road to peace, where men of all nations are involved in common endeavor. For man being what he is, he can rise out of his self-concern and group antagonism only when faced with larger problems and opportunities.

What are the road-blocks and dangers to such a glowing future? First, the inadequacy of management in the prestressed industry. The industry has grown on the enthusiasm and efforts of individuals, challenged by this new material and new concept. But these men are not necessarily trained as managers. By their failure to develop marketing and comptrolling aspects of their business, they risk losing control to less imaginative and enterprising firms. Too many, the author among them, have set their sights more on new techniques and products than on aggressive expansion to fill a need. Both aspects are essential for success of the industry. National organizations, such as the Prestressed Concrete Institute, have recognized the gap and are taking strong steps to overcome it.

A second major road-block is age. Prestressed concrete is a new industry and the men who started it were young in both age and outlook. Now success has been achieved and the industry leaders have aged. There is a discernible trend to be satisfied with present scope, products, techniques. Failure to maintain the spirit behind prestressing will lead to its assuming a passive role.

The Federation Internationale de la Precontrainte has recognized this problem and is encouraging the national groups to assign younger men a more active part in industry and technical association matters. Prestressing has been an emotional enterprise; it would never have achieved its present status, nor opened windows on the future, had the pioneers been motivated by solely practical and materialistic motives.

To look at specific areas of needed development; the first is research. There is much we do not really understand about prestressed concrete and even less about other prestressed materials. High-strength concretes, approaching the properties of aluminum, appear practicable with further applied research. Polymers, stabilized by irradiation or thermal treatment, offer increased strength and impermeability. Artificial aggregates can be developed which possess the desired properties of insulation, thermal expansion, etc. Very high degrees of prestress offer practical applications for increased durability and for composite action with cast-in-place concrete. Use of precast segments of high strength concrete, perhaps coated with epoxy for added protection in view of the thin cover, offer exciting possibilities for long-span bridges, trusses, and space structures.

The application to machinery frames and supports, and eventually to machinery itself, offers a whole new field of utilization as well as improved stability and tolerance control.

The application of prestressed concrete to ship hulls and to offshore floating structures is again a new concept and a new industry. The traditional weight limitations no longer apply to very large structures, and the rigidity, durability, low maintenance, and favorable behavior at low temperature make the oceans the greatest opportunity of all for future utilization of prestressed concrete.

In its industrial organization, the prestressing industry must look more and more to the systems approach to construction, to compatibility and integration of the entire facility. Too much emphasis to date has been placed on a one-to-one replacement of structural steel or reinforced concrete. What is needed is a careful study of the overall function and services required. A structure, whether it be a ship, a building, or housing, is an operating unit, not a static collection of beams and walls. We refer to the "skeleton" of a building and some European languages use the word "carcass." The skeleton is the frame upon which a living body is hung and integrated. So it is not enough to develop a skeleton structure; the goal, rather, should be a complete, living structure, which requires that the systems approach be employed.

However, this does not mean that the "system" must be comprised entirely of prestressed concrete, or that it must be a rigid system. As set forth in Chapter 10, under "Systems (Technical Aspects)," prestressed concrete can develop as a compatible component to many systems.

The industry must also develop a more effective marketing approach, which includes performance criteria. Efforts to contain prestressed concrete within code provisions and engineering structural concepts developed for other materials leads not only to lack of economy but also to poor performance, as for example, at connections. The codes are being modernized and up-dated to include prestressing. However, more effort still has to be expended in engineering education, not only of students but also of practicing engineers. The architectural profession has made striking but not widespread use of prestressing. Familiarity with the potential and practical use of prestressed concrete is as yet not disseminated throughout the profession. There also remain the mechanical engineering, mining and petroleum engineering, naval architectural, etc., professions to which prestressed concrete has made little or no approach. If the industry is to maintain its rate of growth, its marketing efforts must be intensified along the lines indicated and along others as they appear.

Prestressing thus is both a philosophical concept and a highly practical material. Concepts can be transformed into reality only by the properly directed efforts of construction engineers. The most beautiful building or spectacular bridge can be a failure if practical, down-to-earth principles and precautions are ignored or violated by the constructor. Conversely, the constructor has the opportunity to be an essential participant in the creation of these outstanding structures. Thus, construction in prestressed concrete is the translation of ideal and concept into existence and reality, a proper challenge and opportunity for creativity.

Postscript

Since prestressed concrete is a dynamic and vigorous technology, a number of important developments have taken place during the process of publishing this book. These have been denoted by asterisks and footnotes in the appropriate places in the text, with references to the brief descriptive passages in this postscript.

i. Steel Strand for Prestressing

This is generally available now in diameters of 0.6 inch and from at least one mill (Shinko Wire of Japan) in 0.7 inch diameter. Properties are as follows:

Size	Grade	Min. Breaking Strength, Lbs.	Nominal Steel Area	Nominal Wt. Per Ft.
0.6	250 ksi	54,000	.215	.740
0.6	270 ksi	58,600	.215	.740
0.7	250 ksi	74,000	.296	1.032
0.7	270 ksi	79,900	.296	1.032
0.7	Galvanized, Stress-Relieved	74,300	.323	1.110
0.7	Galvanized, Not Stress-Relieved	74,300	.323	1.110

Galvanized strand is also generally available and is increasingly being used for externally wound tanks and for architectural units. In Japan, all wire-wound tanks have used galvanized wire with no reports of any corrosion. Properties are as follows:

Size	Grade	Breaking Strength, Lbs.	Nominal Steel Area, in.	Nominal Wt., Lbs. per ft.
3-8	230k	19,300	0.084	285
	240k	21,400	0.089	303
7-16	230k	26,200	0.114	385
	240k	28,800	0.120	410
1-2	230k	34,500	0.150	506
	240k	38,150	0.159	541

ii. Steel Bars for Pretensioning

High strength steel bars are being rolled in Japan with a spiral external groove, resembling that of strand. Threads can be rolled on the ends to receive nuts for locking anchorages. Deformed bars of similar characteristics are also available from West Europe, especially Germany, for use with both pre- and post-tensioning.

iii. Autoclaving

In East Germany and Japan, considerable study is being directed to the autoclaving of prestressed concrete elements, in sizes up to prestressed pile segments of 15 meters in length and 1 meter in diameter. Autoclaving increases the strength substantially. In Japan, strengths of 10,000 psi (700 K /cm^2) are being obtained with the conventional prestressed concrete mix. Shrinkage is essentially eliminated.

Stress relaxation is increased. Early tests indicate up to 20% stress relaxation during autoclaving.

Creep of the concrete is increased; however, this does not appear too serious due to the short time of high temperature curing.

Corrosion of the tendons is also a matter of concern. The high temperature accelerates the corrosion and the pH is apparently somewhat neutralized. This requires further study; however, early results indicate it may not be too serious a problem.

iv. Driving with "Impact Rammer"

Prestressed concrete piles are now being driven in Japan with vibrators and also with a new type of hammer known as an "Impact Rammer." This transmits directional impulses only, with considerable energy. Stress conditions in the piles appear to be fully satisfactory. The Impact Rammer has been employed on prestressed piles up to 1 meter (39 inches) in diameter and 50 meters (160 feet) in length.

v. Pile Interlocks

Prestressed concrete sheet piles are extensively employed in Japan. Plastic interlocks and galvanized sheet metal interlocks are used to provide sand and water tightness. Various modifications in shape are employed to give maximum bending strength, good shear transfer at the joints, and minimum manufacturing cost.

vi. Use of Precast Concrete for Swimming Pools

In both Japan and the USSR, prestressed concrete is extensively employed for large swimming pools. In the USSR, precast pretensioned planks are laid on a prepared bed and the joints filled with self-stressing cement. In Japan, the base slab is cast-in-place, using shrinkage-compensated cement, and post-tensioned in both directions. Walls are precast units, set in place before pouring the base slab, and thus locked to it by the post-tensioning.

vii. WIRAND Concrete

This is a development of the Battelle Development Corporation. The concrete matrix contains a random dispersion of small metallic filaments which act as crack arrestors. This increases the usable tensile strength by a factor of two or more, decreases the modulus of elasticity, and greatly increases the impact and abrasion resistance, thermal-spall index, durability, and fatigue strength.

viii. Long-Span Bridges

The perfection of construction techniques and engineering control for both precast and cast-in-place construction, the development and availability of launching and supporting gantries, and high-capacity cranes, both land and water, and the development of concentrated tendons (largely as a fall-out from nuclear reactor pressure vessel technology) have combined to produce a significant extension of the economical range for long-span bridges. This, combined with the increased demand or aesthetics, as well as the growing need for reducing the maintenance, has recently (1970) resulted in a strong resurgence in the growth of prestressed concrete bridge construction in many countries.

Certain specific trends have emerged. One of these is the increased utilization of precast concrete segments, assembled with epoxy "perfect fit" joints, erected by means of an overhead launching gantry, and stressed back in progressive cantilever procedure. Among the most notable of such bridges are the viaducts at Lake Chillon, Switzerland, for which each segment was cast against its matching neighbor, with adjustments so as to give a curved soffit profile and to accommodate horizontal curves and superelevation. Alternatively, erection may be performed by two skid A-frame derricks, working progressively out in both directions from the pier. In this latter case, it is usual to provide steps or seats, and erection bolts with which to facilitate the assembly and final positioning.

Another trend also involves the use of precast segments, this time erected on a falsework truss, with the joints being 3 in. (7.5 cm) wide concreted joints. If the falsework truss is under the final bridge, the units must be lifted up by crane or derrick, over the truss and set in position. Each unit is blocked into final position, including adjustments for dead weight deflection of the truss and an allowance for the dead weight deflection of the girder, prior to joining and stressing.

Alternatively, the falsework truss may be an overhead steel bridge, which moves progressively forward, supported by the next pier at the forward end and by the completed bridge structure at the rear end. In this case, a travelling hoist may pick each unit into position, where it is temporarily fixed until the entire span has been erected, joined, and stressed. Such procedures were employed on the Lower Yarra Bridge in Melbourne and the West Avenue Extension in London.

For both of the above precast segment schemes, the trend is to utilize full cross-sections of the bridge, with a length of perhaps 10 to 12 ft. (3 to 4 m). For the Oosterschelde Bridge, however, the length of segments was varied so as to keep the weight of individual lifts approximately uniform despite a changing profile.

Cast-in-place bridges, constructed by progressive cantilevering with travelling forms, continue to be utilized very extensively. Such forms have been fitted with winter protection and heaters to enable concreting, curing and stressing to be carried on through all seasons.

Continuous monitoring is required for variations due to temperature, creep, anchorage set, and shrinkage. In the most sophisticated practice, this is facilitated by computer programming. Minor adjustments in the profile may be made by varying the initial prestress in the subsequent sections or, in the case of cast-in-place construction, by varying the deck concrete thickness. One of the most effective means of adjustment is to provide extra ducts into which additional tendons may be inserted and stressed as necessary to adjust the advancing profile.

For overwater bridges, the use of large pre-assembled spans continues to grow. Spans up to 300 ft. (100 m) in length may be erected as simple spans, then made continuous for live load by a cast-in-place concrete deck.

For the Volga River bridges, major sections of the bridge spans were assembled on the banks from precast elements, joined and stressed; thence, floated on pontoons to position, where they were set directly from the pontoons by flooding or, in some cases, lifted into place by very large floating derricks.

An interesting development is the use of precast pan-type deck units which are erected on falsework. The tendons in rigid ducts are then placed in the joints between the precast units. Concrete is then placed in the joints and on top of the precast units, acting in composite action with them.

Transverse post-tensioning may be beneficially employed in bridge decks to prevent sag due to creep, especially where cantilevered overhangs support a concrete curb. Flat jacks have been utilized very effectively in adjusting the reactions at piers and in decentering of arch bridges.

The above examples illustrate the rapid advances which are being made in the development of equipment and techniques, by which the economical range for prestressed concrete spans is being extended.

ix. Curved Monorail Girders

The precast post-tensioned girders for the Disneyland Monorail in Florida had to accommodate many varying degrees of curvature and superelevation. They were cast within flexible side forms, which were adjusted at close intervals to pre-computed offsets. Extreme care was exercised in mix design, placement, and curing to ensure uniform concrete; that is, with similar moduli of elasticity, shrinkage, and creep characteristics. Similarly, the positioning of tendons and stressing procedures were monitored to ensure uniformity. The resulting girders had riding and guiding surfaces that met the extremely high tolerances necessary for high-speed operation.

References

1. Abeles, P. W. *An Introduction to Prestressed Concrete,* Concrete Publications Ltd., London, Volume I, 1964; Volume II, 1966.

2. American Concrete Institute, *Manual of Concrete Practice,* Detroit, 1968, Volumes I, II, and III, especially:
   (a) Guide for Structural Lightweight Aggregate Concrete;
   (b) Recommended Practice for Hot Weather Concreting;
   (c) Recommended Practice for Cold Weather Concreting;
   (d) Guide for Use of Epoxy Compounds with Concrete;
   (e) Suggested Design of Joints and Connections in Precast Structural Concrete;
   (f) Low Pressure Steam Curing;
   (g) Durability of Concrete in Service.

3. American Concrete Institute, "Tentative Recommendations for Members Prestressed with Unbonded Tendons," *Journal of the American Concrete Institute,* Detroit, February, 1969.

4. American Concrete Institute "Expansive Cement Concretes—Present State of Knowledge," *Journal of the American Concrete Institute,* Detroit, August, 1970.

5. American Concrete Institute, "Concrete Bridge Design," *Proceedings of First International Symposium,* Detroit, 1967.

6. Associated General Contractors of America, *Manual of Accident Prevention in Construction,* Associated General Contractors, Washington, D. C.

7. Bender, M. *Engineering Aspects of Concrete Reactor Pressure Vessels,* Oakridge National Laboratory, ORNL-TM-617, September, 1963.

8. Cement and Concrete Association, *Proceedings of First International Congress on Lightweight Concrete,* London, 1968.

9. Cement and Concrete Association, *Proceedings, Symposium on Design Philosophy and Its Application to Precast Concrete Structures,* Cement and Concrete Association, London, 1967.

10. *CEB/FIP International Recommendations for the Design and Construction of Concrete Structures,* Cement and Concrete Association, London, 1970.

11. Concrete Society, *Safety Precautions for Prestressing Operations,* The Concrete Society, London, 1968.

12. Davidson, I., *Materials Research for Prestressed Concrete Pressure Vessels,* Concrete Society Ltd., London, 1967.

13. de Heer, J. J., *Prestressed Lightweight Aggregate Concrete for Pavements,* FIP Special Report No. 7, Federation Internationale de la Precontrainte, London, 1969.

14. Dikeov, J.; Kukacka, L.; Backstrom, J.; Steinberg, M.; "Polymerization Makes Tougher Concrete," *Journal American Concrete Institute,* October, 1969, Detroit.

15. Federation Internationale de la Precontrainte, London:

    (a)   *Proceedings of I Congress,* London, 1953;
    (b)   *Proceedings of II Congress,* Amsterdam, 1955;
    (c)   *Proceedings of III Congress,* Berlin, 1958;
    (d)   *Proceedings of IV Congress,* Rome-Naples, 1962;
    (e)   *Proceedings of V Congress,* Paris, 1966;
    (f)   *Proceedings of VI Congress,* Prague, 1970.

16. Federation Internationale de la Precontrainte and Permanent International Association of Road Congresses, *Symposium on Prestressed Concrete Roads and Airfield Runways,* Naples, published by FIP, London, 1962.

17. Federation Internationale de la Precontrainte, *Proceedings of Symposium on Mass-Produced Prestressed Concrete (Piles, Poles, Sleepers (ties), and Pipes), Madrid,"* London, 1969.

18. Federation Internationale de la Precontrainte, *Proceedings of Symposium on Steel for Prestressing,* Madrid, London, 1969.

19. FIP-RILEM *International Recommendations for Grout and Grouting of Prestressed Concrete,* Federation Internationale de la Precontrainte, London, 1962.

20. Germain, F., "The Development of the Prestressed Nuclear Reactor Vessel Concept in Europe and the United States," *Journal Prestressed Concrete Institute,* Chicago, August 1967.

21. Gerwick, B., Jr., "Long-Span Prestressed Concrete Bridges Utilizing Precast Elements," *Journal Prestressed Concrete Institute,* Chicago, February 1964.

22. Gerwick B., Jr., *Construction Procedures for Large-Scale Concrete Structures for Ocean Engineering,* American Society of Civil Engineers, New York, 1969.

23. Gerwick, B., Jr., "Prestressed Concrete Underwater Oil Storage System", *Proceedings, First Offshore Technology Conference,* Houston, 1969.

24. Gerwick, B., Jr., and Lloyd, R. R., "Design and Construction Procedures for Proposed Arctic Offshore Structures," *Proceedings, Second Offshore Technology Conference,* Houston, 1970.

25. Guyon, Y., *Constructions en Beton Precontrainte,* Editions Eyrolles, Paris, Volume I 1966, Volume II 1968 (in French).

26. Harris, A. J., et al., "Prestressed Concrete Pressure Vessels for Nuclear Power Stations," *Journal of the Prestressed Concrete Institute,* Chicago, 1965.

27. Harris, A. J., "High Strength Concrete, Manufacture and Properties," *The Structural Engineer,* London, November 1969.

28. Harris, J. D., and Smith, I. C., *Basic Design and Construction in Prestressed Concrete,* Chatto and Windus, London 1963.

29 Harstead, G. A., and Kammerle, E. R., *Grouting of Large-Capacity Tendons for Nuclear Containment Structures*, American Society of Civil Engineers, New York, 1968.

30. Institution of Civil Engineers, *Proceedings, Conference on Prestressed Concrete Pressure Vessels, 1967*, London, 1968.

31. Jackson, G. W. and Sutherland, W. M., *Concrete Boatbuilding*, de Graff, New York, 1969.

32. Leonhardt, F., *Prestressed Concrete, Design and Construction*, second edition, Wilhelm Ernst & Sohn, Berlin, 1964.

33. Libby, J. R., *Prestressed Concrete, Design and Construction*, Ronald Press, New York, 1965.

34. Lin, T. Y., *Design of Prestressed Concrete Structures*, second edition, John Wiley and Sons, 1963.

35. Michailov, K. V., *Development of Concrete and Reinforced Concrete in the U.S.S.R.*, Publication Literature for Structures, Moscow, 1969 (in Russian).

36. Michailov, V. V., "Self-stressed Concrete," *Journal of Prestressed Concrete Institute*, Chicago, (to be published 1971).

37. Mokk, Laszlo, *Prefabricated Concrete for Industrial and Public Structures*, Akademiai Kiado, Budapest, 1964.

38. Moore, D. G., Klodt, D. T., Hensen, R. J., *Protection of Steel in Prestressed Concrete Bridges*, Denver Research Institute, Report No. 2471, Denver, 1968.

39. Portland Cement Association, *Concrete for Railways*, Chicago, 1964.

40. Preston, H. K., *Practical Prestressed Concrete*, McGraw-Hill, New York, 1960.

41. Prestressed Concrete Development Group, *Proceedings, Symposium on the Application of Prestressed Concrete to Machinery Structures, Cement and Concrete Association*, London, 1964.

42. Prestressed Concrete Institute, Chicago:

    (a) *Manual for Quality Control for Plants and Production of Precast Prestressed Concrete Products (Tentative)*, 1966; (Revised edition to be published 1971).

    (b) *Manual for Quality Control for Plants and Production of Architectural Precast Concrete Products (Tentative)*, 1968;

    (c) *Connection Details for Precast-Prestressed Concrete Buildings*, 1963;

    (d) *Summary of Basic Information On Precast Concrete Connections*, 1970;

    (e) *Prestressed Concrete for Long-Span Bridges*, 1968.

    43. Prestressed Concrete Manufacturers' Association of California, Inc., *Connections Manual*, San Francisco, 1969.

43. Prestressed Concrete Manufacturers' Association of California, Inc., *Connections Manual*, San Francisco, 1969.

44. Prestressed Concrete Manufacturers' Association of California, Inc., and Western Concrete Reinforcing Steel Institute, *Recommended Practice for Grouting Post-Tensioning Tendons, Tentative*, Prestressed Concrete Manufacturers' Association of California, Inc., San Francisco, 1967.

45.   Roshore, E. C., "Durability and Behavior of Pretensioned Beams," *Journal of the American Concrete Institute*, **61-47**, Detroit, July, 1964.

46.   Rossnagel, W. E., *Handbook of Rigging in Construction and Industrial Operations*, McGraw-Hill, New York, third edition, 1964.

Schupack, M., "Large Post-Tensioning Tendons," *Journal of Prestressed Concrete Institute*, Chicago, (to be published 1971).

48.   Szilard, R., "Protection of Tendons in Prestressed Concrete Bridges," *Journal of American Concrete Institute*, Detroit, January, 1969.

49.   Tan, Chen Pang, *Prestressed Concrete in Nuclear Pressure Vessels, a Bibliography of Current Literature*, Oakridge National Laboratory, publication ORNL-TM-1675, Rev. 1, February, 1969.

50.   University of California, *Proceedings, Conference on Prestressed Concrete Pressure Vessels*, Berkeley, Calif., 1968.

51.   U.S. Army Engineers, *Safety—General Safety Requirements*, Engineering Manual 385-1-1, 1967.

52.   Weber, John W. "Concrete Crossties in the United States," *Journal of Prestressed Concrete Institute*, Chicago, February, 1969.

53.   Cestelli-Guidi, Carlo, "Cemento Armato Precompresso, Fifth edition, Ulrico Hoepli, Milan, 1970 (in Italian).

54.   Haynes, H. H., and Ross, R. J., "Influence of Length-to-Diameter Ratio on Behavior of Concrete Cylindrical Hulls under Hydrostatic Loading," Naval Civil Engineering Laboratory, Port Hueneme, California, 1970; NCEL Technical Report R696.

55.   Singer, R. H. "Prestressed Concrete Used for Offshore Barges, Storage," Offshore magazine, November 1970.

56.   Ward, D. R. "Rehabilitation Method doubles Platform Life," Offshore Magazine, November 1968.

57.   American Concrete Institute "Recommended Practice for Design, Manufacture, and Installation of Concrete Piles," to be published in Journal of American Concrete Institute, Detroit, 1971.

58.   Gerwick, B. C. Jr., "Prestressed Concrete Developments around the World," Civil Engineering, New York, Dec. 1970.

59.   Anderson, A. R. and Movstafa, S. E., "Ultimate Strength of Prestressed Concrete Piles and Columns," Journal, American Concrete Institute, Detroit, Aug. 1970.

60.   "Post-tensioned Box Girder Bridges, Design and Construction," Western Concrete Reinforcing Steel Institute, Burlingame, California, 1969.

61.   Gerwick, B. C. Jr., and Peters, P. V. *Russian-English Dictionary of Prestressed Concrete and Concrete Construction*, Gordon and Breach, New York, 1966.

62.   Lin, T. Y., "Inter Continental Peace Bridge" (across Bering Straits), ICPB, Inc., San Francisco, California, 1971.

Index